Studies in Computational Intelligence

Volume 652

Series editor

Janusz Kacprzyk, Polish Academy of Sciences, Warsaw, Poland
e-mail: kacprzyk@ibspan.waw.pl

About this Series

The series "Studies in Computational Intelligence" (SCI) publishes new developments and advances in the various areas of computational intelligence—quickly and with a high quality. The intent is to cover the theory, applications, and design methods of computational intelligence, as embedded in the fields of engineering, computer science, physics and life sciences, as well as the methodologies behind them. The series contains monographs, lecture notes and edited volumes in computational intelligence spanning the areas of neural networks, connectionist systems, genetic algorithms, evolutionary computation, artificial intelligence, cellular automata, self-organizing systems, soft computing, fuzzy systems, and hybrid intelligent systems. Of particular value to both the contributors and the readership are the short publication timeframe and the worldwide distribution, which enable both wide and rapid dissemination of research output.

More information about this series at http://www.springer.com/series/7092

Mikhail Z. Zgurovsky · Yuriy P. Zaychenko

The Fundamentals of Computational Intelligence: System Approach

Springer

Mikhail Z. Zgurovsky
National Technical University of Ukraine
Kiev
Ukraine

Yuriy P. Zaychenko
Institute for Applied System Analysis
National Technical University of Ukraine
"Kiev Polytechnic Institute"
Kiev
Ukraine

ISSN 1860-949X ISSN 1860-9503 (electronic)
Studies in Computational Intelligence
ISBN 978-3-319-81739-2 ISBN 978-3-319-35162-9 (eBook)
DOI 10.1007/978-3-319-35162-9

Printed on acid-free paper

This Springer imprint is published by Springer Nature
The registered company is Springer International Publishing AG Switzerland

Preface

The intention of mankind to reveal the essence of brain intelligence and utilize it in computers arose already at the initial stage of cybernetics. These aspirations led to the appearance of new scientific field—artificial intelligence (AI) in the middle of the twentieth century. The main goal of new science became the detection of brain work mechanism and development on this base artificial systems for creative processes automation in information processing and decision-making.

In the past 60 years artificial intelligence has passed complex way of its evolutionary development. On this way were both significant achievements and failures. During the first 40 years of its development in AI several scientific branches were formed such as knowledge bases, expert systems, pattern recognition, neural networks, fuzzy logic systems, learning and self-learning, self-organization, robotics, etc.

In the 1990s by integrating various AI technologies and methods new scientific branch in AI was formed which was called "computational intelligence." This branch appeared as a consequence of practical problems which could not be solved using conventional approaches and methods. In particular these problems have occured while solving the problems of data mining, problems of decision-making under uncertainty, so-called ill-structured problems for which the crisp problem statement was impossible.

The suggested monograph is dedicated to systematic presentation of main trends, approaches, and technologies of computational intelligence (CI). The introduction includes brief review of CI history, the authors' interpretation of CI, the analysis of main CI components: technologies, models, methods, and applications. The interconnections among these components are considered and relations between CI and soft computing are indicated.

Significant attention in the book is paid to analysis of the first CI technology—neural networks. The classical neural network backpropagation (NN BP) is described, the main training algorithms are considered. The important class of neural networks—with radial basic functions is described and its properties are compared with NNBP. The class of neural networks with backfeed—Hopfield and

Hamming are described and their properties and methods of weights adjustment are considered. The results of experimental investigations of these networks in the problem of images recognition under high level of noise are presented and compared. NN with self-organization by Kohonen are considered, its architecture and properties are described and various algorithms of self-organization are analyzed. The application of Kohonen neural networks in the problems of automatic classification and multidimensional visualization is considered.

Great attention in monograph is paid to novel important CI technology—fuzzy logic (FL) systems and fuzzy neural networks (FNN). The general description of fuzzy logic systems is provided, main stages of fuzzy inference process and fuzzy logic algorithms are described. The comparative analysis of fuzzy logic systems properties is presented, their advantages and drawbacks are analyzed. On this base the integration of two CI technologies—NN and FL was performed and as a result the new CI technology was created—fuzzy neural networks. Different FNN are described and their training algorithms are considered and compared.

New class of FNN—cascade neo-fuzzy neural networks (CNFNN) are considered, its architecture, properties, and training algorithms are analyzed. The applications of FNN to the forecast in economy and at stock markets are presented. The most efficient algorithms of fuzzy inference for the problem of forecasting in economy and financial sphere are determined.

The problem of investment portfolio optimization under uncertainty is considered. The classical portfolio optimization problem by Markovitz is described, its advantages and drawbacks are analyzed. The new problem statement of portfolio optimization under uncertainty is considered, which is free of drawbacks of classical model. The novel theory of fuzzy portfolio optimization is presented. For its solution corresponding method based on fuzzy sets approach is suggested. The application of the elaborated theory for investment portfolios determination at Ukrainian, Russian, and American stock exchanges is presented and analyzed.

The problem of corporations bankruptcy risk forecasting under incomplete and fuzzy information is considered. The classical method by Altman is described and analyzed. New methods based on fuzzy sets theory and fuzzy neural networks are suggested. Results of fuzzy methods application for corporations bankruptcy risk forecasting are presented, analyzed, compared with Altman method and the most adequate method was determined.

This approach was extended to the problem of banks bankruptcy risk forecasting under uncertainty. The experimental results of financial state analysis and bankruptcy risk forecasting of Ukrainian and leading European banks with application of fuzzy neural networks ANFIS, TSK, and fuzzy GMDH are presented and analyzed.

The actual problem of creditability analysis of physical persons and corporations is considered. The classical scoring method of creditability analysis is described and its drawbacks are detected. New method of corporations creditability estimation under uncertainty is suggested based on application of FNN. The comparative investigations of corporations creditability forecasting using classical scoring method and FNN are presented and discussed.

The application fuzzy neural networks in the problem of pattern recognition is considered. The applications of FNN for pattern recognition of optical images, handwritten text, are considered.

Great attention in monograph is paid to inductive modeling method, so-called group method of data handling (GMDH). Its main advantage lies therein it enables to construct the structure of forecasting model automatically without participation of an expert. By this possibility GMDH differs from other identification methods. The new fuzzy GMDH method suggested by authors is described which may work under fuzzy and incomplete information. The problem of inductive models adaptation obtained by FGMDH is considered. The results of numerous experimental investigations of GMDH for forecasting at stock exchanges in Ukraine, Russia, and USA are presented and analyzed.

The problems of cluster analysis and automatic classification are considered in detail. The classical and new methods of cluster analysis based on fuzzy sets and FNN are described and compared. The applications of cluster methods for automatic classification of UNO countries by indices of sustainable development are presented and analyzed.

The final chapters are devoted to theory and applications of evolutionary modeling (EM) and genetic algorithms (GA). The general schema and main mechanisms of GA are considered, their properties and advantages are outlined. Special section is devoted to new extended GA. The applications of GA for computer networks structure optimization are considered.

Basic concept and main trends in evolutionary modeling are considered. The evolutionary strategies for artificial intelligence problems solution are presented and discussed. Special attention is paid to the problem of accelerating the convergence of genetic and evolutionary algorithms.

In the conclusion the perspectives of computer intelligence technologies and methods development and implementation are outlined.

The distinguishing features of this monograph are a great number of practical examples of CI technologies and methods and applications for solution of real problems in economy and financial sphere, in particular forecasting, classification, pattern recognition, portfolio optimization, bankruptcy risk prediction of corporations and banks under uncertainty which were developed by the authors and are published in the book for the first time. Just system analysis of presented experimental and practical results enables to estimate the efficiency of the presented methods and technologies of computational intelligence.

All CI methods and algorithms are considered from the general system approach and the system analysis of their properties, advantages, and drawbacks is performed that enables practicians to choose the most adequate method for their own problems solution.

The proposed monograph is oriented first of all to the persons who aspire to make acquaintance with possibilities of current computer intelligence technologies and methods and to implement them in practice. It may also serve as inquiry book on contemporary technologies and methods of CI, it will be useful for students of corresponding specialties.

Contents

Introduction

Human desire to improve the efficiency of own decisions constantly stimulated the development of appropriate computational tools and methods. These tools are characterized by increasing automation and intellectualization of their creative operations, which were previously considered to be the prerogative exclusively of human being with the advent of the computer the question had arisen: Can a machine “think”, is it possible to transfer to a computer a part of human intellectual work? To this sphere refer modeling of certain functions of the human brain, in particular, pattern recognition, fuzzy logic, self-learning, and other creative human activities.

History of work in the field of artificial intelligence counts more than 50 years. The term artificial intelligence (AI) first appeared in 1956, when at Dartmouth College USA the symposium entitled “Artificial Intelligence” (AI) was held, dedicated to solving logic rather than computational problems using computers. This date is the moment of the birth of a new science. Over the years, this branch of science had passed a difficult and instructive way of development. It has formed a number of such fields, as systems based on knowledge, the logical inference, the search for solutions, the pattern recognition systems, machine translation, machine learning, action planning, agents and multi-agent systems, self-organization and self-organizing systems, neural networks, fuzzy logic and fuzzy neural networks, modeling of emotions and psyche, intellectual games, robots and robotic systems, and others.

Currently, there are many definitions of the term Artificial Intelligence. It makes no sense to bring them all in this book. In our view, more important is to highlight the main features and properties that distinguish AI from conventional automation systems for certain technologies. These properties include:

1. the presence of the goal or group of goals of functioning, the ability to set goals;
2. the ability to plan their actions to achieve the goals, finding solutions to problems that arise;
3. the ability to learn and adapt their behavior in the course of work;
4. the ability to work in poorly formalized environment in the face of uncertainty;

5. work with fuzzy instructions;
6. the ability to self-organization and self-development;
7. the ability to understand natural language texts;
8. the ability to generalization and abstraction of accumulated information.

In order to create a machine that would approach in its capabilities to the properties of the human brain, it is necessary, first of all, to understand the nature of human intelligence, to reveal the mechanisms of human thinking. Over the past decade, this issue was the subject of many works. Among the works that have appeared in recent years, it is necessary to appreciate the monograph by Jeff Hawkins and Sandra Blakeslee "On intelligence," translated from English.—M.: Ltd. "P.H. Williams," 2007, in which the authors, in our opinion, had come closest to the understanding the basis of human intelligence. In it the authors, at the intersection of neuroscience, psychology and cybernetics have developed a pioneering theory in which a model of the human brain, was constructed. The main functions of developed model are remembering past experiences and predicting by the brain results of the perception of reality and its own actions. The authors presented many convincing examples of human behavior in different circumstances, which support this idea. So, Jeff Hawkins writes: "… Forecasting, in my opinion, is not just one of the functions of the cerebral cortex, it is the primary function of the neocortex, and the base of intelligence. The cerebral cortex is the organ of vision. If we want to understand what is the intellect, what is creativity, how our brain works, and how to build intelligent machines, we need to understand the nature of forecasts and how the cortex builds them."

Since many years of research work in the field of AI in the early 1990s by integrating a number of intelligent technologies and methods a new direction in the field of AI, called computational intelligence (CI) was created.

There are several definitions of the term computational intelligence. For the first time the term "computational intelligence" (CI) was introduced by Bezdek [1] (1992, 1994), who defined it as: "a system is intelligent in computational sense, if it: operates only with digital data; has a recognition component; does not use knowledge in the sense of artificial intelligence and in addition, when it exhibits:

(a) computing adaptability;
(b) computing fault tolerance;
(c) the error rate that approximates human performance."

Afterward, this definition was clarified and expanded. Marx in 1993 in defining CI focused on technology components of CI [2]: "… neural networks, genetic algorithms, fuzzy systems, evolutionary programming and artificial life are the building blocks of CI."

Another attempt of CI definition was made by Vogel [3]: the technologies of neural networks, fuzzy and evolutionary systems have been integrated under the guise of "computational intelligence"—a relatively new term suggested for the general description of the computational methods that can be used to adapt to new problems solutions and are not based on explicit human knowledge.

Over the years, a large number of works has appeared devoted to various areas in the field of CI, regularly many international conferences and congresses on Computational Intelligence are held, IEEE International Institute publishes a special magazine devoted to the problems of computational intelligence—IEEE Transactions on Computational Intelligence.

The analysis of these works allows us to give the following definition of CI [4].

Under computational intelligence (CI) we'll understand the set of methods, models, technologies and software designed to deal with informal and creative problems in various spheres of human activity, using the apparatus of logic, which identify to some extent mental activity of human, namely, fuzzy reasoning, qualitative and intuitive approach, creativity, fuzzy logic, self-learning, classification, pattern recognition and others.

It is worth noting the relationship between artificial intelligence (AI) and computational intelligence. CI is an integral part of modern AI using special models, methods, and technologies and focused on solving certain classes of problems.

The structure of CI areas and methods is shown in Fig. 1.

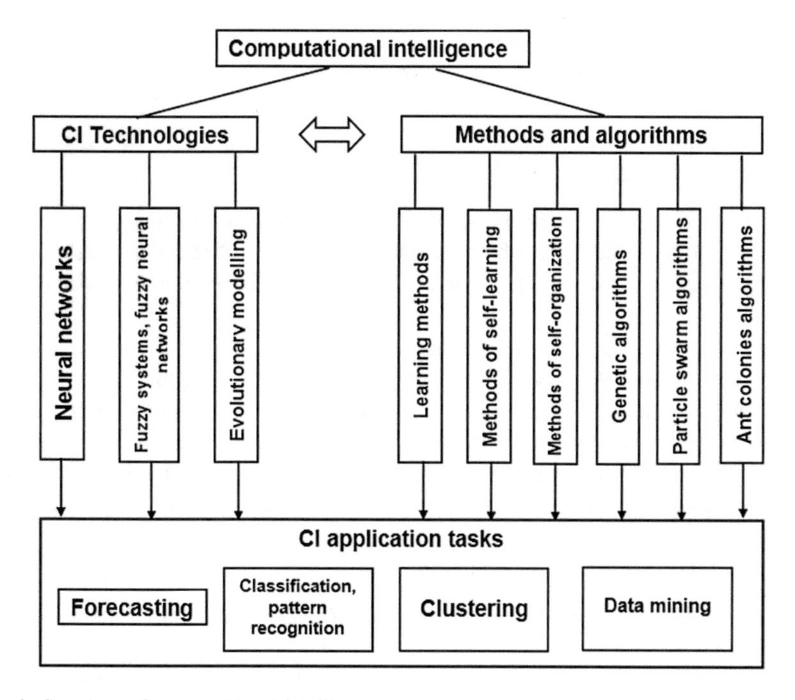

Fig. 1 Structure of computational intelligence

The structure CI includes the following components [4]:

- Technologies;
- Models, methods, and algorithms;
- Applied tasks.

CI Technologies include:

- Neural networks (NN);
- Fuzzy logic systems (FLS) and fuzzy neural networks (FNN);
- Evolutionary modeling (EM).

CI methods and algorithms include:

- Learning methods;
- Methods of self-learning;
- The methods of self-organization;
- Genetic algorithms (GA);
- Particle swarm algorithms;
- Ant colonies algorithms.

CI technologies and techniques are used for solution of relevant applied problems of AI. It is reasonable to distinguish the following basic classes of CI applied problems inherent of human mental activity:

- Forecasting and foresight;
- Classification and pattern recognition;
- Clustering, spontaneous decomposition of the set of objects into classes of similar objects;
- Data Mining;
- Decision-Making.

The term CI is close in meaning, which is widely used in foreign literature the term "soft computing" [5], which is defined as a set of models, methods, and algorithms based on the application of fuzzy mathematics (fuzzy sets and fuzzy logic).

The concept of soft computing was first mentioned in the work of L. Zadeh to analyze soft (soft) data in 1981. Soft computing (SC)—a sophisticated computer methodology based on fuzzy logic (FL), genetic computing, neurocomputing, and probabilistic calculations. Its components do not compete but create synergies. The guiding principle of soft computing is accounting errors, uncertainty, and approximation of partial truth to achieve robustness, low-cost solutions which are more relevant to reality.

Four soft computing components include:

- Fuzzy logic—approximate calculations, information granulation, the calculation in words;
- Neurocomputing—training, adaptation, classification, system modeling, and identification;
- Genetic computing—synthesis, configuration, and optimization using a systematic random search and evolution;
- Probability calculations—management of uncertainty, belief networks, chaotic systems, a prediction.

Traditional computer calculations (hard computing) is too accurate for the real world. There are two classes of problems for soft computing. First, there are problems, for which complete and accurate information is not available and cannot be obtained, and, second, the problems whose definition is not sufficiently complete (uncertain).

Thus, between the computational intelligence (CI) and soft computing are much in common: common paradigms, principles, methods, technologies, and applied problems. Differences between them, in our opinion, consist of approach, SC focuses on methodological, philosophical, and mathematical problems, while CI focuses mainly on the computer algorithms, technology, and the practical implementation of the relevant models and methods.

Technologies and methods of CI are widely used for applications. Thus, neural networks (NN) and fuzzy neural networks are used to predict the nonstationary time processes, particularly in the economy and the financial sector.

NN and GMDH are widely used in problems of classification and pattern recognition problems in diagnostics, including technical and medical spheres.

Neural network self-organization, cluster analysis methods (clear and fuzzy) are used in the automatic classification of objects by their features of similarity—difference.

Evolutionary modeling, genetic algorithms are used for the synthesis of complex systems, pattern recognition, classification, and optimization of computer networks structure. In addition, genetic and particle swarms algorithms are widely used in combinatorial optimization problems, in particular for graphs optimization.

Systems with fuzzy logic and FNN are effectively used for the analysis of the financial state of corporations, prediction of corporations and banks bankruptcy risk, assessment of the borrowers creditworthiness under uncertainty.

Thus, summing up, it should be noted that modern CI technologies and methods closely interact with each other. There is a deep interpenetration of methods and algorithms of computational intelligence into the appropriate technology, and vice versa.

Further development of computational intelligence apparently will go in several directions.

First—it is the extension of CI applications problems, new applications in different tasks and subject areas, such as economy, finance, banking, telecommunication systems, process control, etc.

Second, it is the development and improvement of CI methods themselves, in particular, genetic (GA) and evolutionary algorithms (EA), swarm optimization algorithms, immune algorithms, ant algorithms. One promising area is the adaptation of learning parameters of genetic and evolutionary algorithms in order to accelerate convergence and improve the accuracy of solutions in optimization problems. Relevant is the development and improvement of parallel genetic algorithms.

Third, it is the further integration of various CI technologies, for example, the integration of fuzzy logic and genetic and evolutionary algorithms in problems of decision-making and pattern recognition in conditions of incompleteness of information and uncertainty, genetic and swarm algorithms, as well as algorithms for their learning and self-learning.

The presentation of CI basic techniques and methods, as well as their application to the solution of numerous application problems in various areas: forecasting, classification, pattern recognition, cluster analysis, and risk forecasting in industrial and financial spheres constitute the main content of this monograph.

Chapter 1
Neural Networks

1.1 Introduction

In this chapter the first technology of CI-neural networks is considered.

In the Sect. 1.2 neural network Back propagation (NN BP) is considered which is most widely used, its architecture and functions described. The important theorem on universal approximation for NN BP is considered which grounds its universal applications in forecasting, pattern recognition, approximation etc. gradient method of training neural network BP is described, its properties analyzed in Sect. 1.3. In Sect. 1.4 the extension of gradient algorithm for training NN BP with arbitrary number of layers is considered.

In Sect. 1.5 some modifications of gradient algorithm improving its properties are described. In Sect. 1.6 method of conjugate gradient is considered for network training which has accelerated convergence as compared with conventional gradient method. In Sect. 1.7 genetic algorithm for training NNBP is considered, its properties are analyzed.

In Sect. 1.8 is presented another class of widely used neural networks with radial basis functions. Its structure and properties are described and analyzed. Methods of training weights and parameters of radial functions are considered. The more general class of radial neural networks- so-called Hyper Radial Basis Function networks (HRBF) is considered, its properties are described and analyzed. In Sect. 1.9 efficient algorithm of RBF network training- hybrid algorithm is presented and its properties described. In the Sect. 1.10 some examples of application RBF neural network and methods of selection the number of radial basis functions are presented, comparison of back propagation neural networks and RBF NN is performed.

M.Z. Zgurovsky and Y.P. Zaychenko, *The Fundamentals of Computational Intelligence: System Approach*, Studies in Computational Intelligence 652,
DOI 10.1007/978-3-319-35162-9_1

1.2 Neural Network with Back Propagation. Architecture, Functions

Most widely used and well-known neural networks are the so-called networks with reverse propagation ("back propagation") [1–3].

These networks predict the state of the stock exchange, recognize handwriting, synthesize speech from text, run a motor vehicle. We will see that the back propagation rather refers to the training algorithms, and not to the network architecture. Such a network is more properly described as a feed forward network, with direct transmission of signals.

In Fig. 1.1 Classic three-level architecture of neural network is presented.

Designate: an input vector $X = \{x_1, x_2, \ldots x_N\}$; an output vector $Y = \{y_1, y_2, \ldots y_M\}$, an output vector of the hidden layer $H = \{h_1, h_2, \ldots h_j\}$. A neural network executes functional transformation which can be presented as $Y = F(X)$, where $X = \{x_i\} i = \overline{1, N}$, $Y = \{y_k\} k = \overline{1, M}$, $x_{n+1} = 1$.

The hidden layer actually can consist of several layers, however it is possible to suppose that it is enough to examine only three layers for a this type of networks specification.

For a neural network with N input nodes, J nodes of the hidden layer and M output nodes output y_k is determined so:

$$y_k = g\left(\sum_{j=1}^{J} W_{jk}^{o} h_j\right), k = \overline{1, M} \tag{1.1}$$

where W_{jk}^{o} is output weight of connection from the node j of the hidden layer to the node k of the output layer; g is a function (which will be defined later), executing a mapping $R^1 \rightarrow R'$.

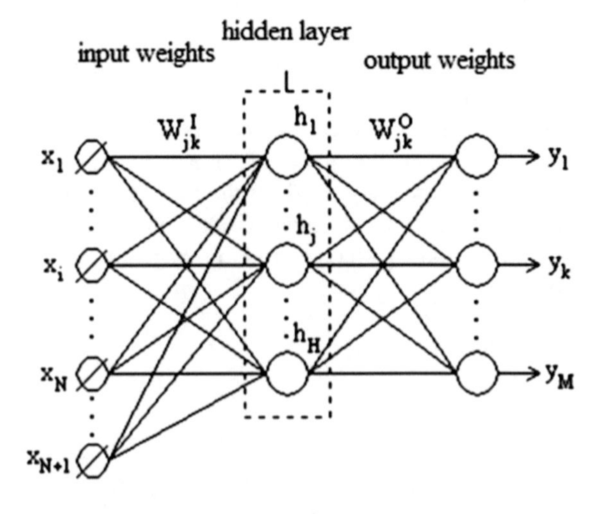

Fig. 1.1 Architecture of back propagation neural network

The output signals of the hidden layer nodes $h_j, j = 1, 2, \ldots J$ are defined by expression:

$$h_j = \sigma\left(\sum_{i=1}^{N} W_{ij}^{I} x_i + W_j^{T}\right), j = \overline{1, J}, \tag{1.2}$$

where W_{ij}^{I} is an input weight of the connection (*i, j*);

W_j^T—threshold value (weight from a node (*n* + 1), which has a permanent signal, equal 1 to node j);

the signal σ at the output of the *j*th input node is the so-called "sigmoid", which is defined as follows

$$\sigma(x) = \frac{1}{1 + e^{-x}} \tag{1.3}$$

The function σ in (1.2) is called the activation function (of a neural network), sometimes called "the function of the ignition of" a neuron.

The function g in Eq. (1.1) may be the same one as $\sigma(x)$ or another. In our presentation we will take g as the function of type σ. It's required that the activation function was nonlinear and had a limited output. The function $\sigma(x)$ is shown in Fig. 1.2.

As activation functions may be also used other non-decreasing monotonous continuous and restricted functions, namely $f(x) = arctg(x),\ f(x) = th(x)$, where $th\,x$ is hyperbolic tangens of x and some other. The advantage of use sigmoidal function $y = \sigma(x)$ is simple computation of derivative: $\sigma'(x) = \sigma(x)(1 - \sigma(x))$.

Feed Forward Neural Network

The behavior of feed forward network is determined by two factors:

- network architecture;
- the values of the weights.

The number of input and output nodes is determined a priori and, in fact, is fixed. The number of hidden nodes is variable and can be set (adjusted) by the user.

To date, the choice of hidden nodes number remains "state of art", though in the literature different methods of setting their number have been proposed.

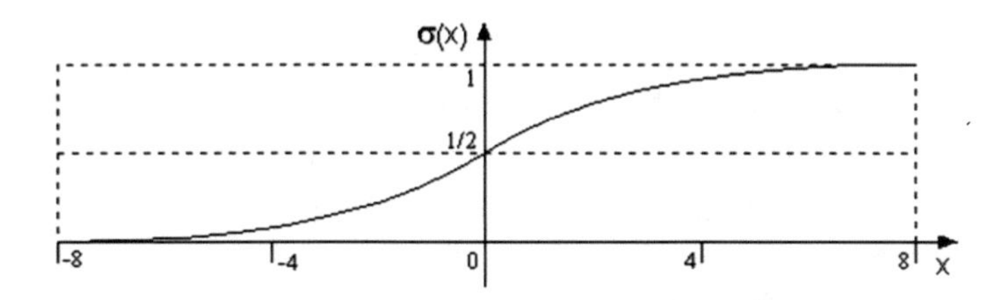

Fig. 1.2 Function $\sigma(x)$

After determining network architecture, just the values of the weights determine its behavior.

It said that the network "trains" if the weights are modified so as to achieve the desired goal. Here it should be borne in mind that the term "training", borrowed from biology, for network means simply adjustment of many parameters (connection weights).

Theorem of Universal Approximation

Neuron networks Back Propagation (NNBP) are applied in various problems of computational intelligence: forecasting, pattern recognition, approximation of multivariate functions etc. Their wide application possibilities are determined by their fundamental property—NNBP are universal approximators. The corresponding theorem of universal approximation was proved by Gornik and others in 1989 year [4]. This theorem is generalization of theorem by Kolmogorov and Arnold and is formulated as follows [2].

Let $\varphi(.)$ be limited, non-constant, monotonously increasing continuous function.

Let I_{m_0} be m_0-dimensional unit hypercube $[0,1]^{m_0}$ and let denote the space of continuous functions on I_{m_0} by symbol $C(I_{m_0})$. Then for any function $f \in C(I_{m_0})$ and arbitrary $\xi > 0$ exists such integer m_1 and a set of real constants α_1, b_i, w_{ij} where $i = 1,\ldots, m_1$, $j = 1,\ldots m_0$, that the function

$$F(x_1, x_2, \ldots, x_{mo}) = \sum\nolimits_{i=1}^{m_1} \alpha_i \varphi\left(\sum\nolimits_{j=1}^{m_0} w_{ij} x_j + b_i\right) \tag{1.4}$$

is approximation of function $f()$, i.e.

$$|F(x_1, x_2, \ldots, x_{m_0}) - f(x_1, x_2, \ldots, x_{m_0})| < \xi \tag{1.5}$$

for all $x_1, x_2, \ldots, x_{m_0}$, belonging to the input space.

This theorem is directly applied to networks BP (multilayered perceptron_) by the following reasons:

1. As an activation function for neurons is used a limited monotonously increasing function "sigmoid" $\varphi(x) = \frac{1}{1+e^{-x}}$
2. Neuron network contains m_0 input nodes and one hidden layer consisting of m_1 neurons.
3. Neuron i of a hidden layer has synaptic weights $w_{i_1}, w_{i_2}, \ldots, w_{i_{m_0}}$ and a threshold b_i.
4. The neuron network output represents linear combination of output signals of hidden layer neurons, weighted with weights $\alpha_1, \alpha_2, \ldots, \alpha_{m_1}$.

So the theorem of universal approximation is the existence theorem, i.e. mathematical proof of possibility of any function approximation by neural network BP even with one hidden layer. But this theorem doesn't give answer how to construct such network. For this problem solution in every practical task for given training sample consisting of inputs X_i and desired outputs $d_i\{(X_1, d_1), (X_2, d_2), \ldots, (X_n, d_n)\}$ it's necessary to carry out the training of the network using one of the training algorithms.

1.3 Gradient Method of Training the Neural Network Back Propagation

The first algorithm that was developed for training Back Propagation network (BP), was a gradient method. Let the criterion of training the network with 3 layers (one hidden layer), be

$$e(w) = \sum_{i=1}^{M} (d_i - y_i(w))^2 \to min \quad (1.6)$$

where d_i is the desired value of the ith output of the neural network, $y_i(w)$—the actual value of the ith output of the neural network BP, for the weight matrix $W = [W^I; W^0]$.

That is, the criterion $e(w)$ is the average squared error of approximation.

Let the activation function for the neurons of the hidden layer be $h_j = \sigma\left(\sum_{i=1}^{N} x_i W_{ij}^I + W_{N+1,j}^I\right)$ and output layer $y_k = \sigma\left(\sum_{j=1}^{I} h_j W_{jk}^0\right)$ be the same and be a function of "SIGMOD"

$$\sigma(x) = \frac{1}{1 + e^{-x}}$$

For such a function the derivative is equal to:

$$\sigma'(x) = \sigma(x) \cdot (1 - \sigma(x))$$

Consider a gradient algorithm for neural network (NN BP) training.

1. Let $W(n)$ be the current values of the weights matrix. The gradient algorithm is as follows:

$$W(n+1) = W(n) - \gamma_{n+1} \nabla_w e(W(n))$$

   where γ_n is the step size at the nth iteration
2. At each iteration, first we train (adjust) input weights.

$$W_{ij}^I(n+1) = W_{ij}^I(n) - \gamma_{n+1} \frac{\partial e(W)}{\partial W_{ij}^I}$$

$$\frac{\partial e(W^I)}{\partial W_{ij}^I} = \sum_{k=1}^{M} -2(d_k - y_k(W^I)) \cdot y_k(W^0) \cdot (1 - y_k(W^0)) \cdot h_j(W) \cdot (1 - h_j(W)) \cdot x_i \quad (1.7)$$

3. Then we train output weights

$$\frac{\partial e(W^0)}{\partial W_{ij}^0} = -2\big(d_k - y_k(W^0)\big)y_k\big(W^0\big)\big(1 - y_k\big(W^0\big)\big)h_j \tag{1.8}$$

$$W_{ij}^0(n+1) = W_{ij}^0(n) - \gamma_{n+1}\frac{\partial e(W)}{\partial W_{ij}^0}$$

where $x_i, \quad i = \overline{1, N+1}$ are inputs of NN BP,
$y_k, \quad k = \overline{1, M}$ are outputs of NN BP,
$h_j, \quad j = \overline{1, J}$—the outputs of the hidden layer.

4. $n := n+1$ and go to the next iteration.
Note: the so-called gradient algorithm with memory is as follows

$$W(n+1) = W(n) - \lambda(1-\alpha)\nabla_W e(W(n)) - \alpha\nabla_W e(W(n-1)), \tag{1.9}$$

where λ—is a training rate, $\alpha \in [0;1]$ is the forgetting coefficient.

The gradient method is the first proposed training algorithm, it is simple to implement, but has the following disadvantages:

- slow convergence;
- finds only a local extremum.

1.4 Gradient Training Algorithm for Networks with an Arbitrary Number of Layers

Recurrent Expression for Calculating of the Error Derivatives of an Arbitrary Layer

Consider the gradient method of training the neural network BP with an arbitrary number of layers N. Let it be required to minimize the criterion:

$$E(w) = \frac{1}{2}\sum_p \sum_j (y_{jp}^{(N)} - d_{jp})^2 \tag{1.10}$$

where $y_{jp}^{(N)}$—is the real output of the jth neuron of the neural network output layer N the at the input of the pth image; d_{jp}—the desired output.

The minimization is performed by stochastic gradient descent, which means the adjustment of the weights as follows:

$$w_{ij}(t) = w_{ij}(t-1) + \Delta w_{ij}$$

$$\Delta w_{ij}^{(n)} = -\eta \frac{\partial E}{\partial w_{ij}} \tag{1.11}$$

where w_{ij} is connection weight of the ith neuron of $(N-1)$- layer to the j—neuron of the Nth layer; $0<\eta<1$—is a step of gradient search, so-called "learning rate".

Further

$$\frac{\partial E}{\partial w_{ij}} = \frac{\partial E}{\partial y_j} \cdot \frac{dy_j}{ds_j} \cdot \frac{\partial s_j}{\partial w_{ij}}, \tag{1.12}$$

where y_j is the output of jth neuron of the nth layer;

s_j—total weighted input signal of the jth neuron.

Obviously

$$s_j = \sum_i w_{ij} y_i^{(n-1)}, \tag{1.13}$$

If the activation function of the jth neuron is $f(\,.\,)$, then

$$y_j = f(s_j) \text{ and } \frac{dy_j}{ds_j} = f'(s_j)$$

In the particular case if f is sigmoid $f = \sigma$, then

$$\frac{dy_j}{ds_j} = f'(s_j) = y_j \cdot (1 - y_j). \tag{1.14}$$

The third factor is $\frac{ds_j}{dw_{ij}} = y_j^{(n-1)}$

As for the first factor $\frac{\partial E}{\partial y_j}$ in (1.12), it can be easily calculated using the outputs of the neurons of the next $(n + 1)$th layer as follows:

$$\frac{\partial E}{\partial y_j} = \sum_k \frac{\partial E}{\partial y_k} \cdot \frac{dy_k}{ds_k} \cdot \frac{\partial s_k}{\partial y_j} = \sum_k \frac{\partial E}{\partial y_k} \cdot \frac{dy_k}{ds_k} \cdot w_{jk}^{(n+1)}, \tag{1.15}$$

Here summation is performed at the neurons of the $(n + 1)$th layer (k).

Introducing a new variable

$$\delta_j^{(n)} = \frac{\partial E}{\partial y_j} \cdot \frac{dy_j}{ds_j}, \tag{1.16}$$

get the recurrence formula for calculating the value $\delta_j^{(n)}$ of the n-th layer using the values $\delta_k^{(n+1)}$ of the next layer (see below—Fig. 1.3)

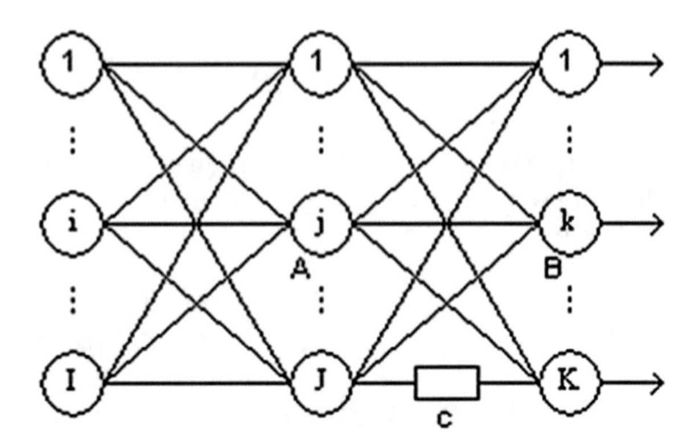

Fig. 1.3 Three successive layers of a neural network (n − 1), n , (n + 1)

$$\delta_j^{(n)} = \sum_{k=1}^{K} \delta_k^{(n+1)} w_{jk}^{(n+1)} \frac{dy_j}{ds_j}. \tag{1.17}$$

For the output layer $n = N$ determine $\delta_i^{(N)}$ directly as

$$\delta_i^{(N)} = \left(y_i^{(N)} - d_i\right) \frac{dy_i}{ds_i}. \tag{1.18}$$

Now the gradient descent algorithm may be written in the following form:

$$\Delta w_{ij}^{(n)} = -\eta \delta_j^{(n)} y_i^{(n-1)}. \tag{1.19}$$

Sometimes to provide to the process of adjusting weights certain inertia to smooth out spikes when moving on the surface the value Δw_{ij} at the previous iteration $t - 1$ is additionally used. In this case the value $\Delta w_{ij}(t)$ is determined:

$$\Delta w_{ij}^{(n)}(t) = \eta(\mu \Delta w_{ij}^{(n)}(t-1) - (1-\mu)\delta_j^{(n)} y_i^{(n-1)}(t)), \tag{1.20}$$

where $\mu \in [0; 1]$

Gradient Training Algorithm for Networks with an Arbitrary Number of Layers
Thus the complete algorithm for training the NN BP using back propagation procedure includes the following steps.

Let one of the possible images. X is entered to NNBP inputs

$$x = \{x_i\}_{i=\overline{1,I}}$$

1. Put $y_i^{(0)} = x_i, i = \overline{1,I}$
2. Calculate sequentially the output values at the nth layer ($n = 1, 2\ldots N$).

$$s_j^{(n)} = \sum_{i=1}^{I} y_i^{(n-1)} w_{ij}^{(n)} \tag{1.21}$$

$$y_j^{(n)} = f(s_j^{(n)}) \tag{1.22}$$

3. Calculate values $\delta_i^{(N)}$ for the neurons of the output layer by the formula (1.18).
4. Define $\Delta w_{jk}^{(N)}$.
5. Using the recurrent formula (1.17), calculate $\delta_k^{(n+1)}$ by $\delta_j^{(n)}$ and $\Delta w_{ij}^{(n)}$ for all the preceding layers $n = N-1, N-2, \ldots, 1$ using formula (1.19).
6. Correct the weights in accordance with the procedure

$$w_{ij}^{(n)}(t) = w_{ij}^{(n)}(t-1) + \Delta w_{ij}^{(n)}(t) \tag{1.23}$$

   In this iteration (t) ends.
7. Calculate $E = E(w(t))$. If $E(w(t)) < \overline{\varepsilon_{giv}^2}$, then STOP. Otherwise go to step 1 of $(t+1)$ iteration.

The algorithm of calculation values $\delta_j^{(n)}$ is illustrated in Figs. 1.4, 1.5, 1.6 and 1.7.

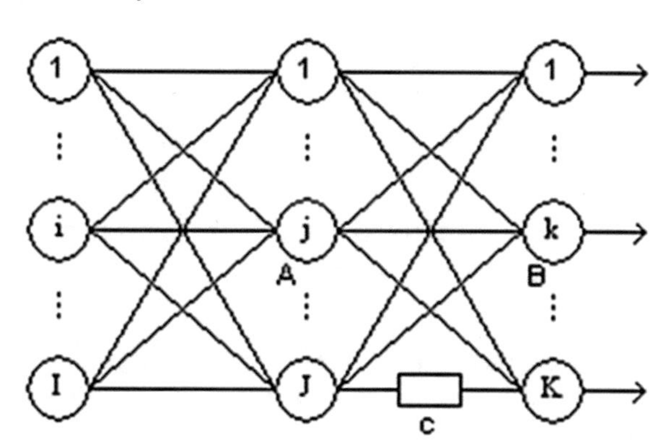

Fig. 1.4 Structure of network BP

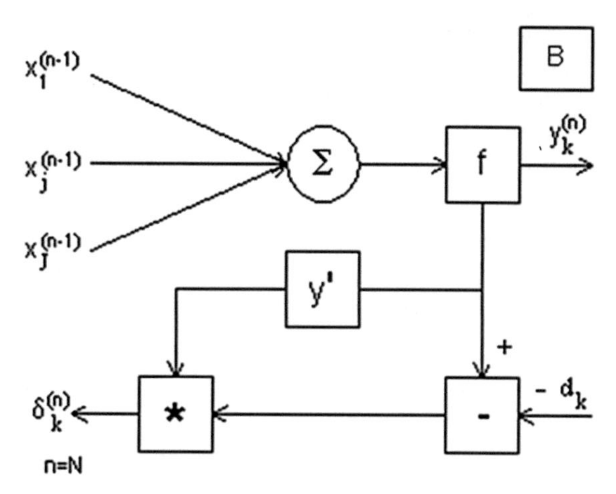

Fig. 1.5 Step 3 of gradient algorithm for last layer n = N

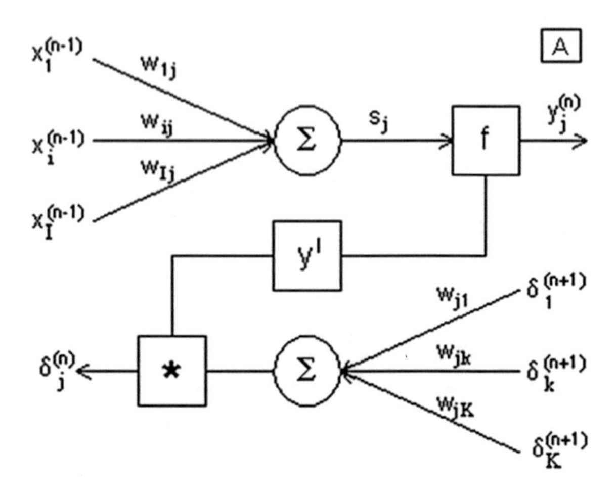

Fig. 1.6 Step 5 of gradient algorithm for layer n

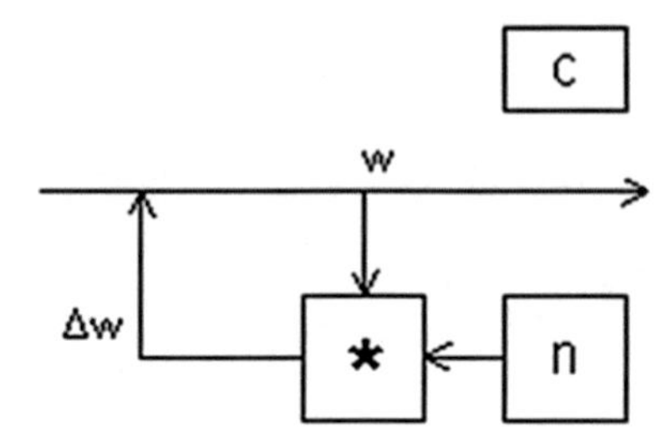

Fig. 1.7 Final step of gradient algorithm calculating w(t)

1.5 The Improved Gradient Training Algorithms for Neural Network BP

During the implementation of the gradient algorithm for training the neural network BP a number of difficulties inherent to gradient-based optimization algorithms may arise:

1. If we are far away from the minimum point of a function $E(w_{ij})$ and move with small step $\eta(t)$, then search process may be delayed. It makes sense to increase the step size. A indicator of this situation is the constant sign $\Delta E(t-1) < 0, \quad \Delta E(t) < 0$ (see Fig. 1.8a).
2. If we are in the neighborhood of the minimum point w_{ij}^*, and the step size η is large, then we may jump over the point w_{ij}^*, after that come back and there the phenomenon of "oscillation" occurs. In this case it is advisable to gradually reduce the step size $\eta(t)$. An indication of this is a change of sign ΔE, i.e. $\Delta E(t-1) < 0, \ \Delta E(t) > 0$ (see Fig. 1.8b).

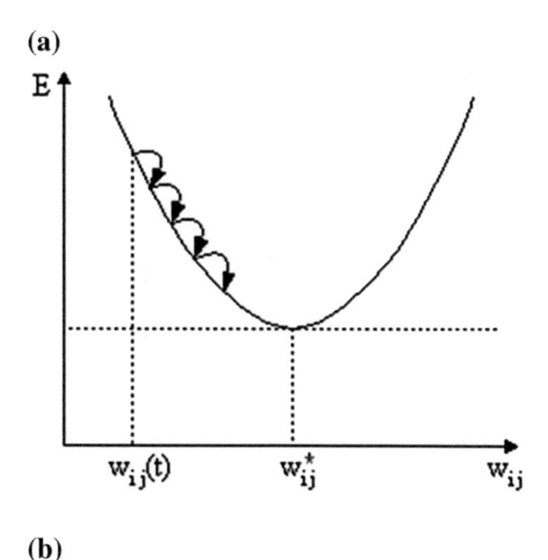

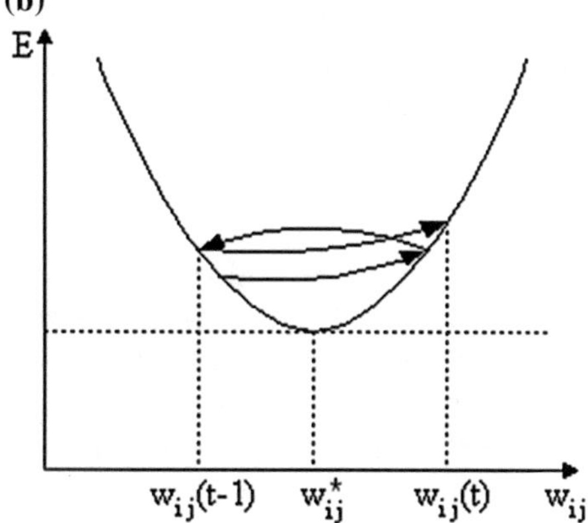

Fig. 1.8 **a** Illustration of algorithm of gradient descent. **b** Illustration of algorithm of gradient descent

Gradient Method with Training Step Correction (Rollback Method)

To overcome the above mentioned difficulties associated with the use of the gradient method, the gradient method with correction of a training step was developed. Here the step size $\eta(t)$ at the $(t+1)$th iteration is described using recurrent expression [1, 2]:

$$\eta(t+1) = \begin{cases} \eta(t) \cdot u, if \;\; \Delta E(w(t))\Delta E(w(t-1)) > 0 \, and \\ \eta(t) \cdot d, otherwise \end{cases} \quad \Delta E(w(t) < 0$$

where $u > 1, \; 0 < d < 1$.

It is recommended to choose $u \cdot d \approx 1$.

The step correction can be performed if several sequential steps were carried out, for example, $t-2$, $t-1$.

Method with the Ejection from Local Minima (Shock BP)

This method is used in the case of a multiextremal dependence $E(w)$ if it's necessary to search the global minimum (see Fig. 1.9).

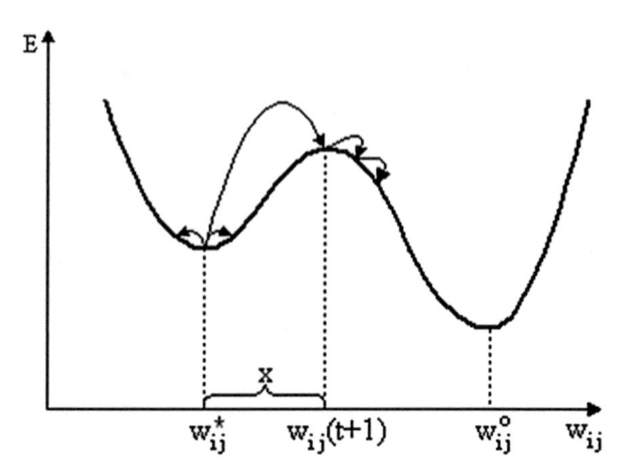

Fig. 1.9 Search of global minimum

If we get stuck in a local minimum w_{ij}^* and the error E for a long time does not change, it makes sense to make a big step in a random direction ξ, to jump out of this shallow and get into the attraction domain of another minimum w_{ij}^*. Then

$$w_{ij}(t+1) = w_{ij}(t) + x\xi$$

where ξ—is random, uniformly distributed in the interval [−1; +1], x is a size of random step. At this point we repeat gradient descend and if we come to another minimum point then make new random step otherwise increase random step as follows: $x_1 = \beta x$, where $\beta > 1$.

Using this procedure we sequentially find local minimal points $\{w_1^*, w_2^*, \ldots, w_m^*\}$ out of which we select the global one.

Training Method with Vector Step (Super SAB)

The main drawback of the classical gradient method is that steps in all directions are the same $\eta(t)$. It does not take into account the fact that different components w_{ij} may be at different distance from the required point minimum (i.e. one component is far away, while others are close).

Therefore Almeida and da Silva [1] have developed a method with vector step, which they called Super SAB. In this method, the search is performed according to expression

$$w_{ij}(t) = w_{ij}(t-1) - \eta_{ij}(t)\delta_j^{(n)} y_j^{(n-1)} \tag{1.24}$$

$$\eta_{ij}(t) = \begin{cases} \eta_{ij}(t-1)\cdot u, \ if \ \frac{\partial E(t)}{\partial w_{ij}} \cdot \frac{\partial E(t-1)}{\partial w_{ij}} > 0. \, and \ \frac{\partial E(t-1)}{\partial w_{ij}} < 0 \\ \eta_{ij}(t-1)\cdot d, \ if \ \frac{\partial E(t)}{\partial w_{ij}} \cdot \frac{\partial E(t-1)}{\partial w_{ij}} < 0 \ and \ \frac{\partial E(t-1)}{\partial w_{ij}} < 0 \ (decrease..of..step) \end{cases}$$

where

$$u > 1, 0 < d < 1, u^* d \cong 1 \tag{1.25}$$

Weights change occurs in accordance with the expression

$$w_{ij}(t) = \begin{cases} w_{ij}(t-1) - \eta_{ij}(t)\delta_j^{(n)} y_j, if \ \frac{\partial E(t)}{\partial w_{ij}} \cdot \frac{\partial E(t-1)}{\partial w_{ij}} > 0 \ and \ \frac{\partial E(t-1)}{\partial w_{ij}} < 0 \\ w_{ij}(t-1), .otherwise \end{cases} \tag{1.26}$$

Autonomous Gradient Method with Approximation of the Topography of a Quadratic Function

Assume that current point is $w(t)$. The value $\nabla E(w(t))$ and antigradient $-\nabla E(w(t))$ are calculated and two trial steps are made:

$$w_1 = w(t) + \nabla E(w)$$

$$w_2 = w(t) - \nabla E(w)$$

Calculate $E(w), E(w_1), E(w_2)$. Further, assuming that $E(w)$ can be approximated by a parabola, find by three points w, w_1, w_2 a minimum point.

The main drawback of the method is that the function $E(w)$ can have a much more complex form, and so this approximation procedure has to be repeated multiple times, which requires large computational costs.

1.6 Acceleration of Convergence of Neural Networks Training Algorithms. The Conjugate Gradient Algorithm

As was shown above, the algorithm for training networks "back propagation" is an implementation of the classical method of steepest descent. This training algorithm is relatively simple to implement and use, which explains its widespread use in the field of neural networks. It has 2 drawbacks:

1. it converges slowly;
2. the method is effective only when finding points of local minimum.

Therefore, were developed more effective training methods, which are alternative to the gradient method: conjugate gradient method and the method based on genetic optimization.

The Method of Conjugate Gradients

The method of conjugate gradients (CG) gives improved convergence rate compared to the method of steepest descent. However, like the method of steepest descent, it is a local optimization method.

In neural networks the objective function (goal) to be minimized is the average error on the whole set of training samples. It is equal to

$$E_{\Sigma}(W) = \sum_{t=1}^{T} \sum_{k=1}^{M} \left(d_{t_k} - y_{t_k}(W) \right)^2 \tag{1.27}$$

where $t = 1, 2, \ldots T$—the set of training samples.

For three-layer network with N input nodes, including the threshold node, J hidden nodes and M output nodes the weight vector W contains $NJ + MJ$ component. In the formula (1.27) M—the number of output nodes; $(d_{t1}, d_{t2}, \ldots d_{tM})$—the desired output for the training sample t, and $y_t(W) = \{y_{t1}(W), y_{t2}(W), \ldots y_{tM}(W)\}$—the response (output signal of the network) on the sample t.

The CG algorithm as well as a more general conjugate directions algorithm is used in the field of optimization with a wide class of problems for which it ensures the convergence to the optimal solution in a finite number of steps. This is a major improvement compared to the method of steepest descent, which requires an infinite number of iterations to find the minimum of the function.

Conjugate Directions

The name of the method comes from the use of conjugate vectors. In a vector space of the dimension D a set of vectors $\{P_1, P_2, \ldots P_D\}$ forms a set of conjugate directions with respect to the matrix A, if [1]

$$P_i A P_j = 0, \text{ for } i \neq j \tag{1.28}$$

where A is a positive denite matrix of the size $D \times D$. Vector satisfying (1.28) are called A-conjugate [1].

The question arises: how does the CG algorithm reach the convergence in a finite number of steps, and on what tasks? Suppose that we need to minimize the function

$$F(W) = (b - AW)^T (b - AW) \tag{1.29}$$

where b and W are D-dimensional vectors, and the matrix $A^{D \times D}$ is positively—defined above. So, we have a quadratic function. Suppose that we seek iteratively optimal vector W^* that minimizes $E(W)$, and start the search from the starting point W_o. Choose a nonzero vector P_1 which serves as the search direction for the next iteration, no matter how W_o and P_1. Define W_1 as the following vector:

$$W_1 = W_0 + \alpha p_1, \tag{1.30}$$

where the scalar α is chosen so as to minimize $E(W_0 + \alpha p_1)$. Now we come to the main idea. The optimal direction in which it is necessary to move at the next iteration is the direction in which you need make only one step straight to the point of optimal solution W^* and it must form A-conjugated pair with the vector p_1.

The optimal direction is therefore $W^* - W_1$, the condition that $(W^* - W_1)$ is A-conjugate direction, is equivalent to the statement that it must satisfy the condition [1]

$$(W^* - W_1)Ap_1 = 0$$

Of course, at this point we do not know the optimal solution W^*, otherwise we wouldn't need any algorithm. at all. However, this condition is important for the following reason. In N-dimensional space, there are exactly N independent vectors that form A conjugate pairs with the vector p_1.

Thus, we need only a finite number of directions to find the optimal solution.

The conjugate direction algorithm systematically constructs the set of A-conjugate vectors. After a maximum of N steps, the algorithm will find the optimal direction, and the convergence will be ensured.

Here we have omitted the important question of determining the one-dimensional problem of minimizing the scalar. For problems in the form Eq. (1.29) such a minimization is performed directly and forms part of the classic SG algorithm, although for more general problems, this task is far from trivial.

In the task of training neural network, Eq. (1.29) does not exist in an explicit form, and in particular we do not have explicit expression for the matrix A, although the gradient of the error ∇E can fulfill this role.

Note that in Eq. (1.29) $-A$ is a factor proportional to the gradient of the function $E(W)$. For such problems in the general form of $E(W)$ the finite steps convergence is no longer guaranteed.

You need to realize that the SG algorithm, similar to the method of gradient descent, finds only locally optimal solutions. Nevertheless, the method gives a significant acceleration of convergence compared with the method of steepest descent.

Description of the Algorithm

1. Set $K = 0$. Initialize the weight vector W and calculate the gradient $G_k = grad\,E(W_k)$. Put an initial direction vector $p_k = -\frac{G_k}{\|G_k\|}$.
2. Find scalar $\alpha > 0$, which minimizes $E(W + \alpha p)$, for which we may use Fibonacci or Golden section algorithms.

$$W(K+1) = W(K) + \alpha p(K)$$

3. If $E(W(K+1)) < \varepsilon$, where ε—is a permissible accuracy of finding the minimum, then STOP. Otherwise compute a new direction:

$$G(k+1) = gradE(W(k+1))$$

4. If $\text{mod}_N(K+1) = 0$, where N is the dimension of the weights space W, then new direction vector is determined as

$$P(k+1) = -\frac{G(k+1)}{\|G(k+1)\|}, \tag{1.31}$$

otherwise put

$$\beta = \frac{G(K+1)^T G(K+1)}{G(K)^T G(K)} \tag{1.32}$$

and calculate the new direction vector $P_{k+1} = \frac{-G(k+1)+\beta p(k)}{\|-G(k+1)+\beta p(k)\|}$

Replace $p(k)$ by $p(k+1)$ and $G(k)$ by $G(k+1)$. Go to step 1 of the next iteration.

1.7 Genetic Algorithm for Neural Network Training

This algorithm is a global optimization algorithm. It uses the following mechanisms [1, 5]:

1. Cross-breeding of parental pairs (cross-over) and generation of descendants;
2. mutation (random disturbances);
3. natural selection of the best descendants.

Let the training goal be minimization of a mean squared error (MSE)

$$E(W) = \frac{1}{M}\sum_{k=1}^{M} (d_k - y_k(w))^2,$$

where $W = [W_I, W_O]$

$$W_1 = \left\| w_{ij}^I \right\|, \quad W_O = \left\| w_{ij}^O \right\|$$

Let the initial population $[W_1(0), \ldots W_i(0), \ldots W_N(0)]$ be given., where N is a population size.

Any specimen i of the N individuals is represented by the corresponding weight matrix W_i.

Calculate the index of suitability (Fitness Index) of the ith specimen and evaluate the quality of prediction (or approximation)

$$FI(W_i) = C - E(W_i) \tag{1.33}$$

where C is a constant.

Crossbreeding of parental pairs. For choosing parents a probabilistic mechanism is used. We determine the probability of selecting the ith parent:

$$P_i = \frac{FI(W_i(0))}{\sum_{i=1}^{N} FI(W_i(0))} \tag{1.34}$$

After selecting the next pair of parents crossing-over occurs. You can apply various mechanisms of crossing. For example: for the first descendant odd components of the vector from the first parent, and the even components of the vector of the second parent are taken and for the second descendant on the contrary—even components of the vector from the first parent, and the odd components of the vector of the second parent. This can be written thus:

$$W_i(0) \oplus W_k(0) = W_i(1) \cup W_k(1)$$

where $W_i = \left[w_{ij}\right]_{j=\overline{1,R}}$.

$$w_{ij}(1) = \begin{cases} w_{ij}(0), & if \;\; j = 2m \\ w_{kj}(0), & if \;\; j = 2m - 1 \end{cases} \quad m = \overline{1, \frac{R}{2}}$$

$$w_{kj}(1) = \begin{cases} w_{kj}(0), & if \;\; j = 2m \\ w_{ij}(0), & if \;\; j = 2m - 1 \end{cases}$$

As a whole, $\frac{N}{2}$ parent pairs are chosen and N descendants are generated.

Then *mutation* noise is acting on descendants as follows:

$$w'_{ij}(n) = w_{ij}(n) + \xi(n),$$

where $\xi(n) = a \cdot e^{-\alpha n}$, $a = const \in [-1; +1]$ is an initial mutation rate, α is a mutation attenuation speed, n is a number of the current iteration.

Selection. Different selection mechanisms can be used, in particular:

1. Replace the old population with the new one.
2. Select N best individuals out of joint population $N_{par} + N_{ch} = 2N$ by the maximum of criterion FI and form the population of the next iteration.

The described iterations are repeated unless the stop condition holds.

As a stop conditions the following variants are used:

1. the number of iterations: $n \geq n_{giv}$ where $n_{giv} = 10^3 - 10^4$ iterations;
2. the achieved value of criterion $\max FI(W_i(n)) \geq FI_{giv}$, where FI_{giv} is a given value.

The main advantage of genetic algorithm (GA) is herein it enables to find global optimum if number of iterations n tends to ∞. But the main disadvantages of the genetic algorithm are following:

1. The number of parameters should be determined experimentally, for example: N—population size; α—attenuation speed of mutations.
2. To implement a genetic algorithm a large amount of computations is demanded.

1.8 Neural Networks with Radial Basis Functions (RBF Networks)

Multilayer neural networks such as Back Propagation carry out the approximation of functions of several variables by converting the set of input variables into the set of output variables [1, 6, 7]. Due to the nature of the sigmoidal activation function this approximation is of global type. As a result of it a neuron, which was once switched on (after exceeding by the total input signal a certain threshold), remains in this state for any value exceeding this threshold. Therefore, every time the value of the function at an arbitrary point of space is achieved by the combined efforts of many neurons, which explains the name of the global approximation.

Another way of mapping input into output is to transform by adapting several single approximating functions to the expected values, and this adaptation is carried out only in a limited area in the multidimensional space. With this approach, the mapping of the entire data set is the sum of local transformations. Taking into account the role played by the hidden neurons, they form the set of basis functions of local type. Since action of each single function (with non-zero values) is recorded only in a limited area of the data space—hence the name is local approximation.

The Essence of Radial Neural Networks

Particular family form a networks with radial basis function, in which hidden neurons implement features, radially varying around the selected center and taking non-zero values only in the vicinity of the center. Such functions defined in the form $\varphi(x)$are called radial basis functions. In such networks, the role of the hidden neuron contains in displaying sphere space around a given single point or a group of points forming a cluster. Superposition of signals from all hidden neurons, which is performed by the output neuron, allows to obtain a mapping of the entire multidimensional space.

The radial network is a natural complement of sigmoidal networks. Sigmoidal neuron is represented in a multidimensional space by a hyper-plane that divides the space into two categories (two classes) in which either of two conditions holds:

$$\text{Either} \quad \sum_i w_{ij} x_i > 0, \quad \text{or} \quad \sum_i w_{ij} x_i < 0.$$

In turn, the radial neuron represents a hyper-sphere, which carries spherical division of the space around a central point. From this point of view it is a natural complement sigmoidal neuron, as in the case of circular symmetry of the data helps reduce the number of neurons required for the separation of different classes.

The simplest neural network radial type operates on the principle of multidimensional interpolation, which consists in the mapping of p different input vectors ($i = 1, 2, \ldots, p$) of the input N-dimensional space into the set of rational numbers p ($p = 1, 2, \ldots, R$). To implement this process, you must use the p hidden neurons of the radial type and ask such a mapping function $F(x)$ for which the condition interpolation fulfills:

$$F(x_i) = d_i \tag{1.35}$$

Use p hidden neurons, connected by links with weights with linear output neurons, means forming output signals of the network by summing the weighted values of the respective basis functions. Consider the radial network with one output and p training pairs $((x_1, d_1), (x_2, d_2), \ldots, (x_p, d_p))$. Let us assume that the coordinates of each of the p centers of the nodes of the network are determined by one of the vectors, i.e. X_i. In this case, the relationship between input and output signals of the network can be determined by the system controls, linear relative weights, which in matrix form looks like

$$w^T \varphi(x) = d. \tag{1.36}$$

Mathematical basis of the radial network work is a *theorem by Cover* [2] about the recognize ability of images. It states that a nonlinear projection of images in a multidimensional space of higher dimension can be linearly separated more likely than their projections in a space with lower dimensionality.

If to denote the vector of radial functions $\varphi(x) = [\varphi_1(x), \varphi_2(x) \ldots \varphi_K(x)]^T$ in N-dimensional input space by $\varphi(x)$ then this space would be split in two spatial class X^+ and X^- if there exists a vector of weights w such that

$$w^T \varphi(x) > 0\,, if\ x \in X^+ \tag{1.37}$$

$$w^T \varphi(x) < 0\,, if\ x \in X^- \tag{1.38}$$

The boundary between these classes is defined by the equation $w^T \varphi(x) = 0$.

In [2] it was proved that every set of images, randomly placed in a multidimensional space, is φ-divided with probability 1, provided correspondingly large dimension k of this space. In practice, this means that the application of a sufficiently large number of hidden neurons that implement the radial function $\varphi_i(x)$ guarantees that the solution of the problem of classification when building only a two-layer network: in which the hidden layer must implement the vector $\varphi(x)$ and the output layer may be composed of a single linear neuron that performs the summation of the output signals from the hidden neurons with weights w.

The simplest neural network of radial type operates on the principle of multi-dimensional interpolation, which consists in the mapping of p different input vectors $(i = 1, 2, \ldots, p)$ of the input N-dimensional space into the set of rational numbers p $(p = 1, 2, \ldots, R)$. To implement this process, you must use the p hidden neurons of the radial type and set such a mapping function $F(x)$ for which the condition interpolation holds

$$F(x_i) = d_i\,. \tag{1.39}$$

Use of p hidden neurons, connected by links with weights with linear output neurons, means forming output signals of the network by summing the weighted values of the respective basis functions. Consider the radial network with one output and p training pairs (x_i, d_i). Let us assume that the coordinates of each of the p centers of the network nodes are defined by one of the vectors x_i, i.e. $c_i = x_i$. In this case, the relationship between input and output signals of the network can be determined by the system of equations, linear related to weights w_i, which in matrix form is written:

$$\begin{bmatrix} \varphi_{11} & \varphi_{12} & \cdots & \varphi_{1p} \\ \varphi_{21} & \varphi_{22} & \cdots & \varphi_{2p} \\ \cdots & \cdots & \cdots & \cdots \\ \varphi_{p1} & \varphi_{p2} & \cdots & \varphi_{pp} \end{bmatrix} \begin{bmatrix} w_1 \\ w_2 \\ \cdots \\ w_p \end{bmatrix} = \begin{bmatrix} d_1 \\ d_2 \\ \cdots \\ d_p \end{bmatrix} \tag{1.40}$$

where $\varphi_{ij} = \left(\left\|x_j - x_i\right\|\right)$ defines a radial function centered at the point x_i, with the input vector x_j. If we denote the matrix elements φ_{ij} as φ and introduce the notation vectors $w = \left[w_1, w_2, \ldots w_p\right]^T$, $d = \left[d_1, d_2, \ldots d_p\right]^T$ then the system of Eq. (1.40) can be represented in the reduced matrix form as follows

$$\varphi w = d \tag{1.41}$$

In [2] it was proved that for the some radial functions in the case $x_1 \neq x_2 \neq \ldots \neq x_p$ quadratic interpolation matrix φ is non-singular and nonnegative definite. Therefore, there exists a solution of Eq. (1.41) in the form

$$w = \varphi^{-1} d \tag{1.42}$$

which allows to obtain the vector of weights of the network output neuron.

The theoretical solution to the problem represented by the expression (1.42) cannot be regarded as absolutely true because of serious limitations in the overall network properties arising from the assumptions done in the beginning. With very large number of training samples and equal number of radial functions the problem from a mathematical point of view becomes infinite (poorly structured), since the number of equations exceeds the number of degrees of freedom of the physical process modeled by Eq. (1.40). This means that the result of such an excessive amount of weight coefficients will be the adaptation of the model to various kinds of noises, or irregularities accompanying the training sample. As a result, interpolating these data hyper-plane will not be smooth, and generalizing capabilities will remain very weak.

That to increase these capabilities, it's ought to decrease the number of radial functions and obtain from excess volume data additional information for the regularization of the problem and improve its conditioning.

Radial Neural Network

Using the decomposition of p basis functions, where p is the number of training samples, it is also unacceptable from a practical point of view, since usually the number of samples is very large, and as a result the computational complexity of the training algorithm becomes excessive. The solution of the system of Eq. (1.40) of dimension $(p \times p)$ for large values of p becomes difficult because a very large matrix (with the exception orthogonal). Therefore, as well as for multi-layer networks, it is necessary to reduce the number of weights, which in this case leads to the reduction of the number of basis functions. So a suboptimal solution in a lower dimension is searched, which accurately approximates the exact solution. If we restrict ourselves to K basic functions, then approximating solution can be represented in the form

$$F(x) = \sum_{i=1}^{K} w_i \varphi(\|x - c_i\|), \tag{1.43}$$

where $K < p$, and $C_i(i = 1, 2, \ldots, K)$ be the set of centers that need to be defined. In special case, if we take $K = p$, we can obtain an exact solution $C_i = X_i$.

The problem of approximation consists in the selection of an appropriate number of radial functions and their parameters, and in the selection of weights w_i $(i = 1, 2, \ldots, K)$ so that the solution of Eq. (1.40) would be the closest to accurate. Therefore the problem of selection of the parameters of the radial functions and the values of the weights w_i of the network can be reduced to minimization of the objective function, which if using the Euclid metric is written in the form

$$E = \sum_{i=1}^{p} \left[\sum_{j=1}^{K} w_j \varphi\left(\|x_i - c_j\|\right) - d_i \right]^2 \tag{1.44}$$

In this equation K is the number of radial neurons, and p is the number of training pairs (x_i, d_i), where x_i is the input vector, d_i—corresponding expected value. Let us denote $d = \left[d_1, d_2, \ldots d_p\right]^T$ the vector of expected values, $w = [w_1, w_2, \ldots w_k]^T$—the vector of network weights, a G—radial matrix is called the matrix of Green [7].

$$G = \begin{bmatrix} \varphi(\|x_1 - c_1\|) & \varphi(\|x_1 - c_2\|) & \cdots & \varphi(\|x_1 - c_K\|) \\ \varphi(\|x_2 - c_1\|) & \varphi(\|x_2 - c_2\|) & \cdots & \varphi(\|x_2 - c_K\|) \\ \cdots & \cdots & \cdots & \cdots \\ \varphi(\|x_p - c_1\|) & \varphi(\|x_p - c_2\|) & \cdots & \varphi(\|x_p - c_K\|) \end{bmatrix} \tag{1.45}$$

When to restrict K basis functions, the matrix G becomes rectangular with the number of rows, usually being much larger than the number of columns ($p \gg K$).

If we assume that the parameters of the radial functions are known, the optimization problem (1.44) reduces to solution systems of equations, linear with respect to the weights w [2]

$$Gw = d. \tag{1.46}$$

Due to the squaredness of the matrix G we can determine the vector of weights w using operations with pseudo-inverse matrix G, i.e.

$$w = G^{+} d, \tag{1.47}$$

where $G^{+} = (G^T G)^{-1} G^T$ denotes pseudoinverse rectangular matrix G. In computational practice, pseudoinverse is calculated by applying the SVD decomposition [2].

In considered up to this point the solution it was used representation of the basis functions with the matrix Green dependent on the Euclidean norm of the vector $\|x\|$. If to take into account the multivariate function may have a different weight on each axis, from a practical point of view, it is useful to specify scaling norm by introducing into the definition of the Euclidean metric weights in a form of matrix Q:

$$\|x\|_Q^2 = (Qx)^T (Qx) = x^T Q^T Qx. \tag{1.48}$$

The scaling matrix for N-dimensional vector x has the form:

$$Q = \begin{bmatrix} Q_{11} & Q_{12} & \cdots & Q_{1N} \\ Q_{21} & Q_{22} & \cdots & Q_{2N} \\ \cdots & \cdots & \cdots & \cdots \\ Q_{N1} & Q_{N2} & \cdots & Q_{NN} \end{bmatrix}$$

In the presentation of matrix $Q^T Q$ by the correlation matrix C in the general case we get:

$$\|x\|_Q^2 = \sum_{i=1}^{N} \sum_{j=1}^{N} C_{ij} x_i x_j. \tag{1.49}$$

If the scaling matrix Q is diagonal, we get $\|x\|_Q^2 = \sum_{i=1}^{N} C_{ij} x_i^2$. This means that the norm of the scaling vector x is calculated according to a standard formula of Euclid, using individual scaling weights for each variable X_i. When $Q = 1$ the weighted Euclid metric reduces to the classical (non-scaled) metric $\|x\|_Q^2 = \|x\|^2$.

Most often as the radial function the Gaussian function is used. When placing its center at the point C_i, it can be specified in abbreviated form as

$$\varphi(x) = \varphi(\|x - c_i\|) = \exp\left(-\frac{\|x - c_i\|^2}{2\sigma_i^2}\right). \tag{1.50}$$

In this expression σ_i—is the parameter, the value of which depends on the width of the function. In the case of Gaussian form of the radial function centered at the point C_i, and weighted scaling matrix Q_i associated with the ith basis function, we obtain a generalized form of Gaussian function [7]

$$\begin{aligned}\varphi(x) = \varphi(\|x - c_i\|_Q) &= \exp\left[-(x - c_i)^T Q_i^T Q_i (x - c_i)\right] \\ &= \exp\left[-\frac{1}{2}(x - c_i)^T C_i (x - c_i)\right],\end{aligned} \tag{1.51}$$

where the matrix $\frac{1}{2}C_i = Q_i^T Q_i$ plays the role of scalar coefficient $\frac{1}{2\sigma_i^2}$ of the standard multivariate Gaussian function, given by (1.50).

A solution is obtained, representing the approximating function in the multidimensional space in the form of a weighted sum of the local radial basis functions (expression (1.50), can be interpreted in a radial neuron network presented in Fig. 1.10 (for simplicity, this network has only one output), which is φ_i determined by the dependence (1.50) or (1.51). The resulting architecture of the radial network has a structure similar to the multilayer sigmoidal networks with one hidden layer. The role of hidden neurons are played by radial basis functions, which differ in their form from the sigmoidal functions. Despite the similarities, these network types are fundamentally different from each other.

The radial network has a fixed structure with one hidden layer and linear output neurons, whereas sigmoidal network may contain a different number of layers and output neurons are linear and nonlinear. Used radial functions may have various

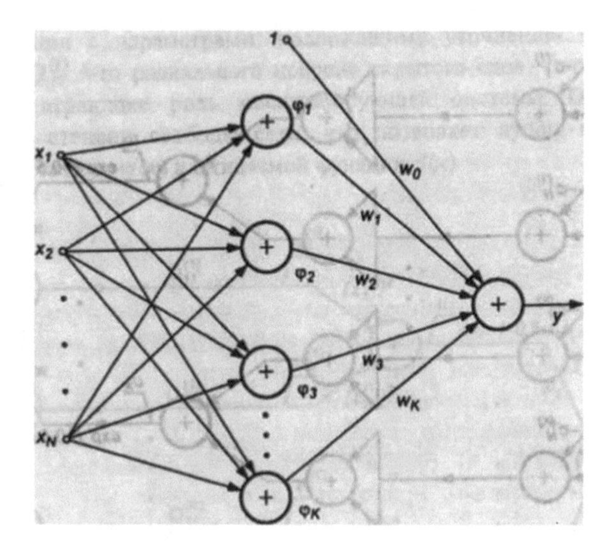

Fig. 1.10 Generalized structure of the RBF networks

structure [2, 7]. Nonlinear radial function of each hidden neuron has its own values for the parameters C_i and σ_i, whereas in sigmoidal networks are used, as a rule, standard activation functions with the same parameter β for all neurons. The argument of the radial function is the Euclidean distance of a sample x from the center C_i, and in a sigmoidal network the scalar product $w^T x$ is used.

Still more large differences between these networks can be noted in a detailed comparison of their structures. Sigmoidal network has a multilayer structure in which an arrangement of neurons is repeated from layer to layer. Every neuron in it sums the signals with subsequent activation. The radial structure of the network is quite different. In Fig. 1.11 shows a detailed diagram of the RBF network structure with the radial function of the form (1.50) in the classical sense of the Euclidean metric. The figure shows that the first layer comprise a non-linear radial function whose parameters (the centers C_i and the coefficients σ_i) are specified in the training process. The first layer does not contain the linear weights unlike sigmoidal network.

Much more complex is the detailed network structure that implements the scaled radial function in the form specified by the expression (1.51). This network is presented in Fig. 1.12, is called HRBF network (Hyper Radial Basis Function) [7]. Radial neuron in it has a particularly complex structure, containing summing signals similar to those used in sigmoidal networks, and exponential activation functions with Parameters to be specified in the training process. The weights $Q_{ij}^{(k)}$ of the Kth radial neuron of the hidden layer are the elements of the matrix $Q(k)$ playing the role of the scaling system. They introduce an additional degree of freedom, so that it allows to bring the output signal of the network $y = f(x)$ closer to the desired function $d(x)$.

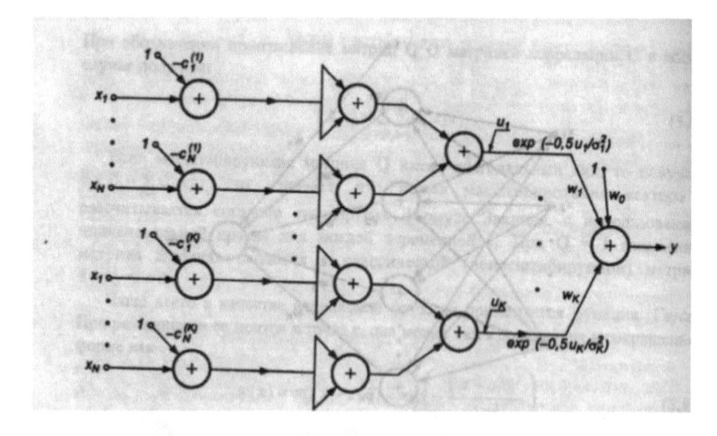

Fig. 1.11 A detailed diagram of the RBF networks structure

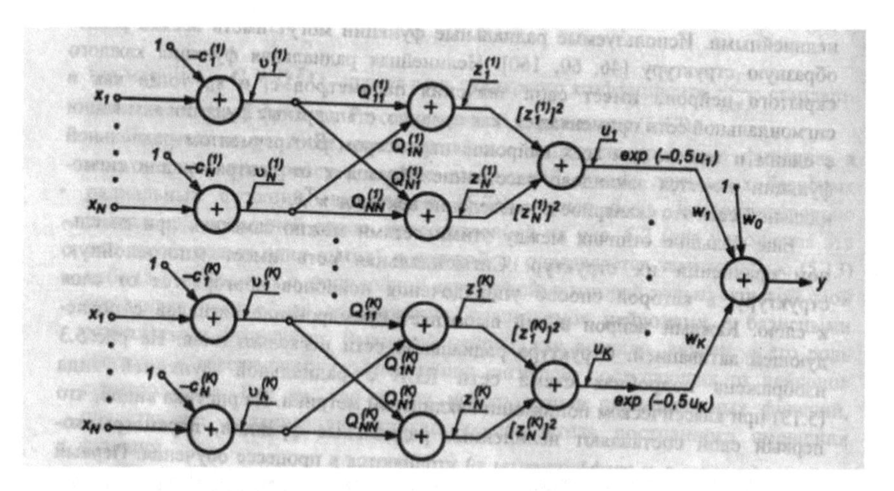

Fig. 1.12 A detailed diagram of HRBF network structure with the scaling matrix Q of arbitrary type

In many practical applications the scaling matrix $Q(k)$ has a diagonal form in which only the elements $Q_{ii}^{(k)}$ have non-zero values. In such a system there is no circular mixing signals corresponding to the various components of the vector x, and the element $Q_{ii}^{(k)}$ plays the role of the individual scaling factor for ith component of the vector X of kth neuron. In Fig. 1.13 a simplified structure of the HRBF network with diagonal matrices $Q(k)$ is presented. It should be noted that in HRBF networks the role of the coefficients σ_i^2 fulfill the elements of the matrix Q which are specified in the training process.

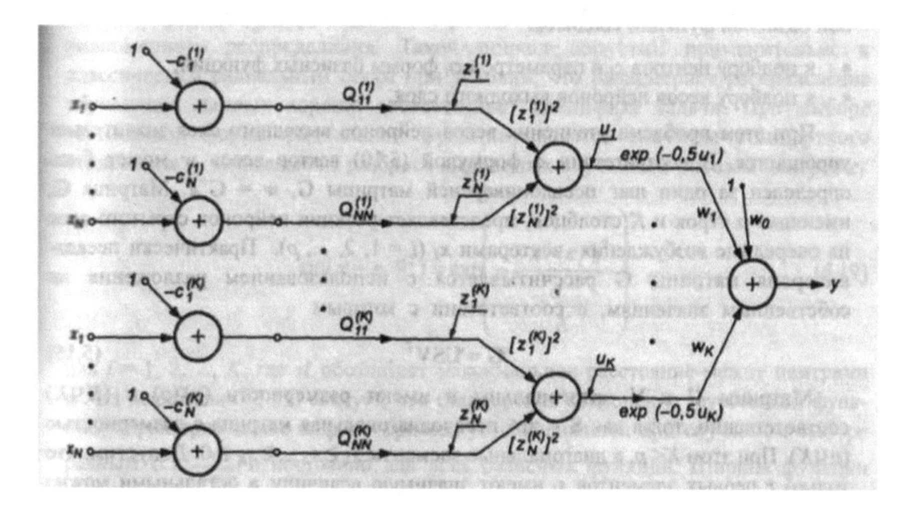

Fig. 1.13 A detailed diagram of of radial HRBF network Q with diagonal scaling matrix structure

Training Methods for Radial Neural Networks

Introduced in the previous subsection, the methods of selection of the weights w_i of the output layer radial RBF network were based on the assumption that the parameters of the basis functions are known, in connection with Green matrix are considered to be defined, and therefore the problem is reduced to the solution of the redundant system of linear equations of the form (1.41). Practically this approach is possible only in a completely unrealistic case, when $K = p$, in which the centers $c_i = x_i$ are known in advance, and the value of the σ_i you can easily determine experimentally with a certain compromise between the monotony and the display precision. In fact, inequality $K \ll p$ always holds, so the training process of RBF network based on the selected type of radial basis function consists of:

- the selection of the centers c_i, and parameters σ_i that form basis functions;
- the selection of the weights of the neurons of the output layer.

The problem of determination the weights of the neurons of the output layer is greatly simplified. In accordance with the formula (1.47) weight vector w can be determined in one step by pseudo-inverse matrix G, $w = G^+ d$.

The matrix G having p rows and K columns, represents the response of neurons in the hidden layer at the next input vector X_i $(i = 1, 2, \ldots, p)$. In practice pseudo-inverse matrix G is calculated using the decomposition by its eigenvalues, according to which [7]

$$G = USV^T, \tag{1.52}$$

Matrices U and V are orthogonal and have the dimensions $(p \times p)$ and $(K \times K)$, respectively, while S is pseudo-diagonal matrix of dimension $(p \times K)$. In this case, $K < p$, and the diagonal elements $s_1 \geq s_2 \geq \ldots \geq s_k \geq 0$. Suppose that only the first r elements of s_i are of significant value, and the others can be neglected. Then the number of columns of the orthogonal matrices U and V can be reduced to r. In this case the obtained reduced matrix U_r and V_r are of the form:

$$U_r = [u_1 u_2 \ldots u_r],$$

$$V_r = [v_1 v_2 \ldots v_r],$$

the matrix s_r becomes diagonal (square $s_r = diag[s_1, s_2, \ldots, s_r]$). This matrix describes the dependence of (1.52) in the form

$$G \cong U_r S_r V_r^T \tag{1.53}$$

Pseudo-inverse to G matrix G^+ is then determined by the expression

$$G^+ = V_r S_r^{-1} U_r^T \tag{1.54}$$

in which $S_r^{-1} = \left[\frac{1}{s_1}, \frac{1}{s_2}, \ldots, \frac{1}{s_r}\right]$, and the vector of network weights, undergoing the training, is defined by the formula

$$w = V_r S^{-1} U_r^T d. \tag{1.55}$$

The advantage of the formula (1.55) is its simplicity. Output network weights are selected in one step by a simple multiplication of the corresponding matrices, some of which (U_r,V_r) are orthogonal and inherently well-ordered. Taking into account the solution of (1.55), which determines the values of the weights of the output layer, the main problem of the training radial networks remains the selection of the parameters of the nonlinear radial functions, especially centers c_i.

One of the simplest, albeit not the most effective way of determining the parameters of the basis functions, is considered to be a random selection. In that decision, the centers c_i of basis functions are chosen randomly based on uniform distribution. This approach is valid as applied to the classical radial networks, provided that a uniform distribution the training data corresponds well to the specifics of the task. When choosing a Gaussian form of the radial function the value of the standard deviations s_i should be given, dependent on location of randomly selected centers c_i

$$\varphi\left(\|x - c_i\|^2\right) = \exp\left(-\frac{\|x - c_i\|^2}{\frac{d^2}{K}}\right) \tag{1.56}$$

for $i = 1, 2, \ldots, K$, where d denotes the maximum distance between the centers c_i. From the expression (1.56) implies that the standard deviation of the Gaussian function, which characterizes the width of the curve is set a random equal to $\sigma = \frac{d}{\sqrt{2K}}$ and constant for all the basis functions. The width function is proportional to the maximum centers divergence and decreases with the increase in their number.

Among the many specialized methods of selection centers some of the most important are the following: self-organizing process of separation into clusters, hybrid algorithm and supervised training algorithms.

The Application of the Self-organization Process for Adaptation the Parameters of Radial Functions

Good results of adjustment the parameters of the radial functions can be obtained by using the algorithm of self-organization. The process of self-organization of the training data automatically splits the space into so-called "Voronoi regions" [7], defining different groups of data. An example of this separation is the two-dimensional space shown in Fig. 1.14. The data are grouped within a cluster represented by the central point, which determines the average value of all its elements. The center of the cluster is to be identified with the center of the corresponding radial functions. For this reason, the number of such functions is equal to the number of clusters and can be adjusted by the algorithm of self-organization.

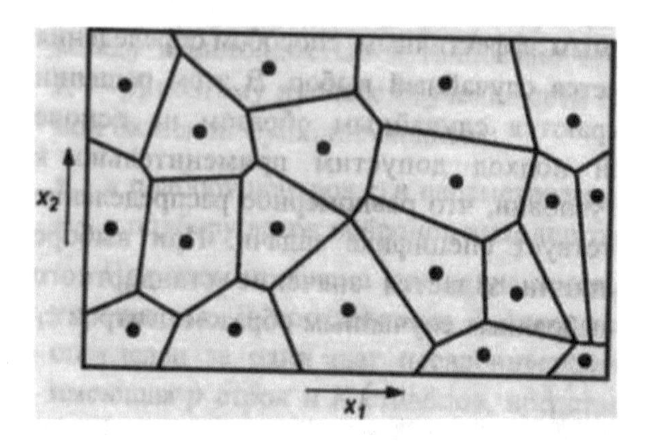

Fig. 1.14 Illustration of the method of dividing the data space into spheres of influence of individual radial functions

Splitting the data into clusters can be done using the version of the algorithm of Linde-Buzo-Grey [2, 7], also called the C-means algorithm (this algorithm is detailly considered in the Chap. 7). In direct (online) version of this algorithm a refinement of the centers is made after the presentation of each next vector X from the set of training data. In the cumulative version (off-line) all centers are specified at the same time after the presentation of all the elements of the set. In both cases, a preliminary selection of centers is often random using a uniform distribution.

If the training data represents a continuous function, the initial values of the centers are primarily placed at points corresponding to all maximum and minimum values of the function. Data about these points and their immediate environment are subsequently removed from the training set and the remaining centers will be uniformly distributed in the area formed by the remaining elements of this set.

In on-line version after the presentation of the kth vector X_k, belonging to the training set, the centre closest to X_k, relatively to the applied metric is selected. This center is subjected to refinement in accordance with the WTA (Winner Takes ALL) algorithm (which is considered in the Chap. 2)

$$c_i(k+1) = c_i(k) + \eta[x_k - c_i(k)], \tag{1.57}$$

where η is a coefficient—learning rate, having a small value (typically $\eta \ll 1$), and decreasing in time. Other centers do not change. All training vectors X are presented several times, usually in a random sequence until the stabilization of their locations.

A variation of the algorithm is also used, according to which the value of the center-the winner is specified in accordance with formula (1.57) while one or more nearest centers are pushed in the opposite direction [2], and this process is implemented according to the expression

$$c_i(k+1) = c_i(k) - \eta[x_k - c_i(k)]. \tag{1.58}$$

This modification of the algorithm allows to alienate centers located close to each other that provides the best survey of the entire data space ($\eta_1 < \eta$).

In cumulative (off-line) versions all training vectors X are presented simultaneously and each of them is mapped to a particular center. The set of vectors assigned to the same center, forms a cluster, a new center which is defined as the average of the corresponding vectors:

$$c_i(k+1) = \frac{1}{N_i} \sum_{j=1}^{N_i} x_j(k). \tag{1.59}$$

In this expression, N is the number of vectors $x(k)$ assigned at the kth cycle to the ith center. The values of all of the centers are determined in parallel. The process of presentation of a set of vectors X and adaptation of centers values is repeated until stabilization of the values of the centers. In practice, most often, we use the direct algorithm, which has slightly better convergence. However, neither of algorithms guarantee convergence to the optimal solution in a global sense, and provides only a local optimization that depends on initial conditions and parameters of the training process.

In poorly chosen initial conditions, some centers may be stuck in the area where the number of training data is negligible small or absent, so the centers modification process will slow down or stop. The way to solve this problem is considered to be simultaneous adjustment of the placement of a large number of centers with fixing the values η for each of them. The center that is closest the to the current vector X is modified most strongly, and the others are modified inversely proportional to their distance to the current vector.

Another approach is to use a weighted distance measure from a center to produce the weighted vector norm which makes the "favorites" those centers that are least probable to be winners. Both approaches do not guarantee a 100 % optimal solution, since they represent actually the procedure, introducing the predefined perturbations into the process of local optimization [7]. The difficulty is also a problem of the selection of training parameter η. When using a constant value η, it should be very small to guarantee the convergence of the algorithm, which excessively increases the training time. Adaptive selection methods allow to make its value dependent on time, i.e. to reduce the growth of the number of iterations. The most known representative of this group is the algorithm of Darken-Moody [7], according to which

$$\eta(k) = \frac{\eta_0}{1 + \frac{k}{T}} \tag{1.60}$$

Coefficient T denotes a time constant, which is adjusted individually for each task. For $k < T$, the value is almost unchanged, but for $k > T$ it gradually decreases to zero. Although adaptive methods of selecting η are more progressive compared to a constant value, they also can not be considered the best solution, especially when modeling dynamic processes.

After fixing the location of centers remains the problem of the selection of the parameter σ_j value corresponding to a particular basis function. Setting σ_j of the radial function affects the shape and size of the coverage area in which the value of this function is non-zero (more precisely, exceeds a certain threshold ε). Selection of σ_j should be performed so that the scope of all radial functions will cover all the space of the input data, and any two zones may overlap only to a small extent. With such selection of σ_j values of the radial network mapping functions will be relatively monotonous.

The easiest way to select the value of σ_j is to accept the Euclidian distance between the *j*th center c_i, and his nearest neighbor [7]. In another algorithm that takes into account the broader neighborhood, on the value of σ_j affects the distance between the *j*th center c_j and its p nearest neighbors. In this case the value of σ_j is determined by the formula

$$\sigma_j = \sqrt{\frac{1}{P}\sum_{k=1}^{P} \left\| c_j - c_k \right\|^2}. \tag{1.61}$$

In practice, the value of p typically lies in the interval [8–10].

When solving any problem the key issue that determines the quality of mapping consists in preliminary selection of a number of radial functions (hidden neurons). As a rule, it's guided by a general principle: the higher is the dimensionality of the vector X, the greater the number of radial functions is necessary to obtain a satisfactory solution. Detailed description of the process of selecting the number of radial functions will be presented in the following subsection.

1.9 Hybrid Training Algorithm for Radial Networks

In the hybrid algorithm, the training process is divided into two stages:

1. selection of a linear network parameters (weights of the output layer) by using the method of pseudo-inverse;
2. adaptation of the nonlinear parameters of the radial functions (center c_i and width σ_i of these functions).

Both stages are closely interconnected. When fixing to current values of the centers and widths of the radial functions (at first this will be the initial value), values of the linear weights of the output layer are selected in one step using SVD decomposition. This fixation of the parameters of the radial functions allows to determine the values of the functions $F_i(x_k)$ $i = 1, 2, \ldots, K$, and $k = 1, 2, \ldots, p$, where i is the number (index) of a radial function, and k is the number of the next training pair (x_k, d_k). Next input signals x_k generate in the hidden layer signals described by vectors $\varphi_k = [1, \varphi_1(x_k), \varphi_2(x_k), \ldots, \varphi_K(x_k)]$, where 1 denotes a unit signal of polarization.

They are accompanied by an output signal of the network y_k, $y_k = \phi_k^T(x)w$, and the vector w contains the weights of the output layer, $w = [w_0, w_1, \ldots, w_K]^T$. In the presence of the p training pairs, we obtain a system of equations

$$\begin{bmatrix} 1 & \varphi_1(x_1) & \ldots & \varphi_K(x_1) \\ 1 & \varphi_1(x_2) & \ldots & \varphi_K(x_2) \\ \ldots & \ldots & \ldots & \ldots \\ 1 & \varphi_1(x_p) & \ldots & \varphi_K(x_p) \end{bmatrix} \begin{bmatrix} w_0 \\ w_1 \\ \ldots \\ w_K \end{bmatrix} = \begin{bmatrix} y_1 \\ y_2 \\ \ldots \\ y_p \end{bmatrix}, \tag{1.62}$$

which in vector form can be written as

$$Gw = y. \tag{1.63}$$

Using a hybrid method at the stage of determination of the output weights vector y is replaced by the vector of expected values $d = [d_1, d_2, \ldots, d_p]^T$ and the system of equations $Gw = d$ is solved in one step using pseudo-inverse

$$w = G^+ d. \tag{1.64}$$

In the algorithm for calculating pseudo-inverse is applied SVD decomposition, which allows to obtain the current value of the vector w in accordance with Eq. (1.64) in one step.

At the second stage, when the values of the output weights are known excitatory signals X are passed through the network until the output layer that enables to calculate the error criterion for sequence of vectors x_k. Then this signal returns back to the hidden layer (back propagation).

For the next consideration assume that there is the network model of HRBF type with a diagonal form of the scaling matrix Q. This means that each radial function is defined in general terms as

$$\varphi_i(x_k) = \exp\left(-\frac{1}{2}u_{ik}\right), \tag{1.65}$$

where the total signal of a neuron u_{ik} is described by the expression

$$u_{ik} = \sum_{j=1}^{N} \frac{(x_{jk} - c_{ij})^2}{\sigma_{ij}^2} \tag{1.66}$$

With availability of p training pairs the goal function can be specified in the form

$$E = \frac{1}{2}\sum_{k=1}^{p} [y_k - d_k]^2 = \frac{1}{2}\sum_{k=1}^{p}\left[\sum_{i=0}^{K} w_i\varphi(x_k) - d_k\right]^2 \tag{1.67}$$

As a result of differentiation of this function we get:

$$\frac{\partial E}{\partial c_{ij}} = \sum_{k=1}^{p} \left[(y_k - d_k) w_i \exp\left(-\frac{1}{2} u_{ik} \right) \frac{(x_{jk} - c_{ij})}{\sigma_{ij}^2} \right], \tag{1.68}$$

$$\frac{\partial E}{\partial \sigma_{ij}} = \sum_{k=1}^{p} \left[(y_k - d_k) w_i \exp\left(-\frac{1}{2} u_{ik} \right) \frac{(x_{jk} - c_{ij})^2}{\sigma_{ij}^3} \right], \tag{1.69}$$

The application of the steepest descent gradient method allows to perform the adaptation of the centers and widths of the radial functions according to the formulae:

$$c_{ij}(n+1) = c_{ij}(n) - \eta \frac{\partial E}{\partial c_{ij}}, \tag{1.70}$$

$$\sigma_{ij}(n+1) = \sigma_{ij}(n) - \eta \frac{\partial E}{\partial \sigma_{ij}}. \tag{1.71}$$

Refinement of non-linear parameters of the radial function completes the second phase (stage) of training. The repetition of both phases leads to a full and quick training network, especially when the initial values of the parameters of the radial functions are chosen close to the optimal.

In practice, the stages in varying degrees affect the adaptation of the parameters. As a rule, algorithm SVD operates faster (it in one step finds a local minimum of the function). To align this imbalance one clarification of linear parameters is usually accompanied by several cycles of adaptation of non-linear parameters.

Example of a Radial Network Application

Neural networks with radial basis functions are used both in solving problems of classification and approximation of functions, in the prediction, i.e. are applied in those areas in which sigmoidal networks have already won solid positions for many years. They perform the same functions as sigmoidal networks, but implement different data processing methods associated with the local mappings. Due to this feature a significant simplification takes place and, therefore, acceleration of the training process.

As an example, consider a three-dimensional function approximation, which is described by the formula [7]

$$\begin{aligned} d = f(x_1, x_2) = & 3(1 - x_1)^2 \exp(-x_1^2 - (x_2 + 1)^2) - \\ & - 10\left(\frac{x_1}{5} - x_1^3 - x_2^5\right) \exp(-x_1^2 - x_2^2) - \\ & - \frac{1}{3} \exp(-(x_1 + 1)^2 - x_2^2) \end{aligned}$$

Let the variables vary within $-3 \le x_1 \le 3$ and $-3 \le x_2 \le 3$. The training data are uniformly distributed in the areas of definitions of variables x_1 and x_2.

In total, for training 625 training samples in the form of data pairs $([x_1, x_2], d)$ were used. To solve the problem a radial network with the structure 2-36-1 was built (2 input x_1 and x_2, respectively, 36 radial neurons of Gaussian type and one output linear neuron, The corresponding d-value functions). The hybrid training algorithm with random initial values of the network parameters was applied. In Fig. 1.15 a chart of the network training (the curve of change of the error with increasing number of iterations) is presented. The chart shows that progress in reducing error is quite large, especially in the early stages of the process (Fig. 1.15).

In Fig. 1.16 a graphical representation of the restored function f(x_1, x_2) and error recovery data are shown(the training network). The maximum approximation error did not exceed the level of 0.06, which is about 1 % of the expected values. Comparing the speed of training and generalizing abilities of the radial network with similar indicators of multilayer perceptron is unanimous in favor of the first one. It trains faster and is much less sensitive to the choice of initial parameter values for basis functions and weights of the output neuron.

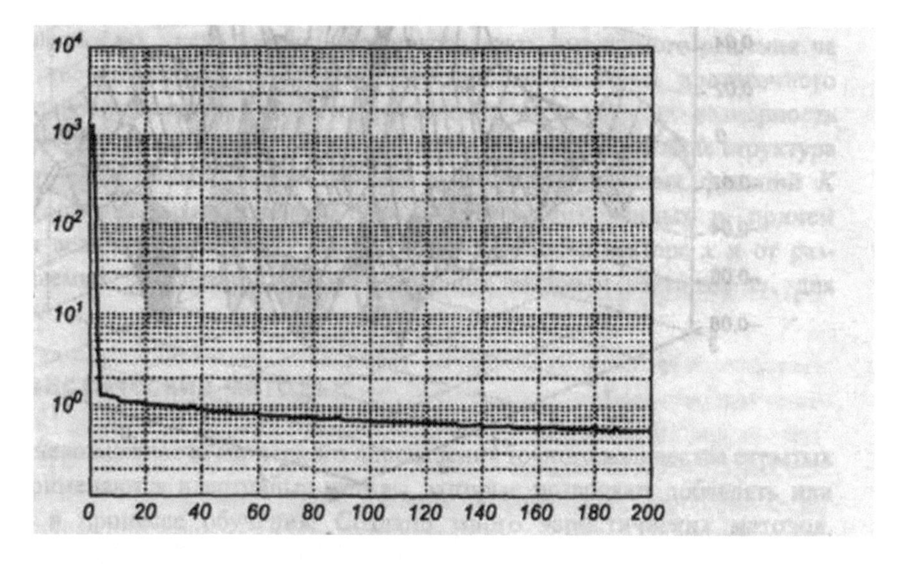

Fig. 1.15 Training curve of radial RBF network for recovery of three-dimensional function

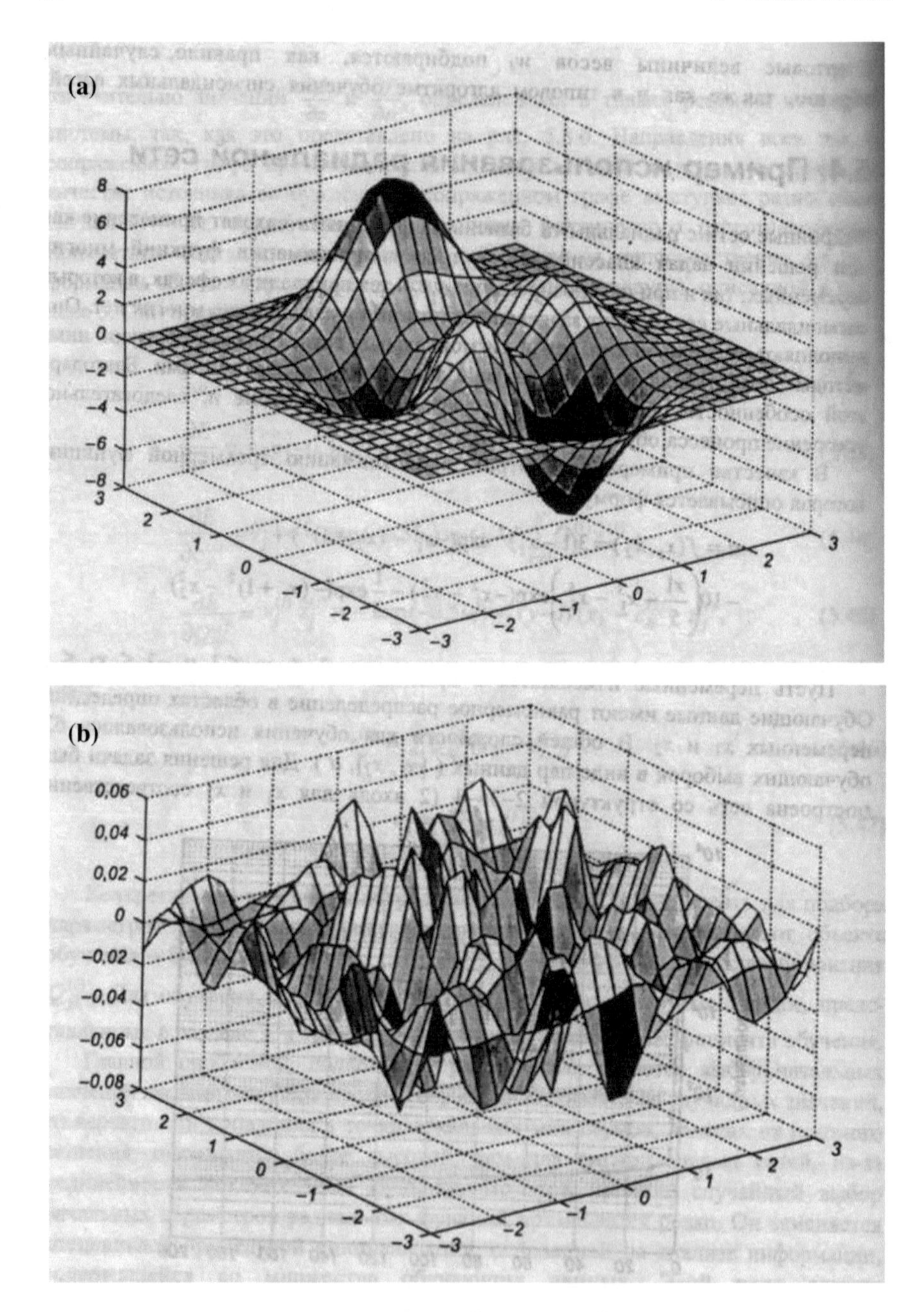

Fig. 1.16 The results of reconstruction of three-dimensional radial functions of the RBF network: **a** the recovered surface; **b** error of recovery

1.10 Methods of Selecting the Number of Radial Basis Functions

The selection of the number of basis functions, each of which corresponds to one hidden neuron, is considered the main problem that occurs when searching solution to the problem of approximation. Like sigmoidal networks, too small number of neurons doesn't allow to reduce sufficiently the generalization error of the training data set, whereas a too large number of them increases the output error on the set of testing data. Selection of the necessary and sufficient number of neurons depends on many factors, including the dimensionality of the problem, the training data size and primarily spatial structure of the approximated function. Typically, the number of basis functions K constitutes a certain percentage of the volume of the training data p, and the actual value of this fraction depends on the dimension of the vector x and the variability in the expected value d_i corresponding to the input vectors for $i = 1, 2, \ldots, p$.

Heuristic Methods

Due to the impossibility of a priori determination of the exact number of hidden neurons adaptive methods are used that allow you to add or delete them in the training process. Many heuristic methods were developed that implement these operations [2, 7]. As a rule, the training of the network starts with any initially chosen the number of neurons, and subsequently the degree of reduction of the RMS error and the change of network parameters values are controlled. If the average change of weight values after a certain number of training cycles is too small: $\sum_i \langle \Delta w_i \rangle < \xi$, two basis functions are added (2 neurons) with centers corresponding to the largest and smallest adaptation error, and then the training of the extended structure continues. At the same time the absolute w_i values of the weights of all the individual neurons are controlled. If they are less than the firstly set threshold δ, the corresponding neurons are subject to removal from the network. Adding neurons and their removal as well begins after run of a certain number of training cycles and can occur throughout the training process until the desired accuracy of the approximation will be achieved.

Another approach to controlling the number of hidden neurons has been proposed by sir Platt [7]. This is a technique that combines elements of self-organization and training. After presentation of each next training sample the Euclidian distance between it and the center of the nearest existing radial functions is determined. If this distance exceeds the threshold $\delta(k)$, a new radial function is generated (i.e. neuron is added), after which the network is subjected to the standard procedure for training using gradient methods (supervised training). The process of adding neurons is continued until the desired level of error is achieved. Crucial for this method is the selection of values $\delta(k)$, according to which a decision is taken to expand the network. Usually $\delta(k)$ decreases exponentially with time (depending on the number of iterations) from the value of $\delta_{\max}$ at the beginning of the process to $\delta_{\min}$ at the end of it. The disadvantage of this approach is the impossibility of reducing the number of neurons in processing information, even when as a result of training some of them degenerate (due to unsuccessful placement centers) or when several neurons begin to duplicate each other, performing the same function. In addition, this method is very sensitive to the selection of the parameters of the training process, especially the values of δ_{max} and δ_{min}.

Comparison of Radial and Sigmoidal Networks

Radial neural networks belong to the same category of networks trained with the teacher as the multilayer perceptron networks. Compared to multi-layer networks with sigmoidal activation functions, they differ in some specific properties which facilitate the approximation characteristics of the simulated process.

Sigmoidal network in which a non-zero value of the sigmoid function extends from some point in space to infinity, solves the problem of the global approximation of the given function. At the same time the radial network based on the functions that have non-zero values only in a certain small area around their centers, implements a local approximation type, the scope of which, as a rule, is more limited. Therefore it is necessary to understand that *generalizing ability of radial networks is slightly worse than sigmoidal networks*, especially on the training data.

Because of the global nature of the sigmoidal function a multi-layer network does not have a built-in mechanism for identifying a data region to which a specific neuron most strongly reacts. Due to the physical impossibility to associate the activity of a neuron with the corresponding area of the training data it's difficult for sigmoidal networks to determine the initial position of the training process. Taking into account poly-modality of an objective function, the achievement of the global minimum in such a situation becomes extremely difficult even with the most advanced training methods.

Radial network solve this problem much better. The most commonly used in practice, the radial functions of the Gaussian type are inherently local in nature and take non-zero values only in the area around a certain center. This makes it easy to determine the dependence between the parameters of the basis functions and the spatial location of the training data in multidimensional space. Therefore it's relatively easy to find a satisfactory initial conditions of the training process with the teacher. The use of such algorithms with initial conditions close to optimal, greatly increases the likelihood of achieving a success with the application of radial networks.

It is believed [2] that the radial networks solve better than sigmoidal networks the classification problems, detection of damages in various systems, pattern recognition tasks, etc.

An important advantage of radial networks is greatly simplified training algorithms. If there is only one hidden layer and the close relationship of the activity of the neuron with the corresponding region of the training data location the starting point of training may be chosen much closer to the optimal solution than in multi-layer networks. It is also possible to separate the stage of selection of parameters of the basis functions from the selection of the values of network weights (hybrid algorithm), which greatly simplifies and speeds up the training process. Time advantage becomes even greater if we take into account the procedure of forming the optimal (from the point of view of ability to generalize) network structure. While for multilayer networks it is a very time consuming task that requires, as a rule, multiple repetition of training. For radial networks, especially based on orthogonalization, the formation of the optimal structure of a network is a natural step in the training process, requiring no additional effort.

References

1. Zaychenko, Yu.P.: Fundamentals of Intellectual Systems Design, p. 352. Kiev Publishing House “Slovo” (2004) (rus)
2. Simon, H.: Neural Networks. Full Course. 2nd edn., p. 1104 (Transl. engl.). Moscow Publishing House “Williams” (2006) (rus)
3. Anderson, J., Rosenfeld, E. (eds.): Neurocomputing: Foundations of Research. MIT Press, Cambridge, MA (1988)
4. Hornik, K., Stinchcombe M., White, H.: Multilayer feed forward networks are universal approximators. Neural Networks **2**, 359–366 (1989)
5. Voronovsky, G.K., Makhotilo, K.V., Petrashev, S.N., Sergeyev, S.A.: Genetic Algorithms-Artificial Neural Networks and Problems of Virtual Reality, p. 212. Kharkov Publishing House “Osnova” (1997) (rus)
6. Korotky, S.: Neural networks: Self-learning. http://www.orc.ru (1999)
7. Osovsky, S.: Neural Networks for Information Processing. Finance and Statistics, p. 344 (transl. from pol.). M.: Publishing House (2002) (rus)
8. Batischev, D.I.: Genetic algorithms for extremal problems solution. In: Teaching Manual, p. 65, Voronezh state Technical University, Voronezh (1995) (rus)
9. Batischev, D.I., Isayev, S.A., Remer, E.K.: Evolutionary-genetic approach for non- convex problem optimization. In: Inter-High School Scientific Works Collection Simulation Optimization in Automated Systems, Voronezh State Technical University, Voronezh, pp. 20–28 (1998) (rus)
10. Batyrshyn, I.Z.: Methods of presentation and processing of fuzzy information in intellectual systems. In: Scientific Review in: News in Artificial Intelligence. Publishing RAAI Anaharsis. №2, pp. 9–65 (1996) (rus)

Chapter 2
Neural Networks with Feedback and Self-organization

In this chapter another classes of neural networks are considered as compared with feed-forward NN—NN with back feed and with self-organization.

In Sect. 2.1 recurrent neural network of Hopfield is considered, its structure and properties are described. The method of calculation of Hopfield network weights is presented and its properties considered and analyzed. The results of experimental investigations for application Hopfield network for letters recognition under high level of noise are described and discussed. In the Sect. 2.2 Hamming neural network is presented, its structure and properties are considered, algorithm of weights adjusting is described. The experimental investigations of Hopfield and Hamming networks in the problem of characters recognition under different level of noise are presented. In the Sect. 2.3 so-called self-organizing networks are considered. At the beginning Hebb learning law for neural networks is described. The essence of competitive learning is considered. NN with self-organization by Kohonen are described. The basic competitive algorithm of Kohonen is considered/its properties are analyzed. Modifications of basic Kohonen algorithm are described and analyzed. The modified competitive algorithm with neighborhood function is described. In the Sect. 2.4 different applications of Kohonen neural networks are considered: algorithm of neural gas, self-organizing feature maps (SOMs), algorithms of their construction and applications.

2.1 Neural Network of Hopfield

Revival of interest in neural networks is connected with Hopfield's work (1982) [1]. This work shed light on that circumstance that neuron networks can be used for the computing purposes. Researchers of many scientific fields received incentive for further researches of these networks; pursuing thus the double aim: the best understanding of how the brain works and application of properties of these networks.

M.Z. Zgurovsky and Y.P. Zaychenko, *The Fundamentals of Computational Intelligence: System Approach*, Studies in Computational Intelligence 652,
DOI 10.1007/978-3-319-35162-9_2

2.1.1 Idea of Reccurrency

The neural network of Hopfield is an example of a network which can be defined as a dynamic system with feedback at which the output of one completely direct operation serves as an input of the following operation of a network, as shown in Fig. 2.1.

Networks which work as systems with feedback, are called "recurrent networks". Each direct operation of a network is called an iteration. Recurrent networks, like any other nonlinear dynamic systems, are capable to show the whole variety of different behavior. In particular, one possible pattern of behavior is that the system can be stable, i.e. it can converge to the only fixed (motionless) point.

When the motionless point is an input to such dynamic system, at the output we will have the same point. Thus the system remains fixed at the same state. Periodic cycles or chaotic behavior are also possible.

It was shown that Hopfield's networks are stable. In general case it may be more than one fixed point. That depends on the starting point chosen for initial iteration to which fixed point a network will converge.

Motionless points are called as **attractors.** The set of points (vectors) which are attracted to a certain attractor in the course of iterations of a network, is called as "attraction area" of this attractor. The set of motionless points of Hopfield's network is its **memory**. In this case the network can work as associative memory. Those input vectors which get to the sphere of an attraction of a separate attractor, are connected (associated) with it.

For example, the attractor can be some desirable image. The area of an attraction can consist of noisy or incomplete versions of this image. There is a hope that images which vaguely remind a desirable image will be remembered by a network as associated with this image.

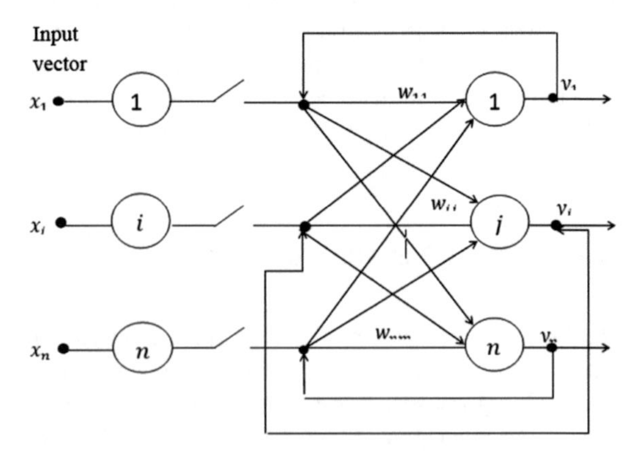

Fig. 2.1 Binary network of Hopfield

2.1.2 *Binary Networks of Hopfield*

In Fig. 2.1 the binary network of Hopfield is represented. Input and output vectors consist of "–1" and "+1" (instead of "–1", "0" can be used). There is a symmetric weight matrix $W = \|w_{ij}\|$ consisting of integers with zeros (or "–1") on a diagonal. The input vector X is multiplied by a weight matrix, using usual matrix and vector multiplication. The input vector X is fed to corresponding neurons and the output vector is determined. However, only 1 component of an output vector $Y = [y_j]$ is used on each iteration. This procedure is known as **"asynchronous correction"**. This component which can be chosen incidentally or by turn enters to a threshold element, whose output is –1, or +1). Corresponding component of an input vector is replaced with this value and, thus, forms an input vector for the following iteration. Process proceeds until input and output vectors become identical (that is, the motionless point will be reached) This algorithm is described below.

2.1.3 *Description of the Algorithm of Asynchronous Correction*

At the first moment a key k is closed (see Fig. 2.1) so an input vector x is fed with weight $\|w_{ij}\|$ to input neurons and the total signal at the input of *j*th neuron $S_j(x)$ is defined. Further, the key k is disconnected and outputs of neurons are fed to their inputs. The following operations are to be made:

- Calculate components of an output vector y_j, $j = 1, 2, \ldots$, n, using the formula

$$y_j = T\left(\sum_{i=1}^{n} w_{ij}x_i\right) \tag{2.1}$$

  where

$$T(x) = \begin{cases} -1, & if\ x<0 \\ 1, & if\ x>0 \\ y\ is\ not\ changed, & if\ x=0 \end{cases}$$

- To execute asynchronous correction, i.e. [2–4]:

Step 1: start with an input vector $(\boldsymbol{x}_1, \boldsymbol{x}_2, \ldots, \boldsymbol{x}_n)$.
Step 2: find y_j according to the formula (2.1).
Step 3: replace $(\boldsymbol{x}_1, \boldsymbol{x}_2, \ldots, \boldsymbol{x}_n)$ with $(\boldsymbol{y}_1, \boldsymbol{x}_1, \boldsymbol{x}_2, \ldots, \boldsymbol{x}_n) = Y$ and a feed Y back to input X.
Step 4: repeat process to find y_2, y_3, etc. and replace the corresponding inputs.

Repeat steps 2–3 until the vector: $Y = (y_1, y_2, \ldots, y_n)$ ceases to change. It was proved each such step reduces the value of communications energy E if at least one of outputs has changed:

$$E' = \frac{1}{2}\sum_{i=1}^{n}\sum_{j=1}^{n} w_{ij}x_i x_j, \tag{2.2}$$

so convergence to a motionless point (attractor) is provided.

Asynchronous correction and zeros on a diagonal of a matrix W guarantee that power function (2.2) will decrease with each iteration [2, 5]. Asynchronous correction is especially essential to ensuring convergence to a motionless point. If we allow whole vector to be corrected on each iteration, it is possible to receive a network with periodic cycles as terminal states of an attractor, but not with motionless points.

2.1.4 Patterns of Behavior of Hopfield's Network

Weight matrix distinguishes behavior of one Hopfield's network from another so there is a question: "How to define this weight matrix?"

The answer is it should be given a set of certain weight vectors which are called **etalons**. There is a hope that these etalons will be the fixed points of a resultant Hopfield's network, though it is not always so. In order to ensure these etalons to be attractors, the weight matrix $W = \|w_{ij}\|$ should be calculated so [5]:

$$w_{ij} = \begin{matrix} \sum_{k=1}^{N} (x_{ki} - 1)(x_{kj} - 1), \ \mathit{if}\ i \neq j \\ 0, \ \mathit{if}\ i = j, \end{matrix} \tag{2.3}$$

where N is a number of the etalons, X_k is the kth etalon.

If etalon vectors form a set of orthogonal vectors, it is possible to guarantee that if the weight matrix is determined as shown above in formula (2.3), each etalon vector will be a motionless point. However, generally in order that etalons become motionless points, orthogonality isn't obligatory.

It should be noted that Holfield network weights aren't trained like BP or RBF networks but are calculated in accordance with formula (2.3).

2.1.5 Application of Hopfield's Network

Hopfield's network can be used in particular for images recognition. But the number of the recognizable images isn't too great owing to limitation of memory of Hopfield's networks. Some results of its work are presented below in the experiment of adjusting network to recognize letters of the Russian alphabet [2].

Initial images are:

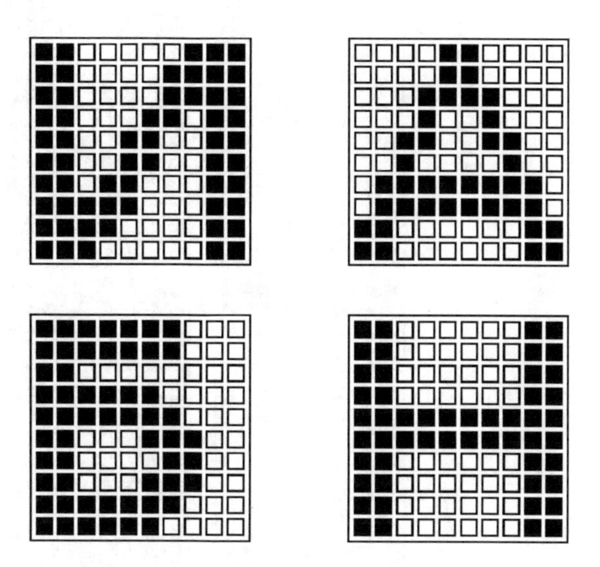

Research was conducted thus: consistently increasing noise level each of 4 images, they are fed to inputs of Hopfield's network. Results of network functioning are given in Table 2.1.

Thus, Hopfield's neural network perfectly copes with a problem of recognition of images(pattern) in experiments with distortion of 0–40 %. In this range all images(pattern) were recognized without mistakes (sometimes there are insignificant distortions for 40 % level of noise).

At 45–60 % level of noise images(patterns) are recognized unstably, often there is "entangling" and at the neural network output appears absolutely other image (pattern) or its negative.

Beginning from 60 % noise level at the system output the negative of the tested image(pattern) appears which sometimes is partially distorted (starts appearing at 60–70 %).

Table 2.1 Experiments with Hopfield network

The tested image	Percent of image distortion	View of the distorted image	Result of recognition
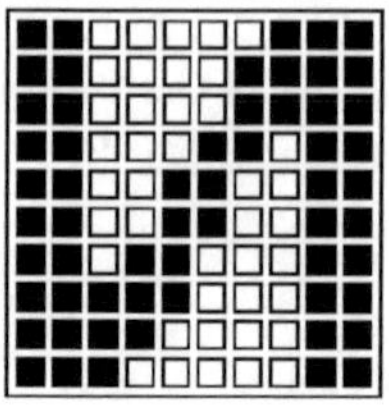	10%	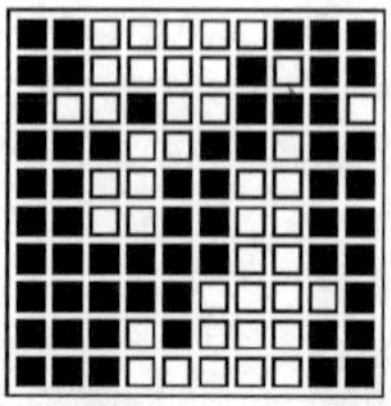	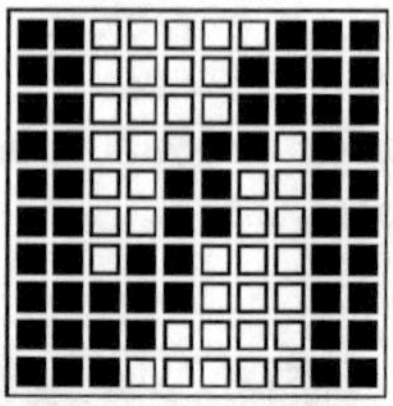
	20%	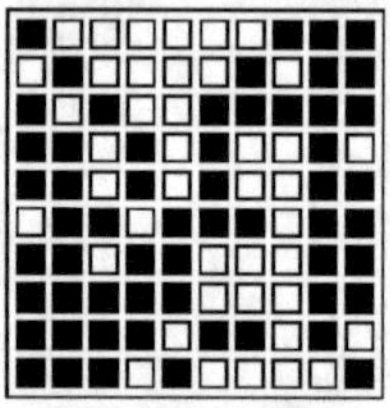	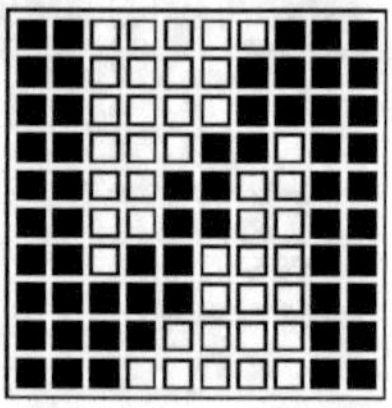
	30%	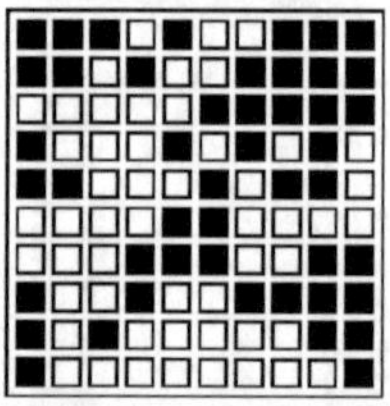	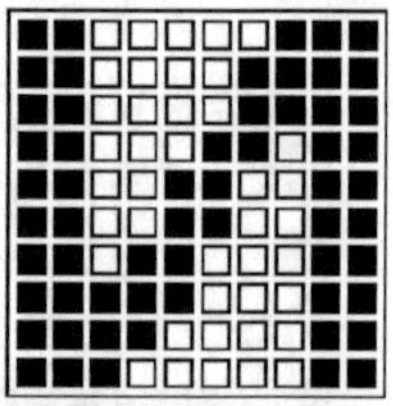
	35%	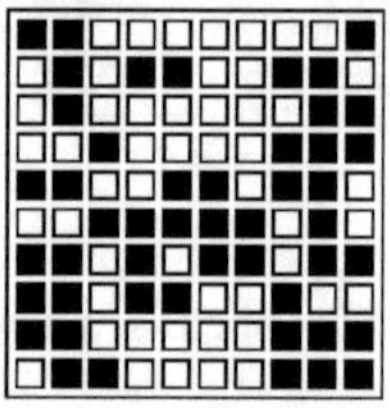	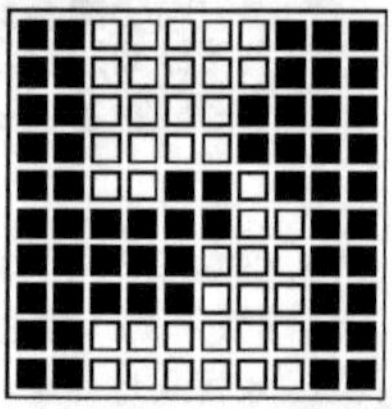
	40%	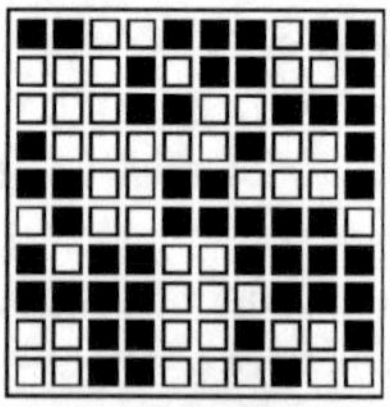	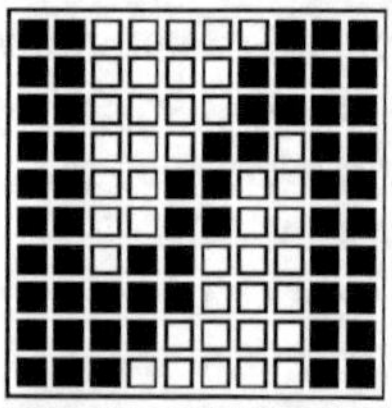
	45%	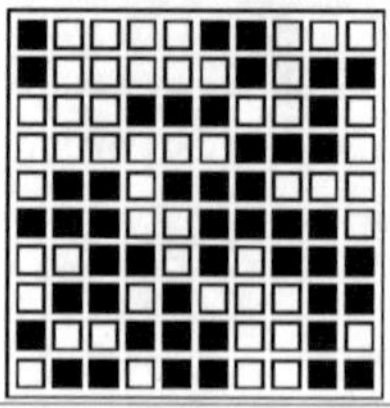	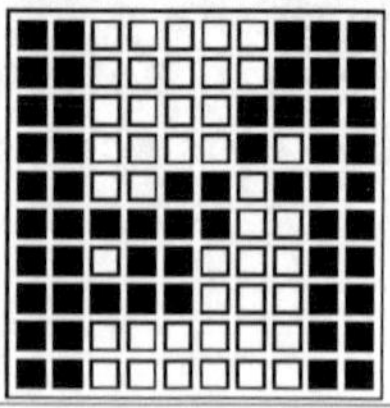

(continued)

Table 2.1 (continued)

50%	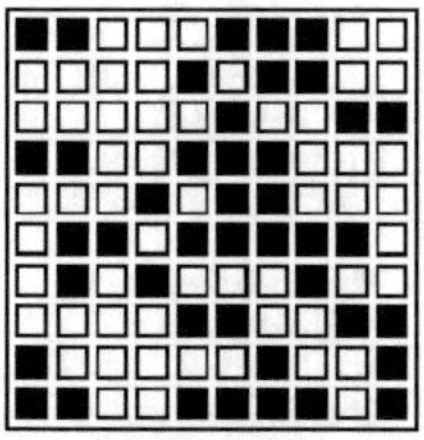	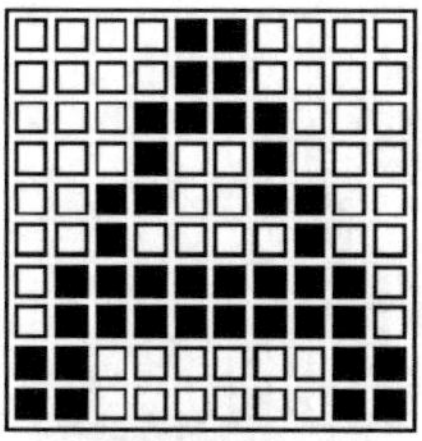
60%	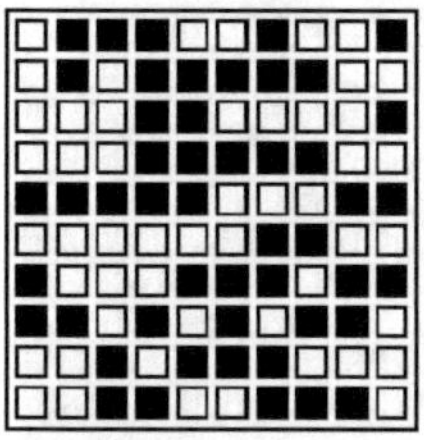	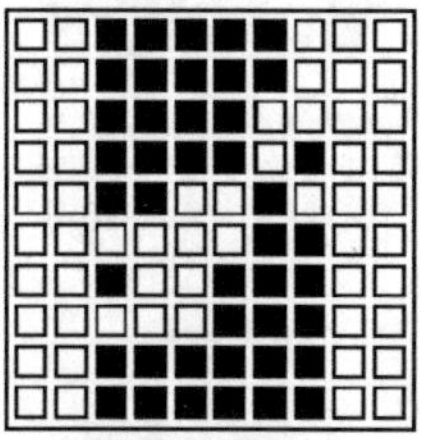
70%	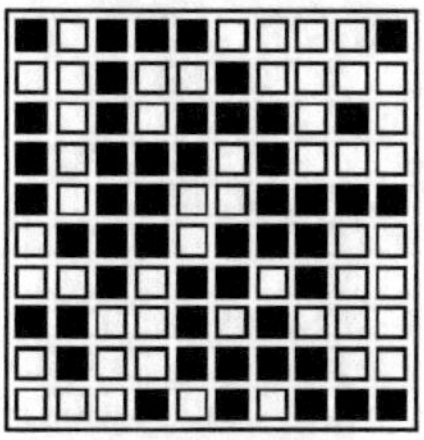	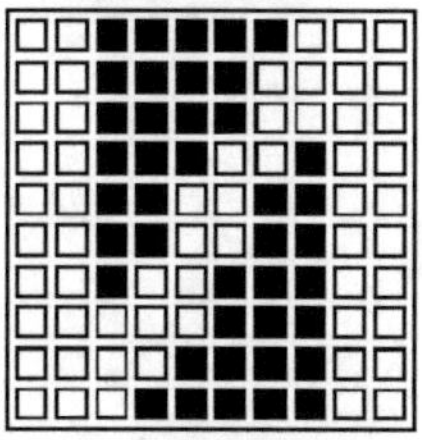
80%	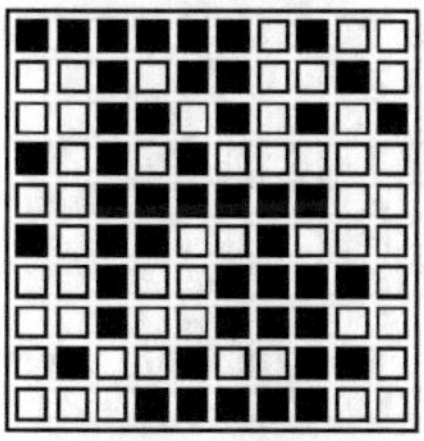	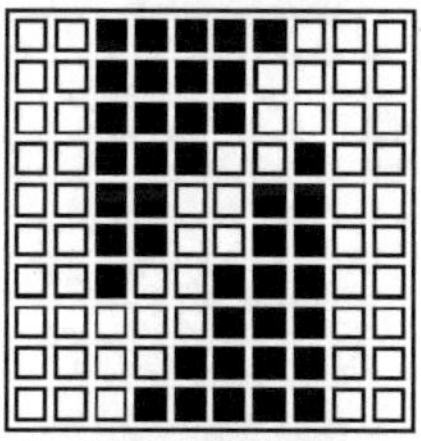
90%	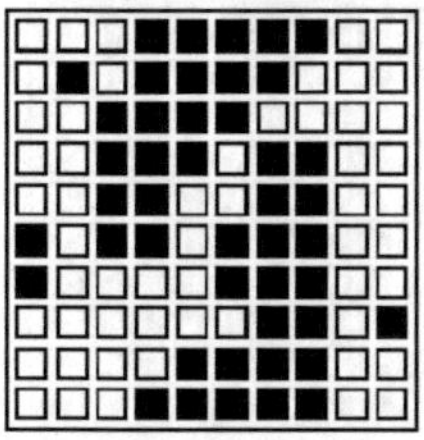	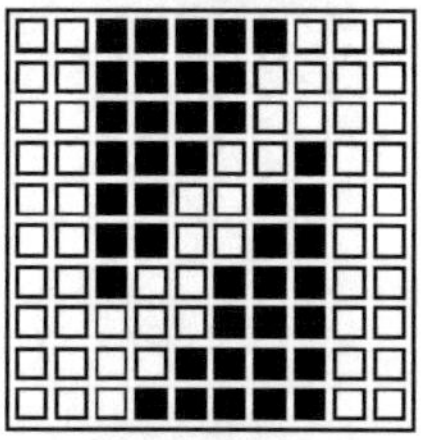
100%	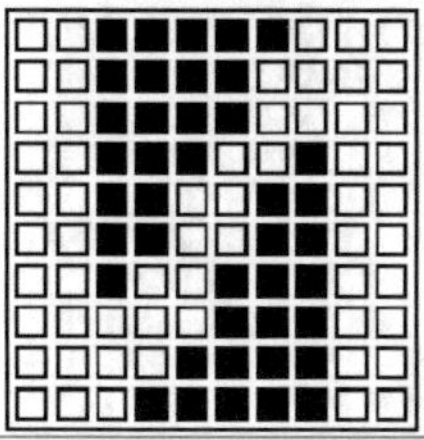	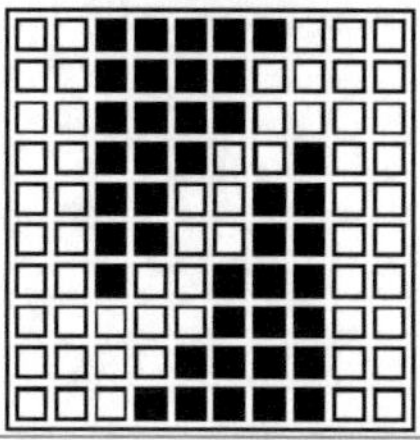

(continued)

Table 2.1 (continued)

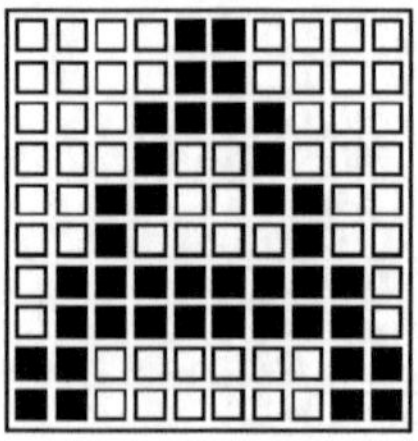	10%	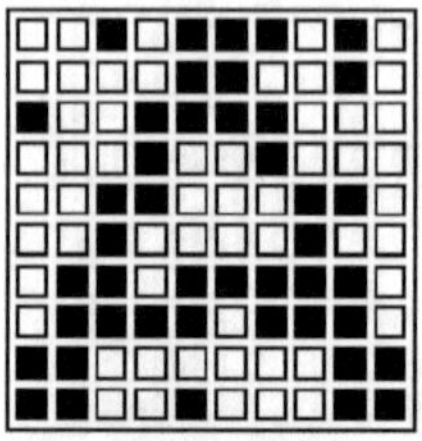	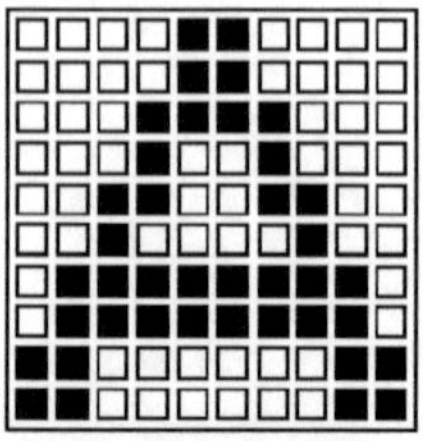
	20%	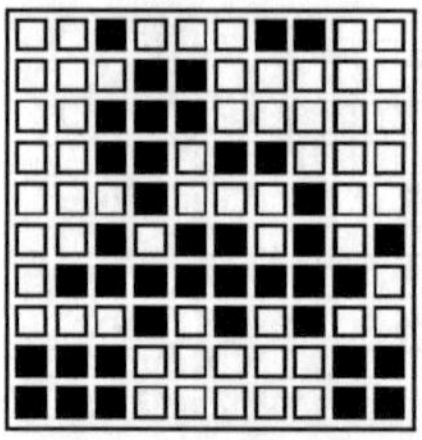	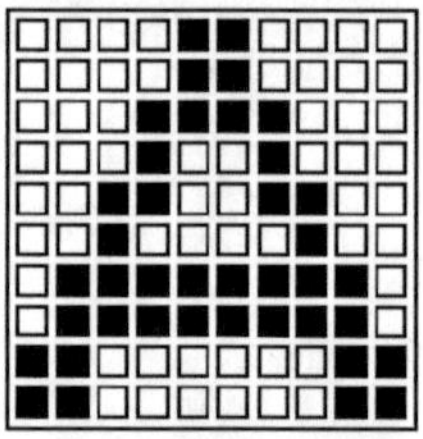
	30%	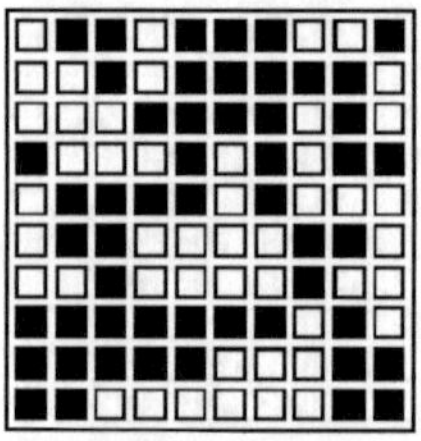	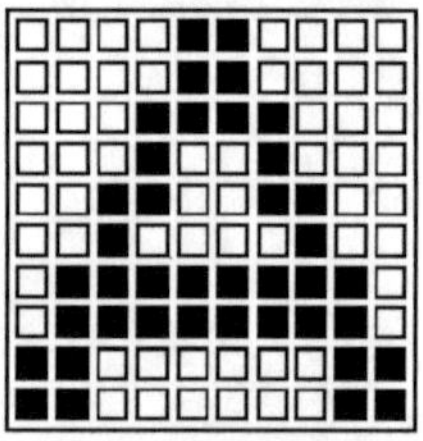
	40%	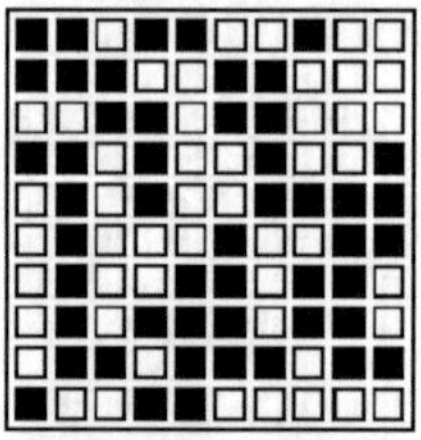	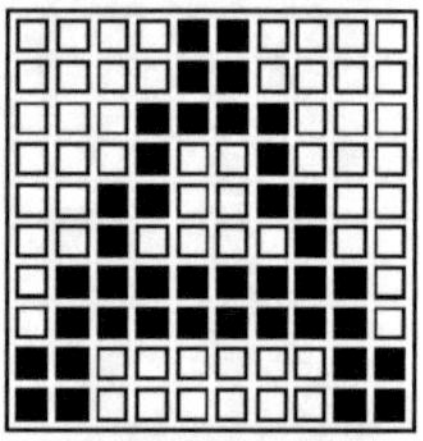
	45%	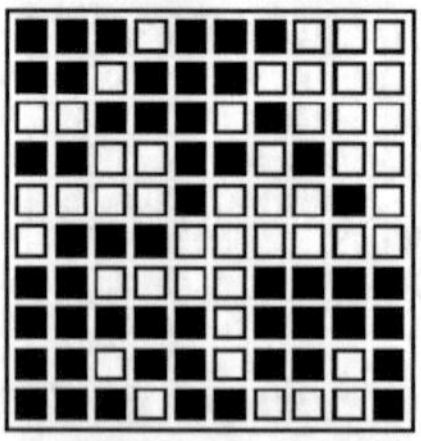	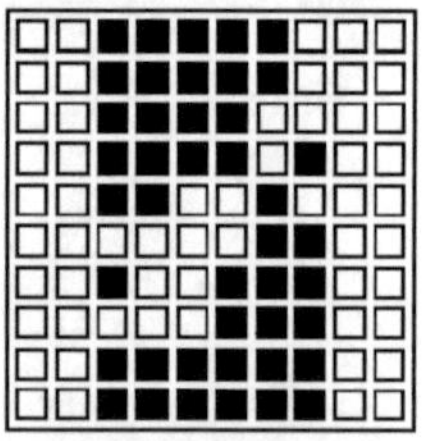
	50%	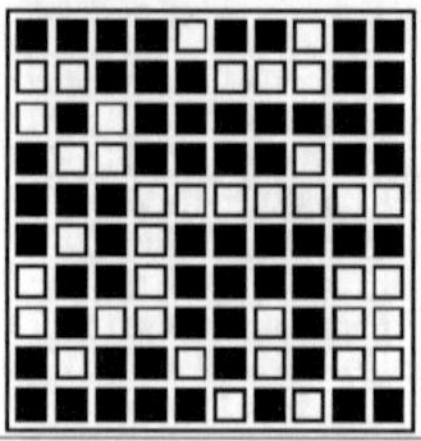	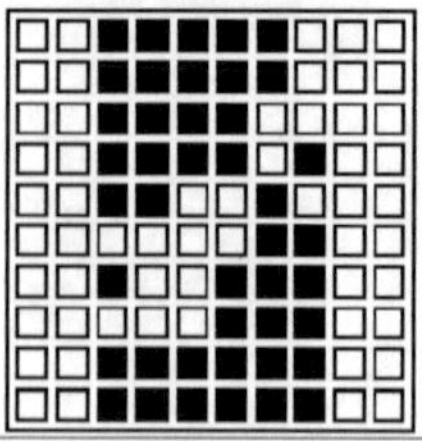

(continued)

Table 2.1 (continued)

60%	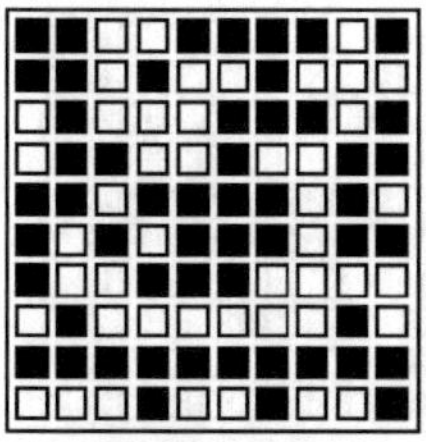	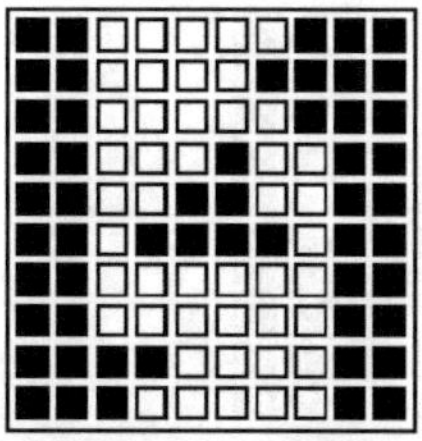
65%	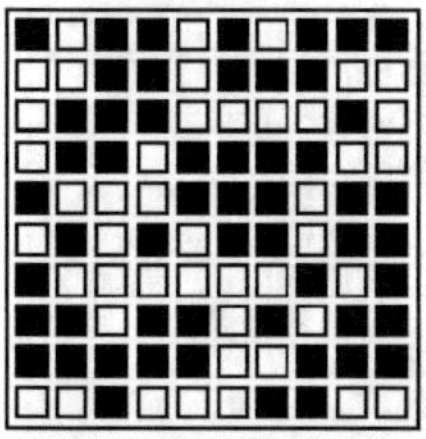	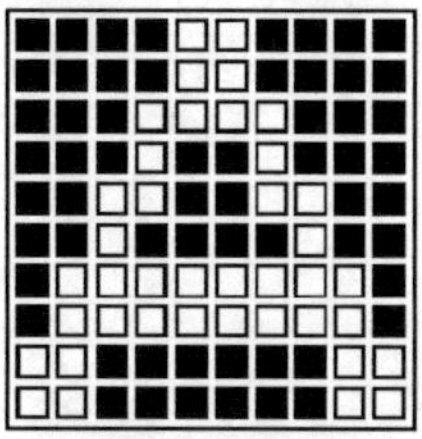
70%	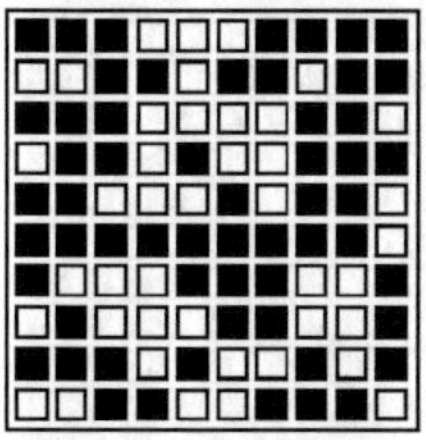	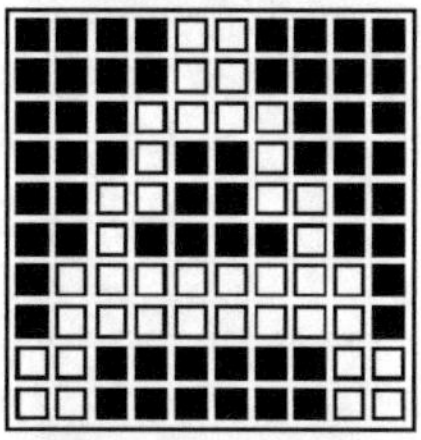
80%	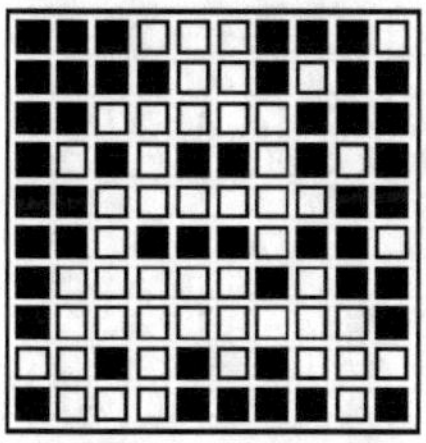	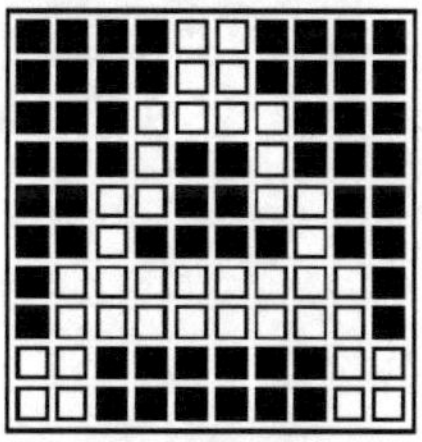
90%	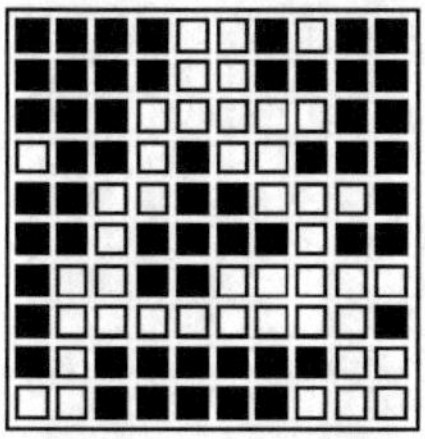	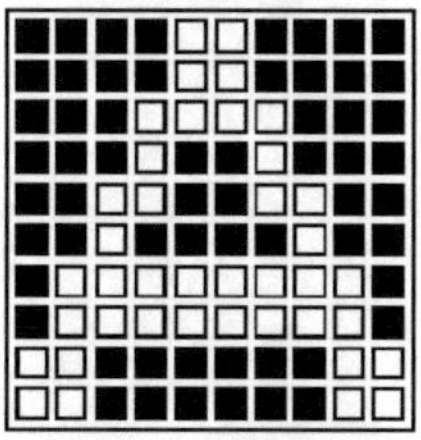
100%	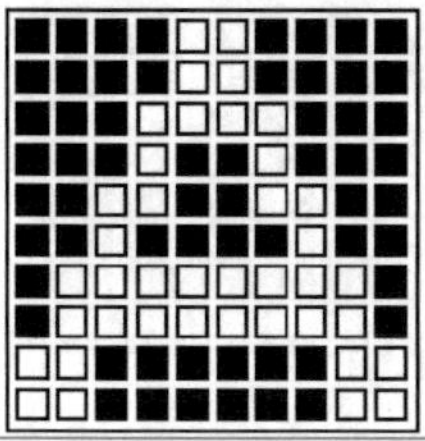	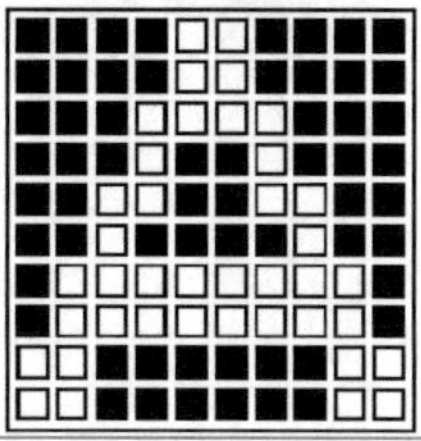

(continued)

Table 2.1 (continued)

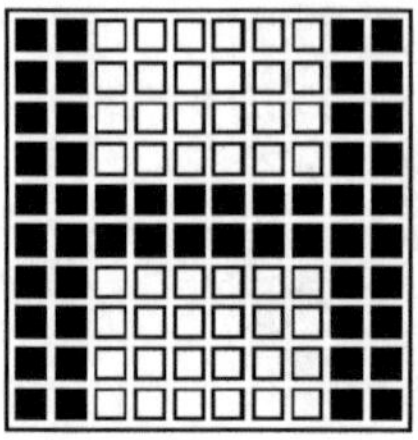	10%	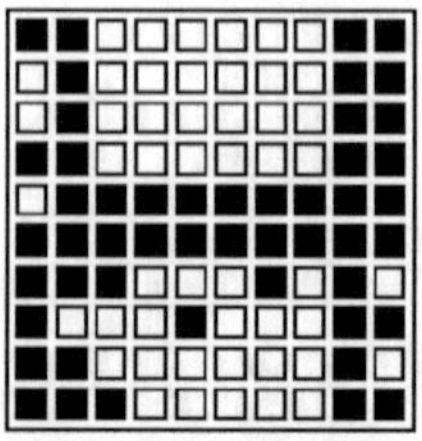	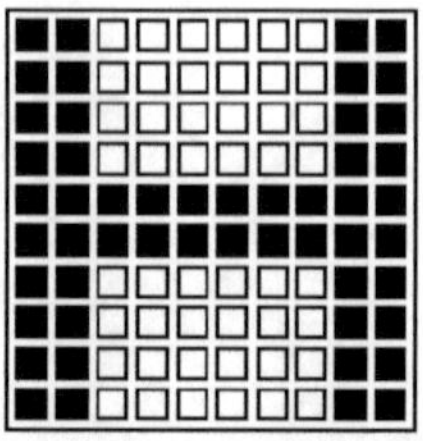
	20%	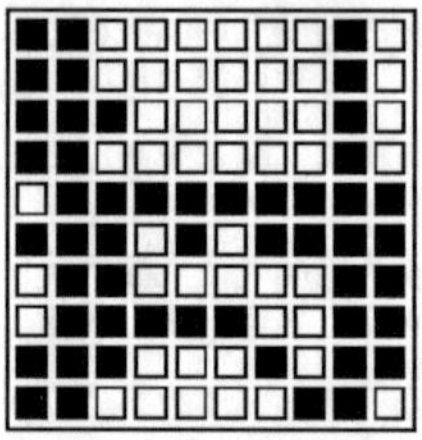	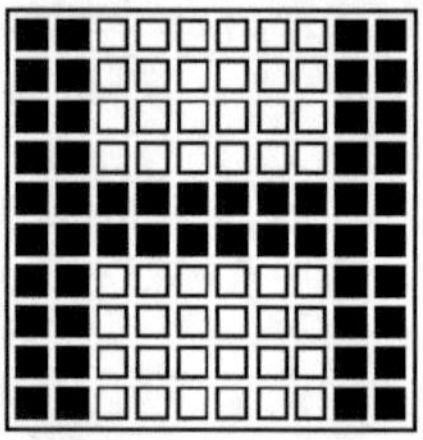
	30%	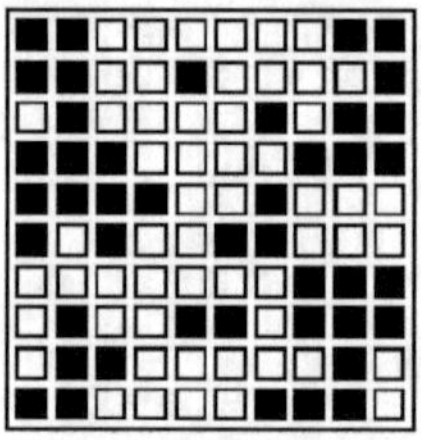	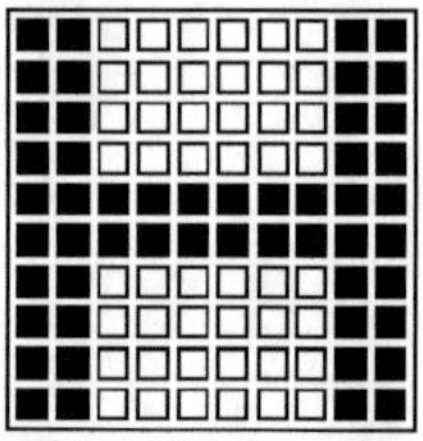
	40%	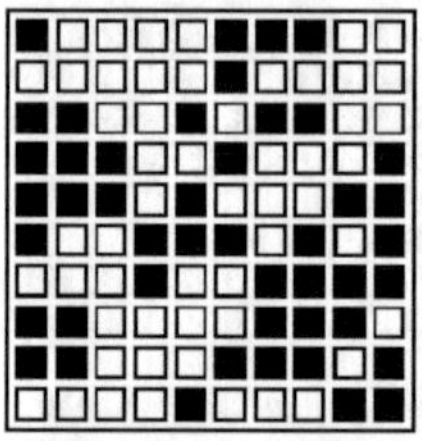	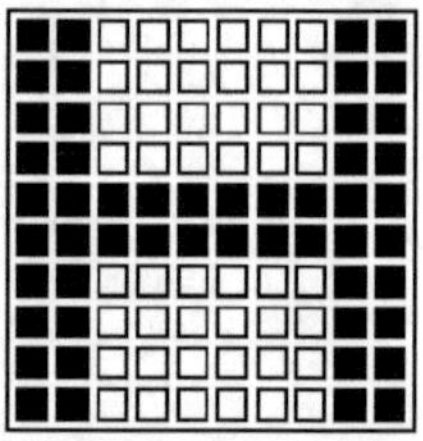
	45%	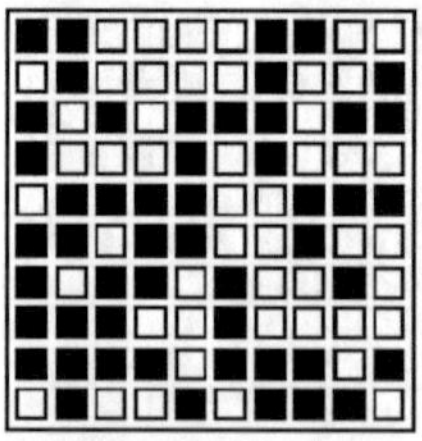	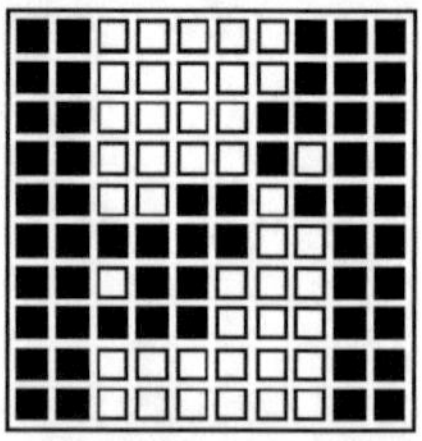
	50%	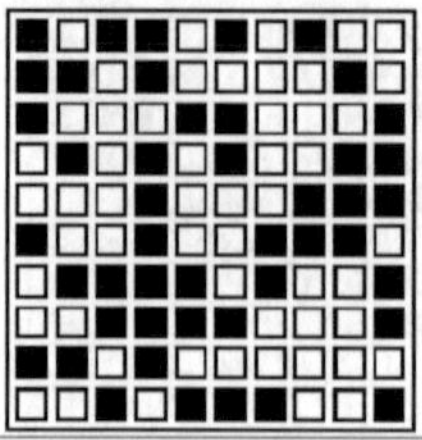	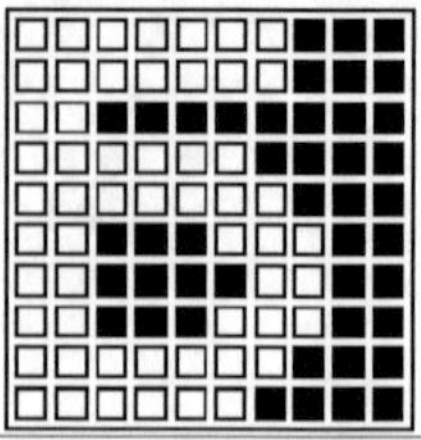

(continued)

Table 2.1 (continued)

55%

60%

70%

80%

90%

100%

2.1.6 The Effect of "Cross–Associations"

Let's complicate a task and train our neural network with one more pattern:

The letter "П" is very similar to letters "И" and "H" which already exist in the memory (Fig. 2.2). Now Hopfield's neural network can't distinguish any of these letters even in an undistorted state. Instead of correctly recognizable letter it displays the following image (for distortion of an pattern from 0 to 50 %):

It looks like to each of letters "И", "H", "П" but isn't the correct interpretation of any of them (Fig. 2.3).

From 50 to 60 % noise level at the output of neural network at first appears an image presented above (Fig. 2.3) in slightly distorted form, and then its negative.

Since 65 % of noise level, at the neural network output steadily there is an image negative to shown in Fig. 2.3.

The described behavior of a neural network is known as effect of "cross associations" [2]. Thus "A" and "Б" letters are recognized unmistakably at noise level up to 40 %.

At 45–65 % noise level at the network output appear, their slightly noisy interpretations, the image similar to a negative of a letter "B" (but very distorted), or a negative of the tested image (pattern). At distortion level of 70 % and more neural network steadily displays its negative of the tested image.

The experimental investigations had revealed the following shortcomings of Hopfield's neural network:

1. existence of cross-associations when some images(patterns) are similar to each other (like the experiments with letter П);
2. due to storing capacity restrictions the number of the remembered attractors (patterns) is only (0, 15–0, 2) n, where n is dimension of a weight matrix W.

These circumstances significantly limit possibilities of practical use of Hopfield's network.

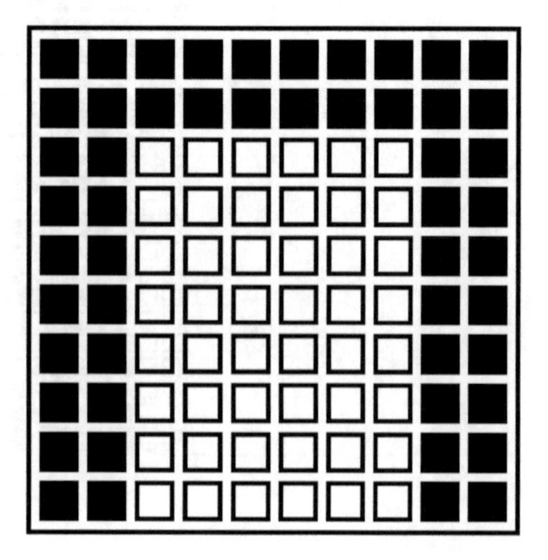

Fig. 2.2 New test symbol

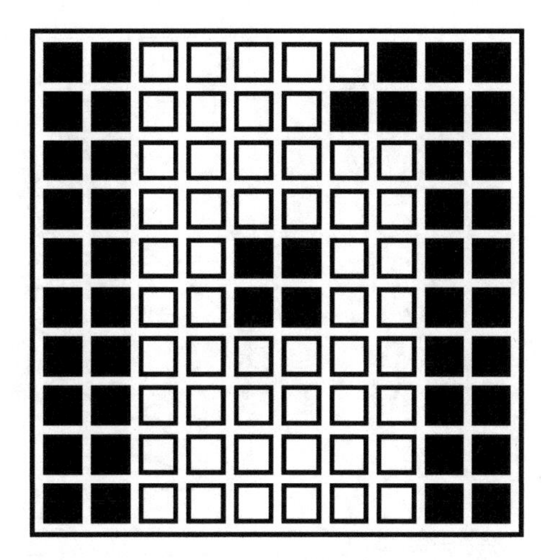

Fig. 2.3 Output symbol for tested symbols И, Н, П

2.2 Neural Network of Hamming. Architecture and Algorithm of Work

When there is no need that the network would display an etalon pattern in an explicit form, that is it is enough to define, say, a class of a pattern, associative memory is realized successfully by Hamming's network. This network is characterized, in comparison with Hopfield's network, by smaller costs of memory and volume of calculations that becomes obvious of its structure and work (Fig. 2.4) [2, 6].

The network consists of two layers. The first and second layers have m neurons, where m is a number of patterns. Neurons of the first layer have n synapses connected to the network inputs (forming a fictitious zero layer). Neurons of the second layer are connected among themselves by synaptic connections. The only synopsis with positive feedback for each neuron is connected to his axon.

The idea of network functioning consists in finding of Hamming distance from the tested pattern to all patterns represented by their weights. The number of different bits in two binary vectors is called as Hamming distance. The network has to choose a pattern with the minimum of Hamming distance to an unknown input signal therefore the only one of a network outputs corresponding to this pattern will be made active.

At an initialization stage the following values are assigned to weight coefficients of the first layer and a threshold of activation function:

$$w = \frac{x_i^k}{2}, i = \overline{1,n}; k = \overline{1,m} \tag{2.5}$$

$$T_k = \frac{n}{2}; \quad \overline{1,m} \tag{2.6}$$

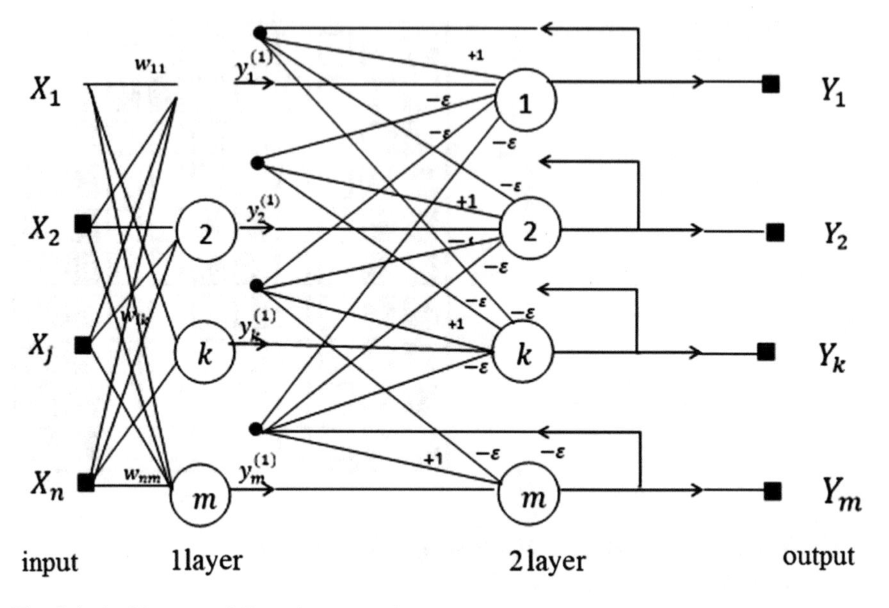

Fig. 2.4 Architecture of Hamming network

Here x_i^k is the ith an element of the kth pattern.

Weight coefficients of the braking synapses in the second layer are set equal to some value $-\varepsilon$, where $0<\varepsilon<\frac{1}{m}$, where m is a number of classes.

The neuron synapse connected with its own axon has a weight (+1).

2.2.1 *Algorithm of a Hamming Network Functioning*

1. Enter the unknown vector $X = \{x_i : i = \overline{1,n}\}$ to a network input and determine outputs of the first layer neurons (the top index in brackets in formula (2.7) specifies number of a layer):

$$y_j^{(1)} = f(s_j^{(1)}) = f\left(\sum_{i=1}^{n} w_{ij}x_i + T_j\right), \ j = \overline{1,m} \tag{2.7}$$

After that initialize states of axons of the second layer with received values:

$$y_j^{(2)} = y_j^{(1)} \ j = \overline{1,m} \tag{2.8}$$

2. Calculate new states of the second layer neurons

$$s_j^{(2)}(p+1) = y_j(p) - \varepsilon \sum_{k=1}^{m} y_k^{(2)}(p), \; j = \overline{1,m} \tag{2.9}$$

And values of their axons:

$$y_j^{(2)}(p+1) = f\left[s_j^{(2)}(p+1)\right], \; j = \overline{1,m} \tag{2.10}$$

Activation function f has a threshold, thus the size of a threshold should be rather big so that any possible values arguments won't lead to saturation.

3. Check, whether output of the second layer neurons has changed since the last iteration. If yes, then pass to a step 2 of the next iteration, otherwise—the end.

From the description of algorithm it is evident that the role of the first layer is very conditional, having used once on a step 1 value of its weight coefficients, the network doesn't come back to it any more, so that the first layer may in general be excluded from a network and replaced with a matrix of weight coefficients.

Note advantages of neural network of Hamming:

- small costs of memory;
- the network works quickly;
- algorithm of work is extremely simple;
- capacity of a network doesn't depend on dimension of an input signal (as in Hopfield's network) and equals exactly to a number of neurons.

2.2.2 Experimental Studies of Hopfield's and Hamming's Networks

Comparative experimental researches of Hopfield's and Hamming's neural networks in a problem of symbols recognition were carried out. For learning of a network input sample of symbols (1, 7, e, q, p) was used. Then generated noisy patterns from this sample were entered and their recognition was performed. Level of noise changed from 0 to 50 %. Results of recognition of the specified symbols are presented in Fig. 2.5. On the screen 4 images (patterns) are presented (from left to right, from top to down): the initial image—a etalon, the noisy image, result of Hamming network, result of Hopfield's network (Figs. 2.6, 2.7, 2.8, 2.9).

2.2.3 Analysis of Results

Results of experiments with Hopfield's and Hamming's networks are presented in the Table 2.2. A corresponding table element is a result of recognition by a network

Symbol 1

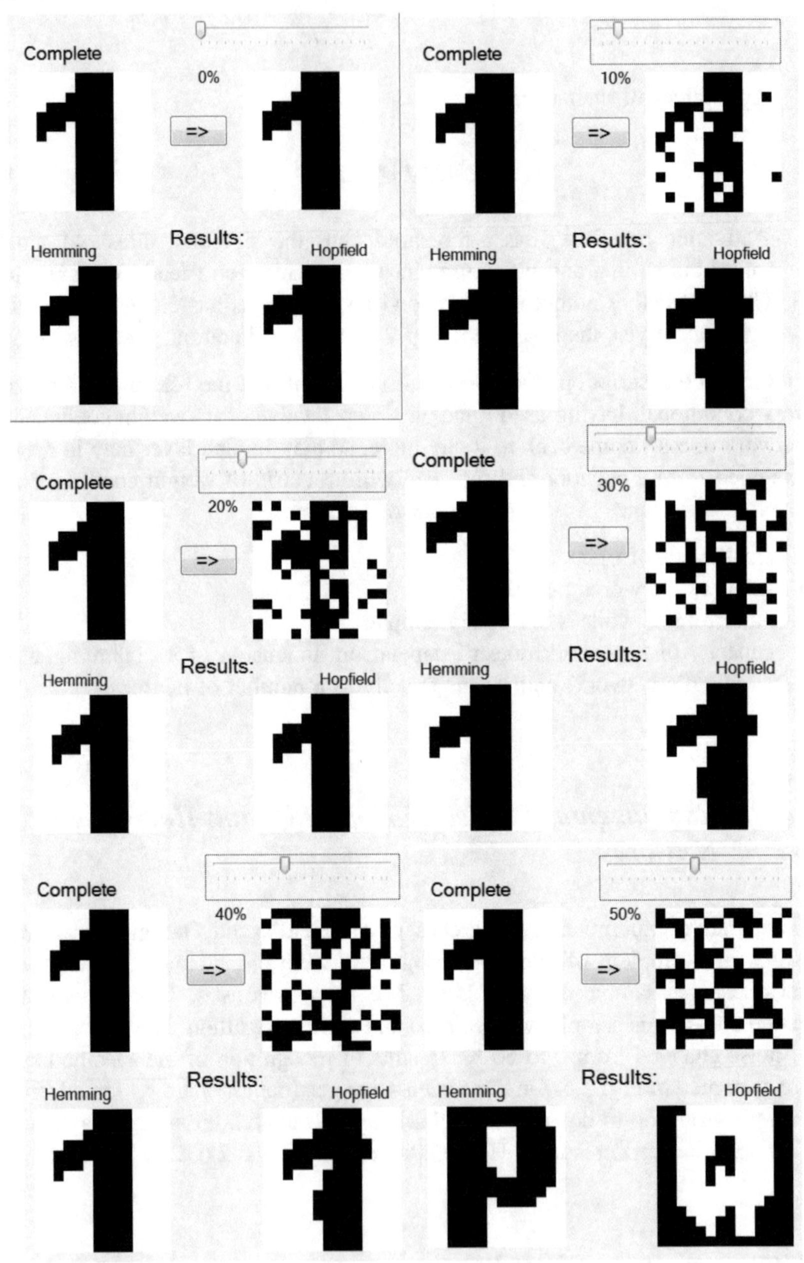

Fig. 2.5 Recognition of the symbol 1

Symbol 7

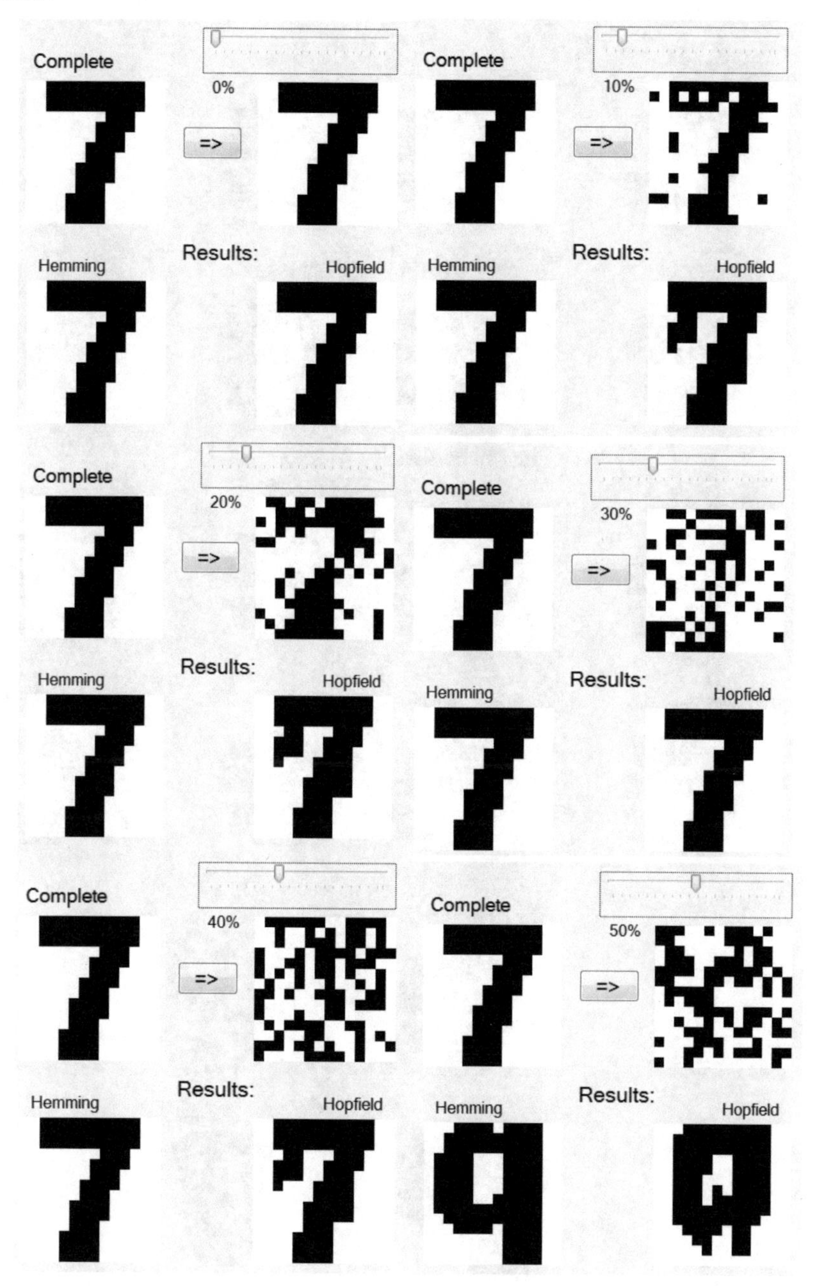

Fig. 2.6 Recognition of the symbol 7

Symbol E

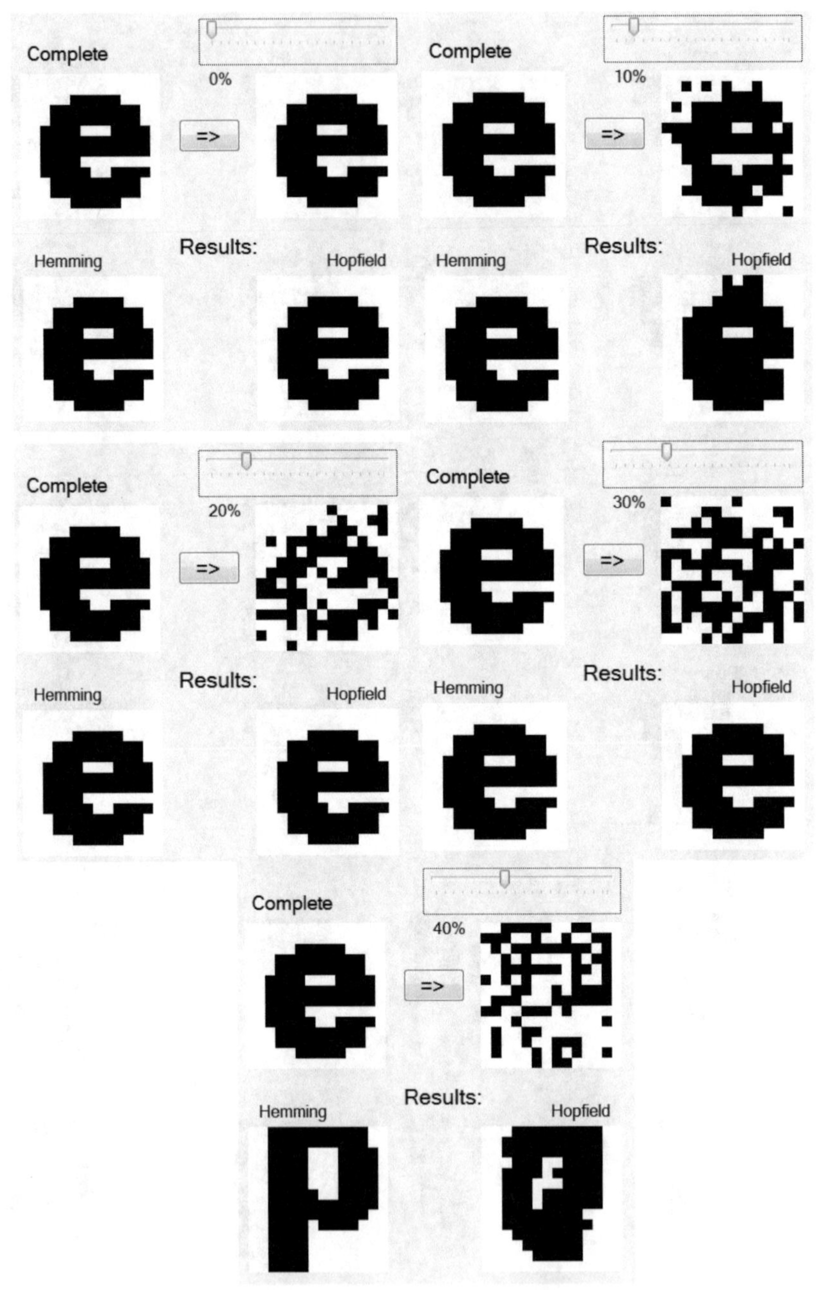

Fig. 2.7 Recognition of the symbol e

Symbol Q

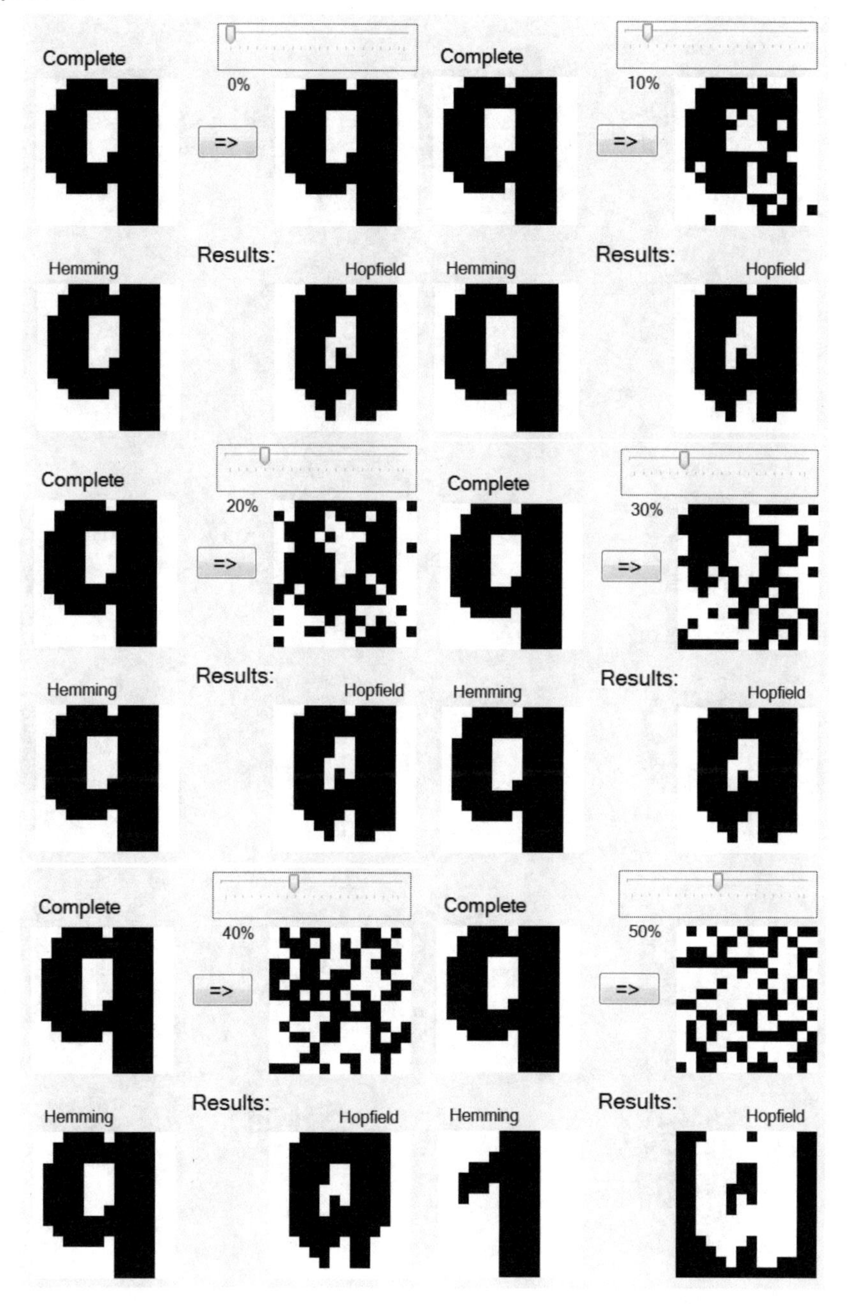

Fig. 2.8 Recognition of the symbol q

Symbol P

Fig. 2.9 Recognition of a p symbol

Table 2.2 Comparative results of experiments with Hopfield and Hamming networks

Recognition (Hamm, Hop)	0 %	10 %	20 %	30 %	40 %	50 %
1	(1; 1)	(1; 0.5)	(1; 1)	(1; 0.5)	(1, 0.5)	(0; 0)
7	(1; 1)	(1; 0.5)	(1; 0.5)	(1; 1)	(1, 0.5)	(0; 0)
E	(1; 1)	(1; 0.5)	(1; 1)	(1; 1)	(0; 0)	(0; 0)
Q	(1; 0)	(1; 0)	(1; 0)	(1; 0)	(1; 0)	(0; 0)
P	(1; 1)	(1; 1)	(1; 0)	(1; 0)	(1; 0)	(0; 0)

(Hamming; Hopfield) of a symbol at the specified noise level, and namely: 0—it isn't recognized; 0.5—it is recognized with defects; 1—it is recognized correctly.

Hamming'snetwork in general performed well (except for "e" symbol) and recognized correctly up to the level of noise 40 %), while Hopfield's network results of recognition are much worse, at recognition of symbols with a similar elements (e, p, q) there were difficulties—recognition level was less than 30 %, it is the effect of cross associations.

2.3 Self-organizing Neural Networks. Algorithms of Kohonen Learning

2.3.1 Learning on the Basis of Coincidence. Law of Hebb Learning

In 1949 the Canadian psychologist D. Hebb published the book "Organization of Behaviour" in which he postulated the plausible mechanism of learning at the cellular level in a brain [2, 7].

The main idea of Hebb consisted therein when the input signal of neuron arriving through synaptic communications causes activation operation of neuron, efficiency of such input in terms of its ability to cause operation of neuron in the future has to increase.

Hebb assumed that change of efficiency has to happen in a synapse which transmits this signal to an neuron input. The latest researches confirmed this guess of Hebb. Though recently other mechanisms of biological learning at the cellular level were detected, but in recognition of merits of Hebb this law of learning was called in his honor [7].

The law of Hebb learning belongs to a class of laws of learning by competition.

Linear Associative Elements

In Fig. 2.10 the architecture of the neural network (NN) consisting of m neurons which are called as "linear associators" is presented.

The input vector in the linear associator is a vector, $X = \{x_i\}$, $i = \overline{1, n}$, which is taken out of space R_n according to some distribution $\rho(x)$.

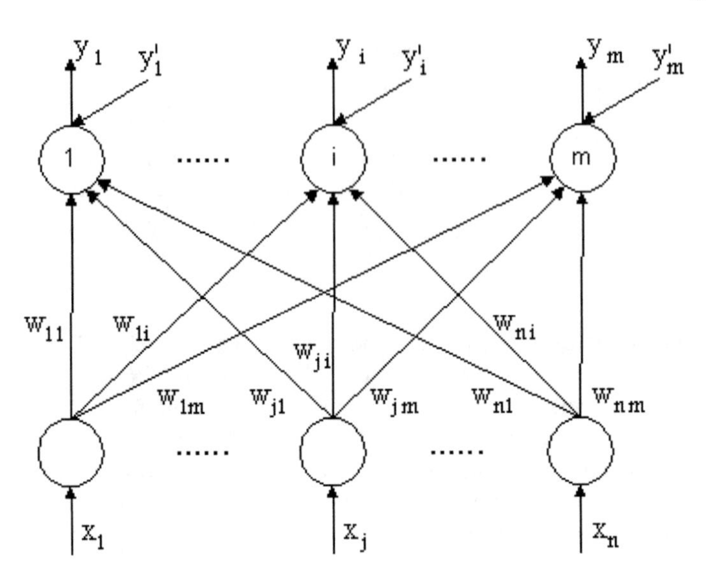

Fig. 2.10 Network with linear associators–neurons

The output vector is obtained from input X by the following formula:

$$Y = WX \tag{2.11}$$

where $W = \left\| w_{ij} \right\|$ is a weight matrix $n \times m$; $W = (W_1, W_2, \ldots, W_m)$, W_j is W-matrix column, $W_j = (W_{1j}, W_{2j}, \ldots, W_{nj})^T$ is a weight vector.

We will designate through $Y' = \{y_j\}$ a desirable output. The main idea of a linear associative neural network consists that the network has to learn on pairs input-output:

$$(X_1, Y_1), (X_2, Y_2), \ldots, (X_L, Y_L)$$

When to an neural network input the signal X_k is given, desirable output Y' has to be equal Y_k. If on an network input the vector $X_k + \varepsilon$ is given (where ε—rather small), the output has to be equal $Y_k + \varepsilon$ (i.e. at the output have to receive the vector close to Y_k).

The law of Hebb's learning is as follows [2, 7]:

$$w_{ij}^{new} = w_{ij}^{old} + y_{kj} x_{ki}, \tag{2.12}$$

Where x_{ki}—is the ith vector component X_k; y_{kj}—jth vector component Y_k.

In a vector form the expression (2.12) is written so:

$$W^{new} = W^{old} + X_k Y_k^T = W^{old} + Y_k X_k^T, \tag{2.13}$$

To realize this law in the course of learning the corresponding components $Y_k = \lfloor y_{kj} \rfloor$, which are shown by dashed lines (arrows) in Fig. 2.10 are entered.

It is supposed that before learning, all $w_{ij}^{(0)} = 0$. Then as a result of display of the learning sample $(X_1, Y_1), \ldots, (X_L, Y_L)$ the final state of a matrix of W is defined so:

$$W = Y_1 X_1^T + Y_2 X_2^T + \cdots + Y_L X_L^T. \tag{2.14}$$

This Eq. (2.14) is called "a formula of the sum of external product" for W. This name comes from that fact that $Y_k X_k^T$—is an external product.

The reformulation of Hebb law in the form of the sum of external product allows to conduct additional researches of opportunities of this law to provide associations of pairs of vectors $(X_k Y_k)$.

The first conclusion consists that if vectors $\{X_1, X_2, \ldots, X_L\}$ are ortogonal and have unit length, i.e. are orthonormalized, then

$$Y_k = W X_k, \tag{2.15}$$

In other words, the linear associative neural network will produce desirable transformation "input-output".

This is the consequence of an orthonormalization property:

$$X_i^T X_j = \delta_{ij} = \begin{cases} 0, & i \neq j \\ 1, & i = j \end{cases} \tag{2.16}$$

Then

$$W X_k = \sum_{r=1}^{L} Y_r X_r^T X_k = Y_k. \tag{2.17}$$

But the problem consists that an orthonormalization condition is very rigid (first of all it is necessary, that $L \leq n$).

Further we are restricted by the requirement that $\|X_i\| = 1$. It would be much more useful if it was succeeded to lift this restriction. These goal can be achieved, but not in a linear associative neural network. Here, if vector s X_k aren't orthonormalized, then a reproduction error Y_k appears at the output:

$$W X_k = \sum_{r=1}^{L} Y_r X_r^T X_k = Y_k + \sum_{r \neq k} Y_r X_r^T X_k = Y_k + \eta. \tag{2.18}$$

It is desirable to achieve that η be minimum. To provide $\eta = \min$ or $\eta = 0$, it is necessary to pass to a nonlinear associative network with nonlinear elements.

2.3.2 *Competitive Learning*

Competitive learning is used in problems of self-learning, when there is no classification of the teacher.

The laws of learning relating to category competitive, possess that property that there arises a competitive process between some or all processing elements of a neural network. Those elements which appear winners of competition, get the right to change their weights, while the rest the of weights don't change (or change by another rule).

Competitive learning is known as "Kohonen's learning". Kohonen's learning significantly differs from Hebb learning and BP algorithm by therein the principle of self-organization is used (as opposed to the principle of controlled learning with the teacher).

The competitive law of learning has long and remarkable history [2]. In the late sixties—the beginning of the 70th Stephen Grossberg suggested the whole set of competitive learning schemes for neural networks. Another researcher who dealt with problems of competitive learning was van der Malsburg. The learning law of van der Malsburg was based on idea that the sum of the weights of one input element connected with various processing neurons has to remain a constant in the course of learning i.e. if one of weights (or some) increases, the others have to decrease.

After considerable researches and studying works of Grossberg, van der Malsburg and others Toivo Kohonen came to the conclusion that the main goal of competitive learning has to consist in designing of a set of vectors which form a set of equiprobable representatives of some fixed function of distribution density $\rho(x)$ of input vectors. And though learning laws of this type were independently received by many researchers, T. Kohonen was the first who paid attention to a question of equiprobability. Exactly thanks to this idea and the world distribution of T. Kohonen book "Self-organization and associative memory" [8] his name began to associate with this law of learning.

2.3.3 *Kohonen's Learning Law*

The basic structure of a layer of Kohonen neurons is given in Fig. 2.11. The layer consists of N processing elements, each of which receives n input signals $x_1, x_2, \ldots x_n$ from a lower layer which is the direct transmitter of signals. To an input x_i and communication *(i, j)* we will attribute weight w_{ij}.

Each processing element of a layer of Kohonen counts the input intensity I_j in compliance with a formula [2, 8]:

$$I_j = D(W_j, X), \tag{2.19}$$

where $W_j = (w_{1j}, w_{2j}, \ldots, w_{nj})^T$ and $X = (x_1, x_2, \ldots, x_n)$; $D(W_j, X)$—some measure (metrics) of distance between W_j and X.

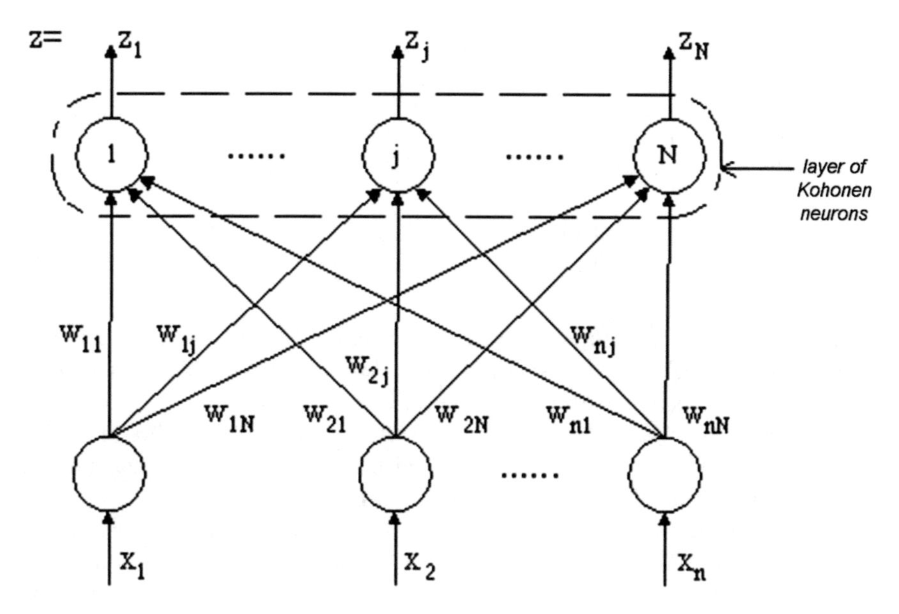

Fig. 2.11 Architecture of Kohonen network

We will define two most general forms of function $D(W_j, X)$:

1. Euclide distance: $d(W, X) = \|W - X\|$;
2. Spherical arc distance:

$$\Delta(W, X) = 1 - W^T X = 1 - \cos\theta \tag{2.20}$$

where $W^T X$—the scalar product, and is supposed that $\|W\| = \|X\| = 1$.

In this statement, unless otherwise stated, we'll use Euclidean distance $d(W, X)$. At implementation of the Kohonen law as soon as each processing element (neuron) counted the function I_j, a competition between them takes place is, whose purpose is to find an element with the smallest value I_j (i.e. $I_{j_{\min}}$). As soon as the winner of such competition is found, his output z is put equal to 1. Output signals of all other elements remain equal to 0.

At this moment a learning by Kohonen takes place.

The learning data for Kohonen's layer assumed to consist of sequence of input vectors $\{X\}$, which are taken randomly with the fixed density of probabilities distribution $\rho(x)$. As soon as next vector X it is entered into a network, the processing Kohonen's neurons start competing to find the winner for whom $\min_j d(X, W_j)$ is reached. Then for the winner neuron j^* output is established $z_j* = 1$, and for all others $z_j = 0,\ j \neq j^*$.

At this moment a change of weights according to Kohonen learning law is performed:

$$W_j^{new} = W_j^{old} + \alpha\left(X - W_j^{old}\right) z_j, \tag{2.21}$$

where $0 < \alpha < 1$.

This law can be written in the following form:

$$W_j^{new} = \begin{cases} (1-\alpha)W_j^{old} + \alpha X, \, for\, winner\; j = j^* \\ W_j^{old}, \; \forall\, j \neq j^* \end{cases} \tag{2.22}$$

It is evident that at such learning law the weight vector W_j moves to an input vector X. At the beginning of learning process $\alpha = 1$ and then in process of learning it monotonously decreases up to the value $\alpha = 0, 1$.

This algorithm realizes the principle "The winner takes all" therefore in foreign literature it is called WTA. Further it should be noted similarity of learning by Kohonen and statistical process of finding of "k-means".

K-means for the fixed set of vectors $\{X_1, X_2, \ldots, X_L\}$, which are chosen randomly from some population with the fixed density of probabilities distribution $\rho(x)$, make a set of k vectors $W = (W_1, W_2, \ldots, W_k)$ such that the following functional is minimized:

$$\min_{\{w_i\}} \sum_{i=1}^{L} D^2(X_i, W(X_i)) \tag{2.23}$$

Where $W(X_i)$ is a vector W, closest to X_i.

In summary, it is necessary to emphasize that the learning by Kohonen's algorithm generally doesn't generate a set of equiprobable weight vectors, that is a set of such vectors that X which is chosen randomly, according to density of probabilities distribution ρ will have equal probability to be the closest to each of weight vectors W_j.

2.3.4 Modified Competitive Learning Algorithms

As it was already noted above, we seek to get vectors W_j, which would be approximately equally probable in the sense of being the closest to the vectors of X taken from $\Re^n$ with some density of probability distribution In other words, for any vector of X taken from $\Re^n$ with probability $\rho(x)$, it is desirable that the probability of that X will appear to be the closest to W_i, has to be approximately equal to $\frac{1}{N}$ for all $i = \overline{1, N}$.

There are some approaches for the solution of the problems arising at implementation of a basic learning law of Kohonen [2, 9].

1. The first approach is called as Radial Sprouting. It is the best for Euclidean metrics and metrics (distances) similar to it. All weight vectors $W_i = \lfloor w_{ij} \rfloor$ are originally set equal to $w_{ij}(0) = 0$. All input vectors X at first are multiplied by some small positive scalar β. Process begins with β, close to 0. It provides proximity of input vectors of X to vectors W_j. In process development β slowly increases, until reaches the value $\beta = 1$. As soon as it occurs, weight vectors "are pushed out" from the initial values and follow input vectors. This scheme works quite well, but usually some weight vectors will lag behind process and as a result will be not involved in the competion process that slows down learning process.
2. Other approach ("noise addition") consists in adding randomly distributed noise to data vectors X that facilitates effect of achievement $\rho(X) > 0$, in all area Ω_x Level of noise is chosen at first rather big so that noise vector be much greater, than a data vector X. But in the process of learning noise level gradually decreases. This approach works correctly, but it appears even more slowly, than approach of "Radial Sprouting". Therefore approaches of "Radial Sprouting" and "noise addition" solve a problem of presentation of badly representable laws with small probability of distribution in some area, but they don't solve a problem of equiprobable positioning of vectors W_j.

In general, the basic Kohonen's learning law will bring to a surplus at placement of vectors W_j in those areas where probabilities distribution density $\rho(X)$, is large, and to shortage of vectors W_j in areas where density of probabilities distribution $\rho(X)$ is small.

3. The third approach which was offered by Duane Desieno, is to build-in "consciousness" (or memory) in each element k to carry out monitoring (control) of history of successful results (victories) of each neuron. If the processing Kohonen element wins competition significantly more often than $\frac{1}{N}$ times (time), then his "consciousness" excludes this element from competition for some time, thereby giving the chance to elements from the oversaturated area to move to the next non-saturated areas. Such approach often works very well and is able to generate good set of equiprobable weight vectors.

The main idea of the consciousness mechanism is a tracking of a share of time during which the processing element j wins competition. This value can be calculated locally by each processing element by formula:

$$f_j(t+1) = f_j(t) + \beta \left(z_j - f_j(\mathrm{t}) \right) \tag{2.24}$$

When competition is finished and the current value z_j (0 or 1) is defined, the constant β takes a small positive value (typical value $\beta = 10^{-4} = 0,0001$) and the share f_j is calculated. Right after it the current shifts value b_j is defined

$$b_j = \gamma(\frac{1}{N} - f_j) \tag{2.25}$$

where γ—positive constant ($\gamma \approx 10$).

Further the correction of weights is carried out. However, unlike a usual situation in which weight are adjusted only for one processing element-winner with $z_i = 1$, here separate competition is being held for finding of the processing element which has the smallest value of

$$D(W_j, X) - b_j \tag{2.26}$$

The winner element corrects further the of weight according to the usual law of Kohonen's learning.

The role of the shift b_j is as follows. For often winning elements j the value $f_j > \frac{1}{N}$ and $b_j < 0$ therefore for them value $D(W_j, X) - b_j$ increases in comparison with $D(W_j, X)$ for seldom winning elements $f_j \ll \frac{1}{N}$, $b_j > 0$ and $D(W_j, X) - b_j$ decreases that increases their chances to win competiton. Such algorithm realizes the consciousness mechanism in work of self-organizing neuron networks and therefore it is called as CWTA (Conscience Winner Takes All) [9].

2.3.5 Development of Kohonen Algorithm

In 1982 T. Kohonen suggested to introduce into the basic rule of competitive learning information on an arrangement of neurons in an output layer [2, 8, 10]. For this purpose neurons of an output layer are ordered, forming a one-dimensional or two-dimensional lattice. The arrangement of neurons in such lattice is marked by a vector index $i = (i_1, i_2)$. Such ordering naturally enters distance between neurons $|i - j|$.

The modified rule of competitive of Kohonen's learning considers distance of neurons from winner neuron [2, 8, 9]:

$$W_j(t+1) = W_j(t) + \alpha\,(X - W_j)\Lambda(d(i, j^*)) \tag{2.27}$$

where Λ—function of the neighborhood. $\Lambda(d(i, j^*))$ is equal 1 for winner neuron with an index j^*, and gradually decreases in process of increase in distance d, for example, by function

$$\Lambda(d) = e^{-d^2/R^2} \tag{2.28}$$

Both rate of learning α, and radius of interaction R gradually decreases in the course of learning so at a final stage of learning we come back to the basic law of weights adaptation only of winner neurons $\alpha(t) = a_0 e^{-kt}$.

As we can see in this algorithm the principle 2 is realized: winner takes away not all but maximum (income) therefore in foreign literature it is called as WTM (Winner Takes MOST).

Learning by Kohonen modified algorithm WTM reminds a tension of an elastic grid of prototypes on a data file from the learning sample. In process of learning the elasticity of a network gradually decreases and weights changes also decrease except winner neuron.

As a result we receive not only quantization of input's, but also we order input information in the form of a one-dimensional or two-dimensional topographic map of Kohonen. On this grid each multidimensional vector has the coordinate, and the closer are coordinates of two vectors on the grid, the closer they are in the initial space.

2.3.6 *Algorithm of Neural Gas*

Acceleration of convergence of the modified Kohonen WTM'S algorithm and the best self-organization of a network gas molecules motion can be received with application of the method offered by M. Martinez, S. Berkovich and K. Shulten [9] and called by authors "algorithm of neural gas" owing to similarity of its dynamics. In this algorithm on each iteration all neurons are sorted depending on their distance to a vector X. After sorting neurons are marked in the sequence corresponding to increase in their remoteness:

$$d_0 < d_1 < d_2 < \cdots < d_{n-1}, \tag{2.29}$$

where $d_i = \left\| X - W_{m(i)} \right\|$ designates remoteness from a vector X of the ith neuron taking as a result of sorting position m in the sequence begun with neuron winner to which remoteness d_0 is put into compliance. Value of the neighborhood function for ith neuron is determined by a formula [9]:

$$\Lambda(i,x) = \exp\left\{ -\frac{m(i)}{\sigma(t)} \right\}, \tag{2.30}$$

where $m(i)$ defines the sequence received as a result of sorting of $(m(i) = 0, 1, 2, \ldots, n-1)$, and $\sigma(t)$ is the parameter similar to R neighbourhood level in Kohonen WTM's algorithm decreasing eventually with t.

At $\sigma(t) = 0$ adaptation only of the winner neuron occurs and the algorithm turns into usual (basic) Kohonen's algorithm, and at $\sigma(t) \neq 0$ of adaptation are subject the weights of many neurons neighbors, and the level of change of scales depends on size $\Lambda(i,x)$.

For achievement of good results of self-organization, process of learning has to begin with rather great value of σ, but eventually its size decreases to zero. Change of $\sigma(t)$ can be linear or exponential. In work [9] it was offered to change value according to expression

$$\sigma(t) = \sigma_{\max}\left(\frac{\sigma_{\min}}{\sigma_{\max}}\right)^{\frac{t}{T_{\max}}}, \tag{2.31}$$

where $\sigma(t)$—value on iteration of t; $\sigma_{\min}$ and $\sigma_{\max}$—the accepted minimum and maximum values of σ. Value $T_{\max}$ defines the maximum number of iterations. The learning coefficient of the *i*th neuron $\alpha_i(t)$ can also change both linearly, and exponentially, and its exponential dependence is defined by expression

$$\alpha_i(t) = \alpha_i(0)\left(\frac{\alpha_{\min}}{\alpha_i(0)}\right)^{\frac{t}{T_{\max}}}, \tag{2.32}$$

where $\alpha_i(0)$ is an initial value α, and $\alpha_{\min}$ is a priori set minimum value corresponding to $t = \mathrm{T}_{\max}$. In practice the best results of self-organization are reached at linear change of $\alpha(t)$ [9]. For reduction of calculations volume at realization of neural gas algorithm it is possible to use the simplification consisting that adaptations of weights happen only for the first k of neurons neighbors in the ordered sequence of $d_0 < d_1 < d_2 < \ldots < d_k$.

Algorithm of neural gas together with Kohonen's algorithm (CWTA) with memory considering a share of victories of each neuron of f_j are the most effective tools of neurons self-organization in Kohonen's network.

Comparative Analysis of Algorithms of Self-organization

Above considered algorithms were compared at the solution of a problem of recovery of the two-dimensional learned data of difficult structure which are presented in Fig. 2.12 [9]. For recovery of data 2 sets of the neurons including 40 and 200 elements were used which after ordering positions of neurons will reflect data distribution. They have to locate in areas of the maximum concentration of data.

Results of self-organization of 40 neurons when using three algorithms are presented in Fig. 2.13 algorithm with memory (CWTA) (Fig. 2.13a), neural gas (Fig. 2.13b) and basic algorithm of Kohonen (Fig. 2.13c). For comparison in Fig. 2.15 similar pictures are obtained by Kohonen's network consisting of 200 neurons (Fig. 2.14).

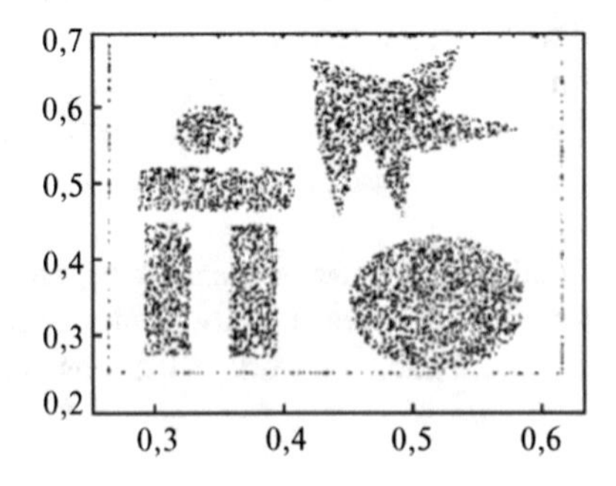

Fig. 2.12 Data structures to be simulated

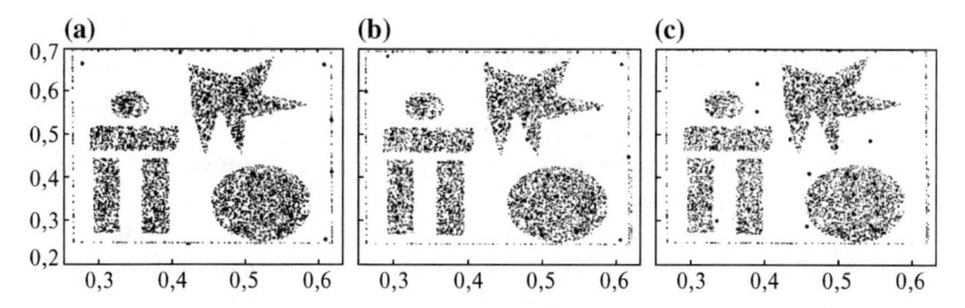

Fig. 2.13 Results of self-organization of different algorithms with 40 neurons

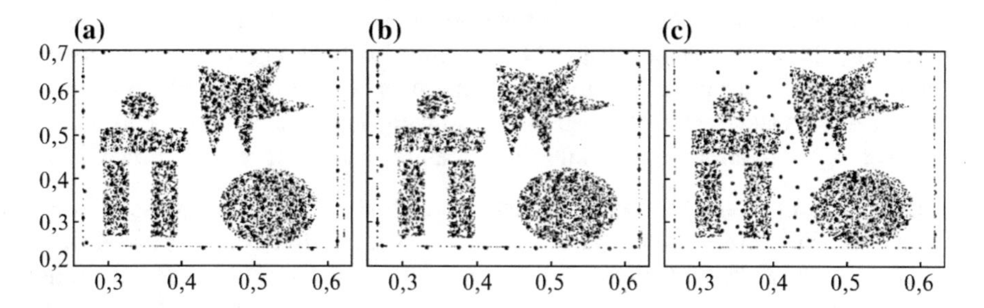

Fig. 2.14 Results of self-organization for 200 neurons

As follows from the given results, irrespective of number of neurons the best results of self-organization were received with use of algorithms of the self-organization with memory and neural gas.

For quantitative comparison of results it is possible to use criterion "a quantization error":

$$E_q = \sum_{i=1}^{n} \left\| X_i - W_i^* \right\|, \tag{2.33}$$

where W_i^* is winner neuron weight at presentation of a vector X_i.

At 200 neurons the following criterion values were received [9]: E_q = 0,007139 for CWTA; E_q = 0,007050 for algorithm of neural gas and E_q = 0,02539 for basic Kohonen's algorithm.

2.4 Application of Kohonen Neural Networks

Neural networks with self-organization are used in two main directions.

1. For automatic classification of objects.
2. For visual display of properties of multidimensional space (representation of multidimensional vectors of features).

In the first case the problem of automatic splitting a set of the objects represented by multidimensional vectors in some feature space on similarity—to difference of feature values is solved. Such task sometimes is called as a task of the cluster analysis and it is in detail considered in the 7th chapter.

In the second case the problem of visualization of properties of multidimensional feature vectors on the two-dimensional plane is solved.

So-called "Self-organizing Maps" (SOM) are used for this purpose [10].

For this in multidimensional space the spatial lattice is stretched in which nodes are processing neurons (a layer of Kohonen's neurons). Further the next points to each of these neurons are defined. They define area of an attraction of this neuron. Average value of each feature of neurons in the attraction area is defined $-X_{i\ \mathrm{cp}}$.

Further nodes of a lattice are mapped on the plane in the form of squares or hexagons, the corresponding value of a feature for this neuron is painted in the certain color: from blue (the minimum value of a feature) to red (the maximum value). As a result we receive the self-organizing feature map similar to a geographical map. Such maps are built on all features and we receive a certain atlas.

Self-organizing Maps (SOMs) represents a method of design of N-dimensional input space in discrete output space which makes an effective compression of input space in a set of the coded (weight) vectors. The output space usually represents itself a two-dimensional lattice. SOM uses a lattice for approximation of probability density function of an input space, thus keeping its structure i.e. if two vectors are close to each other in the input space, they have to be close and on the map as well. During self-organization process SOM carries out an effective clustering of input vectors, keeping structure of initial space [10].

Stochastic Algorithm of Learning

Learning of SOM is based on strategy of competitive learning. We will consider N-dimensional input vectors X_p where the index p designates one learning pattern. The first step of process of learning is definition of structure of the maps, usually two-dimensional lattice. Map is usually square, but may be rectangular. The number of elements (neurons of an output layer) on the map is less, than the number of the learning patterns (samples). The number of neurons has to be equal in an ideal to the number of the independent learning patterns. The structure of SOM is given in Fig. 2.15. Each neuron on the Map is connected with a N-dimensional weight vector which forms the center of one cluster. Big cluster groups are formed by grouping together of "similar" next neurons.

Initialization of Weight Vectors Can Be Carried Out in Various Ways

1. To each weight $W_{kj} = \left\{W_{kj1}, W_{kj2}, \ldots, W_{kjN}\right\}$, where k is a number of rows and J is a number of columns, random values are attributed. Initial values are limited to the range of the corresponding input parameter (variable). Though it is simple to realize random initialization of weight vectors, such way of initialization gives a big variation of components into SOM that increases learning time.

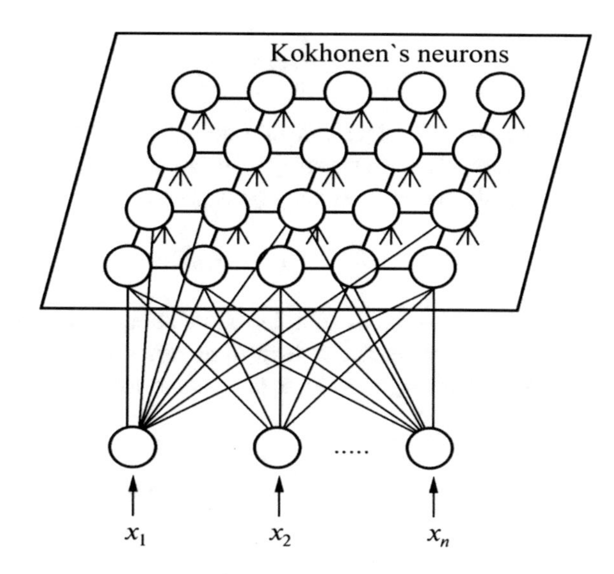

Fig. 2.15 Structure of SOM

2. To weight vectors randomly chosen input patterns are attributed, i.e. $W_{kj} = X_p$, where p is a pattern index. Such approach can result in premature convergence until weight aren't disturbed with small casual values.
3. To find the main components of vectors of input space and to initialize weight vectors with these values.
4. Other technology of initialization of scales consists in definition of a hyper cube of rather big size covering all learning patterns. The algorithm begins work with finding of four extreme points by definition of four extreme learning patterns. At first find two patterns with the greatest distance from each other in an Euclidean metrics. The third pattern is placed in the most remote point from these two patterns, and the fourth pattern—with the greatest distance from these three patterns according to Euclid. These four patterns form lattice corners on the Map (SOM: (1, 1); (1, J); (K, 1); (K, J). Weight vectors of other neurons are defined by interpolation of four chosen patterns as follows. The weight of boundary neurons are initialized so [10, 11]:
5. $W_{1j} = \frac{w_{1J} - w_{11}}{J-1}(j-1) + W_{11},$
6. $W_{Kj} = \frac{w_{KJ} - w_{K1}}{J-1}(j-1) + W_{K1},$
7. $W_{kj} = \frac{w_{KJ} - w_{11}}{K-1}(k-1) + W_{11},$ (2.34)
8. $W_{KJ} = \frac{w_{KJ} - w_{1J}}{K-1}(k-1) + W_{1J},$
9. For all $j = 2, 3, \ldots, J-1$ and $k = 2, 3, \ldots, K-1$.
10. other weight vectors are initialized so:

$$W_{Kj} = \frac{w_{KJ} - w_{K1}}{J - 1}(j - 1) + W_{K1}. \quad (2.35)$$

The standard algorithm of learning for SOM is stochastic in which weight vectors adapt after display of each pattern of a network. For each neuron the related code vector (weight) adapts so:

$$W_{kj}(t+1) = W_{kj}(t) + h_{mn,kj}(t)\left\lfloor X_p - W_{kj}(t)\right\rfloor, \quad (2.36)$$

where *m, n*—index of a line and column of winner neuron correspondingly. The winner neuron is defined, as usual, by calculation of Euclidean distance from each weight vector to an input vector and a choice of neuron, the closest to an input vector, i.e.

$$\left\|W_{mn} - X_p\right\|^2 = \min_{(k,j)}\left\{\left\|W_{kj} - X_p\right\|^2\right\}, \quad (2.37)$$

Function $h_{mn,kj}(t)$ in the Eq. (2.36) is considered as function of the neighborhood. Thus, only those neurons which are in the vicinity (in the neighborhood) of *(m, n)* winner neuron, change the weights. It is necessary for ensuring convergence, that $h_{mn,kj}(t) \to 0$ at $t \to \infty$.
Function of the neighborhood usually is function of distance between coordinates of the neurons presented on the map i.e.

$$h_{mn,kj}(t) = h\left(\left\|c_{mn} - c_{kj}\right\|^2, t\right), \quad (2.38)$$

where $c_{mn}, c_{kj} \in R$, and with increase in distance $\left\|c_{mn} - c_{kj}\right\|^2, h_{mn,kj}(t) \to 0$. The neighbourhood can be determined by a square or a hexagon. However are most often used a smooth Gaussian kernel:

$$h_{mn,kj}(t) = \alpha(t)\exp\left(-\frac{\left\|c_{mn} - c_{kj}\right\|^2}{2\sigma^2(t)}\right). \quad (2.39)$$

Here $\alpha(t)$ is the learning speed, and $\sigma(t)$-kernel width. Both functions $\alpha(t)$ and $\sigma(t)$ are monotonously decreasing functions with increase i***Creation of SOM in Batch Mode***

Stochastic learning algorithm of SOM is too slow owing to need of weights adaptation of all neurons after each display of a pattern. The version of SOM learning algorithm in batch mode was developed. The first package SOM learning algorithm was developed by Kohonen and is described below [11].

1. To initialize weight vectors by purpose of the first KJ of the learning patterns where KJ—total number of patterns on the map.

Until stop condition won't be satisfied,
for each neuron of kj do
make the list of all patterns of X_p which are the closest to a weight vector of this neuron;
end.

2. For each weight vector of w_{kj} calculate new value of a weight vector as an average of the corresponding list of patterns.

Also the accelerated version of package SOM learning algorithm was developed. One of the design problems arising at creation of SOM—determination of the map sizes. Too many neurons can cause glut when each learning pattern is attributed to various neurons. Or alternately, final SOM can successfully create good clusters from similar patterns, but many neurons will be with zero or close to the zero frequency of use where the frequency of neuron means number of patterns for which the neuron became the winner. Too small number of neurons, on the other hand, leads to clusters with big intra cluster dispersion.

The Accelerated Package Algorithm of Creation of SOM

1. To initialize weight vectors of w_{kj}, using any way of initialization.
2. Yet until stop condition(s) won't be satisfied

for each neuron k_j do
calculate an average value for all patterns for which this neuron was the winner;
designate average as $\bar{w}_{kj}$;
end.

3. To adapt weight value for each neuron, using expression

$$W_{kj} = \frac{\sum_n \sum_m N_{nm} h_{nm,kj} \bar{W}_{nm}}{\sum_n \sum_m N_{nm} h_{nm,kj}}, \tag{2.40}$$

where indexes m, n are summarized according to all numbers of rows and columns; N_{nm} is the number of patterns for which the neuron appeared to be the winner, and $h_{nm,kj}$—function of the neighborhood which specifies, whether neuron of (m, n) gets into area of the neighborhood of (k, j) neuron and in what degree;
end.
The method of search close to optimum structure of SOM consists in beginning with small structure and to increase the Maps sizes when the increase in number of neurons is required. We will notice that development of the map takes place along with learning process. Consider one of algorithms of SOM structure development for the rectangular Map [11].

Algorithm of SOM Structure Development

To initialize weight vectors for small SOM,
yet until the stop condition(s) won't be satisfied do;
until a condition of the Map growth won't be true, do;
to learn SOM for t displays of patterns, using any method of SOM learning;
end.

If the condition of the Map growth is satisfied,

1. find *(k, j)* neuron with the greatest error of quantization (intercluster dispersion);
2. find the most distant direct neighbor of *(m, n)* in rows of the Map;
3. find the most remote neuron in Map columns;
4. insert a column between neurons of *(k, j)* and *(r, s)*, and a line between neurons of *(k, j)* and *(m, n)* (this step keeps rectangular structure of the Map);
5. for each neuron *(a, b)* in a new column initialize the corresponding vector of W_{ab}, using expression

$$W_{ab} = \gamma\left(W_{a,b-1} + W_{a,b+1}\right), \tag{2.41}$$

   and for each neuron in a new row calculate

$$W_{ab} = \gamma\left(W_{a-1,b} + W_{a+1,b}\right) \tag{2.42}$$

   where $\gamma \in (0, 1)$;
   end.
6. To adjust the weights of the final structure of SOM, using additional learning iterations, until convergence will be reached.

The increase in the sizes of the map needs to be stopped when one of the following criteria is satisfied:

- the maximum size of the Map is reached;
- the greatest error of quantization for neuron will become less than threshold ε determined by the user;
- the error of the map quantization converged to a preset value.

Some aspects of this algorithm demand the explanation.
There are constants ε, γ and the maximum SOM size, and also various stop conditions. For parameter γ a good choice is $\gamma = 0.5$. The idea of an interpolation step consists in assigning a weight vector to new neuron so that it will take patterns from *(k, j)* neuron with the greatest error of quantization to reduce an error of this neuron. Value $\gamma < 0.5$ will locate neuron *(a, b)* closer to *(k, j)* that, perhaps, will lead to that more patterns will be taken by it at *kj* neuron, value $\gamma > 0.5$ will cause a boomerang effect.

The threshold of an quantization error ε is important to provide the sufficient size of the SOM Map. Small value ε leads to too big SOM size whereas too great value of ε can lead to increase in learning time to reach rather big size of structure.

It is easy to define the upper bound of the Map size, it is equal simply to the number of learning patterns p_T. However it, naturally, is undesirable. The maximum size of the Map can be presented as βp_T, where $\beta \in (0, 1)$. Optimum value β depends on the solved task, and it is necessary to take measures to provide not too small value β if the increase in the SOM sizes isn't provided.

Process of learning of SOM is too slow owing to a large number of iterations of scales correction. For reduction of computing complexity and acceleration of convergence of learning some mechanisms are offered. One of them is batch mode of designing of SOM. Two other mechanisms include control of functions of the neighborhood and learning speed.

If Gaussian neighborhood function is used, all neurons drop to the winner neuron neighborhood area, but with different degree. Therefore introducing a certain threshold θ, it is possible to limit number of neurons which will get to this area and by that to reduce computational costs of correction of their weights. Besides, the width of the neighborhood function σ can be changed dynamically during learning process. For example, it is possible to choose this function so:

$$\sigma(t) = \sigma(0)e^{-\frac{t}{\tau_1}}, \tag{2.43}$$

where $\tau_1 > 0$, is some constant; $\sigma(0)$ is the initial rather big variation.

Similarly it is possible to use the learning speed $\alpha(t)$ decreasing with increase in time:

$$\alpha(t) = \alpha(0)e^{-\frac{t}{\tau_2}}, \tag{2.44}$$

where $\tau_2 > 0$, is some constant; $\alpha(0)$ is initial, rather big variation.

Clustering and Visualization. Application of SOM

Implementation of SOM learning process consists in a clustering (grouping) of similar patterns, keeping thus topology of input space. After learning the set of the learned weights is obtained without obvious borders between clusters.

The additional step for finding of borders between clusters is required.

One of the ways to define and visualize these borders between clusters is to calculate the unified matrix of distances (U-matrix [11]) which includes geometrical approximation of distribution of weight vectors on the Map. The U-matrix expresses for each neuron distance to weight vectors of the next neurons. Great values in a matrix of distances of U indicate location of borders between clusters.

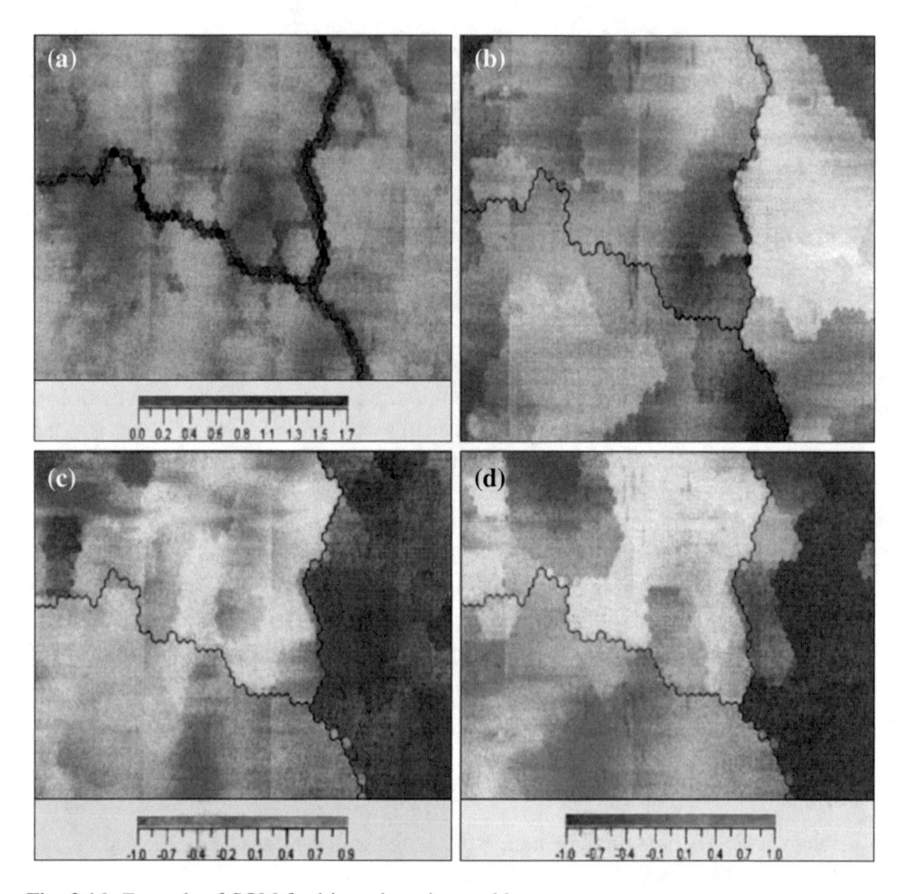

Fig. 2.16 Example of SOM for irises clustering problem

As an example consider a problem of a clustering of irises. Its statement and the description are given in Chap. 4. For example, in Fig. 2.16, the U-matrix for a problem of irises clustering with use of scaling of the SOM Map in shades of the grey is presented. Thus borders between clusters are marked in more dark color.

For the same problem in Fig. 2.16b clusters on a full map are visualized. Borders between them can usually be found by one of clustering methods, for example, Ward's method (see Chap. 7). The clustering by Ward's method uses approach "from down-up" in which each neuron originally forms an own cluster. On the subsequent iterations two next clusters merge in one until the optimum or certain number of clusters is designed. The end result of a clustering is the set of clusters with the minimum intra cluster dispersion and big inter-cluster dispersion.

For definition, what clusters need to be united, the metrics (distances) according to Ward is used (see Chap. 7). The metrics of distance is defined so:

$$d_{rs} = \frac{n_r n_s}{n_r + n_s} \|W_r - W_s\|^2, \tag{2.45}$$

where r and s are indexes of clusters; n_r and n_s are number of patterns in the corresponding clusters; W_s and W_r are vectors of the gravity centers of these clusters. Two clusters s and r unite (merge) if their metrics of d_{rs} is the smallest. For the new created cluster q its weight is defined thus:

$$W_q = \frac{1}{n_r + n_s}(n_r W_r + n_s W_s), \tag{2.45}$$

and $n_q = n_r + n_s$.

It's worth to note that for preservation of topological structure two clusters can be integrated only if they are adjacent (neighbors).

The main advantage of SOM consists in easy visualization and interpretation of the clusters created on the Map. In addition to visualization of the full map presented in Fig. 2.17b, separate components of vectors can be visualized, i.e. for each input feature the separate Map for visualization (display) of values distribution of this feature in space can be constructed, using the corresponding color gamut. Such Map and the planes of features can be used for research and the analysis of data. For example, the marked region on the visualized map can be designed on the feature plane to find distribution of values of the corresponding input parameters (features) for this region. In Fig. 2.16c, d SOM's for the third and fourth features for a problem of the irises clustering are presented.

The learned SOM can be also used as the classifier [11] as information on clusters is inaccessible during learning process, (SOM) it is necessary to investigate the clusters created on the Map manually and to assign appropriate tags of classes. Further the vector of input data is entered into the map and the winner neuron is defined. The corresponding tag of a cluster to which the entered input vector belongs, is used as a name (a number) of a class.

In the mode of a recall SOM can be used for interpolation of the missed values in a pattern. When entering the input of such pattern, ignoring inputs with the

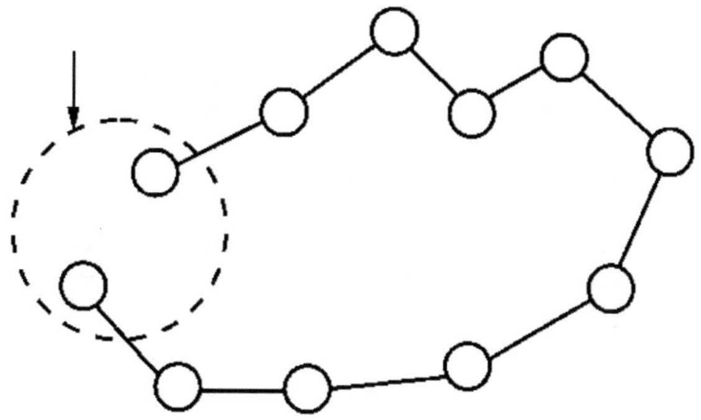

Fig. 2.17 Area of violation of a continuity of mapping using SOMs

missed values the winner neuron is defined. Further the missed value is determined by the corresponding feature value of a winner neuron, or by interpolation of values of the neighbor neurons.
The described topographic maps give an evident presentation of structure of data in multidimensional input space, which geometry we aren't able to imagine otherwise [10]. Visualization of multidimensional information is the main use of Kohonen's maps.
Note that in consent with the general everyday principle "free lunches don't happen" topographic maps keep the proximity relation only locally, i.e. neighbors on the map of area are close and in the initial space, but not on the contrary (Fig. 2.17). Generally there is no mapping cutting the dimension and keeping the relation of proximity globally.
In Fig. 2.17 the arrow shows the area of violation of a continuity of mapping, neighbors points on the plane are displayed on the opposite ends of the map.
The convenient instrument of visualization is the coloring of topographic maps how it used on usual maps. Each feature generates the corresponding coloring of the map by average value of this feature at the data which got to this cell [10].
Having collected maps of all of the interesting features, we'll receive the topographical atlas giving an integrated presentation of the structure of multi-dimensional data. Self-learning Kohonen's networks are widely used for data preprocessing at pattern recognition in space of very big dimension. In this case, that procedure to be effective, it is required to compress at first input information by one or another way:

1. or to lower dimension, having defined significant features;
2. or to make quantization of data.

SOM are applied to the solution of a wide range of real problems, including the analysis of images, recognition of the speech, the analysis of musical patterns, processing of signals, robotics, telecommunications, data mining of the hidden knowledge and the analysis of time series [9].

References

1. Hopfield, J.J.: Neural Networks and physical systems with emergent collective computational abilities. Proc. Natl. Acad. Sci. USA, **79**, 2554—2558 (1982)
2. Zaychenko, Y.P.: Fundamentals of intellectual systems design. Kiev. Publishing House, "Slovo", pp. 352 (2004) (rus)
3. Chung, F.L., Lee, T.: Fuzzy competitive learning. Neural Netw. 7, 539–552 (1994)
4. Deb, K., Joshi, D., Anand. A.: Real-coded evolutionary algorithms with parent-centric recombination. In: Proceedings of the IEEE Congress on Evolutionary Computation, pp. 61–66, (2002)
5. Hopfield, J.J.: Neurons, dynamics and computation. Phys. Today, **47**, 40–46 (1994)

6. Heykin, S.: Neural networks. Full course. (2nd edn). Transl. engl. Moscow.-Publishing House "Williams". pp. 1104 (2006). (rus)
7. Hebb, D.O.: The Organization of Behavior: A Neuropsychological Theory. Wiley, New York (1949)
8. Kohonen, T.: Self-Organization and Associative Memory. (3rd edn). Springer, New York (1988)
9. Osovsky, S.: Neural networks for information processing, transl. from pol.—M.: Publishing house Finance and Statistics. pp. 344 (2002). (rus)
10. Kohonen, T.: Self-organized formation of topologically correct feature maps. Biol. Cybern. **43**, 59–69 (1982)
11. Engelbrecht, A.: Computational Intelligence. An Introduction (2nd edn). John Wiley & Sons, Ltd., pp. 630 (2007)

Chapter 3
Fuzzy Inference Systems and Fuzzy Neural Networks

3.1 Introduction

In recent years, the attention of many researchers in the field of artificial intelligence systems attracts the problem of decision making under uncertainty, the incompleteness of the initial data and quality criteria.

There is a new trend in the theory of complex decision-making, which is rapidly developing—making decisions under uncertainty. A promising approach for solving many decision-making problems under uncertainty and incomplete information is based on fuzzy sets and systems theory created by Zadeh [1].

The introduction by Zadeh of the concept of linguistic variables described by fuzzy sets [2] gave rise to a new class of systems—fuzzy logic systems (FLS), which allows to formalize fuzzy expert knowledge. The use of fuzzy inference systems (FIS) and built on the their basis fuzzy neural networks (FNN) has allowed to solve many problems of decision-making under uncertainty, incompleteness and qualitative information—forecasting, classification, cluster analysis, pattern recognition.

Chapter 3 is devoted to the detailed consideration of FL systems. It discusses the basic algorithms of fuzzy inference Mamdani, Tsukamoto, Larsen and Sugeno (Sect. 3.2). In Sect. 3.3 the methods of defuzzification are described.

In the Sect. 3.4 the important Fuzzy approximation theorem (FAT-theorem) is considered which is theoretical ground for wide applications of FNN.

Further fuzzy controller (FC) Mamdani and Tsukamoto and classical learning algorithm on the basis of back-propagation are detailly considered. A new learning algorithm of FC Mamdani and Tsukamoto for Gaussian membership functions (MF) of gradient type is described (Sect. 3.6).

Next FNN ANFIS is considered, its architecture and gradient learning algorithm are presented (Sect. 3.7). Then FNN TSK, the development of FNN ANFIS, is described and its hybrid training algorithm is reviewed (Sect. 3.8). In the Sect. 3.9 adaptive wavelet-neuro-fuzzy networks are considered and different learning

M.Z. Zgurovsky and Y.P. Zaychenko, *The Fundamentals of Computational Intelligence: System Approach*, Studies in Computational Intelligence 652,
DOI 10.1007/978-3-319-35162-9_3

algorithms in batch and on-line mode are presented. Cascade neo-fuzzy neural networks (CNFNN) are considered, training algorithms are presented and GMDH method for its structure synthesis is described and analyzed (Sect. 3.10).

3.2 Algorithms of Fuzzy Inference

Used in different expert and control systems fuzzy inference mechanism is grounded on a knowledge base, formed by experts in the subject area as a set of fuzzy rules of the predicate form [3–5]:

$$\begin{aligned} P_1 &: \text{ if } x \text{ is } A_1, \text{ then } y \text{ is } B_1 \\ P_2 &: \text{ if } x \text{ is } A_2, \text{ then } y \text{ is } B_2 \\ &\ldots \\ P_n &: \text{ if } x \text{ is } A_n, \text{ then } y \text{ is } B_n \end{aligned}$$

where x, $x \in X$—input variable (the name for the known data values);
y, $y \in Y$—output variable (name for a data value to be calculated);
A_i and B_i—membership functions defined on the sets X and Y.

Let's provide a more detailed explanation. Expert knowledge $A \to B$ reflects fuzzy causal relationship between premises and conclusions, so it can be called a fuzzy relation and denoted as R:

$$R : A \to B,$$

where "$\to$" is called fuzzy implication.

Relation R can be regarded as a fuzzy subset of the direct product $X \times Y$ complete set of premises X and conclusions Y. Thus, the process of obtaining a (fuzzy) output results B' from the use of this observation A' and knowledge $A \to B$ can be represented as the composite fuzzy rule "modus ponens":

$$B' = A' \cdot R = A' \cdot (A \to B),$$

where "$\cdot$"—convolution operation.

Both the operation of the composition and the implication operation in the algebra of fuzzy sets may be implemented in different ways (in this case will be different and the resultant outcome), but in any case, the total fuzzy inference is carried out in the following four steps [4].

1. **Introduction of fuzziness (fuzzification)**. Membership functions defined on the input variables are applied to their actual values to determine the degree of truth of each premise(antecedent) of each rule.

2. **The logical inference**. The calculated value of the truth of the antecedent for each rule applies to the conclusion (consequent) of each rule. This results in one fuzzy subset that will be assigned to each variable output for each rule. The rules of inference typically use only operations min (minimum) or prod (multiplication). The inference MINIMUM output membership function " is cut off" at a height corresponding to the calculated degree of truth of the rule antecedents (fuzzy logic "AND"). At the inference by product output membership function is scaled by calculating the degree of truth of the rule antecedents.
3. **Composition**. All the fuzzy subsets assigned to each output variable (all rules) are combined together to form a single fuzzy subset for all output variables. With such a merge operation normally is used max (maximum) or sum (SUM). When the composition MAXIMUM is used the combined output of fuzzy rules is constructed as a point-wise maximum over all fuzzy sets (fuzzy logic "OR"). When the composition SUM is used combined output of fuzzy rules is formed as the point-wise sum of all fuzzy subsets assigned to the output variable by inference rules.
4. **Transformation to non-fuzziness (defuzzification)**. It's used when you need to convert the fuzzy set of conclusions in a crisp number. There is a considerable number of methods of transferring to non-fuzziness, some of which are discussed below.

Example 3.1 Suppose that a system is described by the following fuzzy rules:

$$P_1: \text{ if } x \text{ is } A\text{, then w is } D$$
$$P_2: \text{ if } y \text{ is } B\text{, then w is } E$$
$$P_3: \text{ if } z \text{ is } C\text{, then } w \text{ is } F$$

where x, y and z—the names of the input variables; w—the name of the output variable, A, B, C, D, E, F—set of membership function (e.g. of triangular shape).

The procedure for obtaining inference is illustrated by Fig. 3.1. It is assumed that the specific (clear) values of the input variables are: x_0, y_0, z_0.

At the first step, based on these values and on the basis of membership functions A, B, C, degrees of truth $\alpha(x_0), \alpha(y_0)$ и $\alpha(z_0)$ of prerequisites (antecedents) for each of the three aforementioned rules are determined. At the second stage, a "cut-off" membership functions of rules consequence (D, E, F) on levels $\alpha(x_0), \alpha(y_0)$ and $\alpha(z_0)$ are determined. At the third stage, the output membership functions, truncated at the previous stage are considered and their composition (unification) with operations max is produced, resulting in a combined fuzzy subset described by the membership function $\mu_{\Sigma}(w)$ corresponding to the output variable w. Finally, in the fourth step is, if necessary, the crisp value of the output variable, for example, using

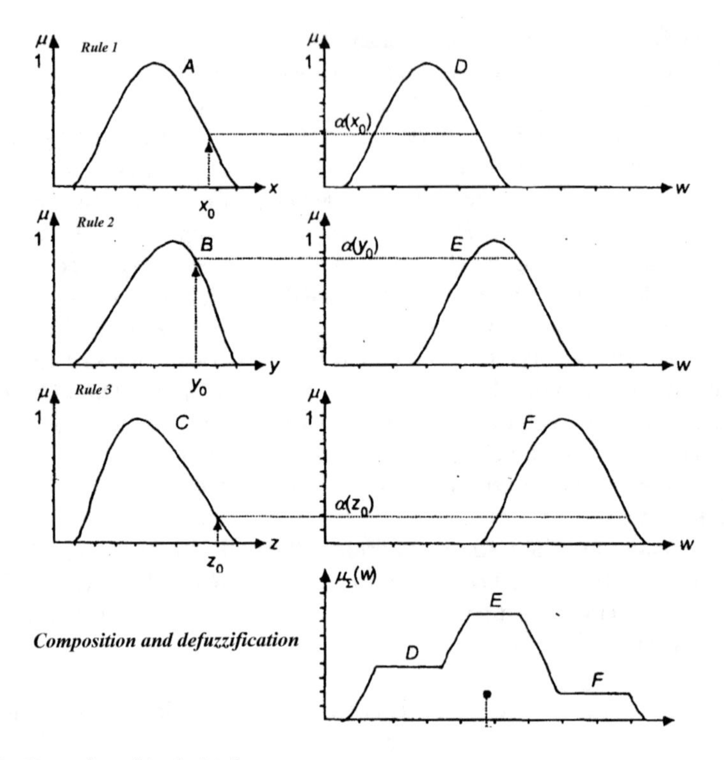

Fig. 3.1 Illustration of logical inference

centroid method: is defined as the center of gravity for the curve $\mu_\Sigma(w)$: $w_0 = \frac{\int_\Omega w \cdot \mu_\Sigma(w)dw}{\int_\Omega \mu_\Sigma(w)dw}$.

Consider the following modification of the most commonly used fuzzy inference algorithm assuming for simplicity that the knowledge base contains two fuzzy rules of the form:

$$P_1: \text{ if } x \text{ is } A_1 \text{ and y is } B_1, \text{ then } z \text{ is } C_1$$
$$P_2: \text{ if } x \text{ is } A_2 \text{ and y is } B_2, \text{ then } z \text{ is } C_2$$

where x and y—the names of the input variables, z—the name of the output variable, $A_1, B_1, C_1, A_2, B_2, C_2$—some membership functions. At the same time a crisp value z_0 must be determined on the basis of the above information and crisp values x_0 and y_0.

3.2.1 Mamdani Fuzzy Inference

This algorithm corresponds to the example considered in the Fig. 3.1. In this situation, it can be formally described as follows [3–5]:

1. Introduction of fuzziness. the truth degrees for premises (antecedents) of each rule are determined: $A_1(x_0), A_2(x_0), B_1(x_0), B_2(x_0)$.
2. The logical inference. The cut-off levels for the premises of each rule are determined (using operation minimum):

$$\alpha_1 = A_1(x_0) \wedge B_1(y_0);$$
$$\alpha_2 = A_2(x_0) \wedge B_2(y_0);$$

where by "∧" denotes the operation of the logical intersection (min).

Then the membership functions (MF) of consequent of each rule are calculated (operation of fuzzy implication):

$$C_1' = (\alpha_1 \wedge C_1(z));$$
$$C_2' = (\alpha_2 \wedge C_2(z)).$$

3. Composition. Unification of found truncated membership functions for consequences using MAX operation is performed (this operation is designated hereinafter as "∨"), resulting in a fuzzy subset of the final output with membership function:

$$\mu_\Sigma(z) = C_1'(z) \vee C_2'(z) = (\alpha_1 \wedge C_1(z)) \vee (\alpha_2 \wedge C_2(z)) \tag{3.1}$$

4. Defuzzification It's performed to find crisp z_0, using for example, centroid method.

3.2.2 Tsukamoto Fuzzy Inference Algorithm

Assumptions are the same as in the previous algorithm, but here it is assumed that the function $C_1(z)$, $C_2(z)$ are monotonous (see. Fig. 3.2) [3–5]:

1. Introduction of fuzziness (like Mamdani algorithm).
2. Fuzzy inference. First, "cut-off" levels" α_1 and α_2 (as in Mamdani algorithm) are determined and then the following equations are to be solved:

$$\alpha_1 = C_1(z_1) \quad \text{and} \quad \alpha_2 = C_2(z_2)$$

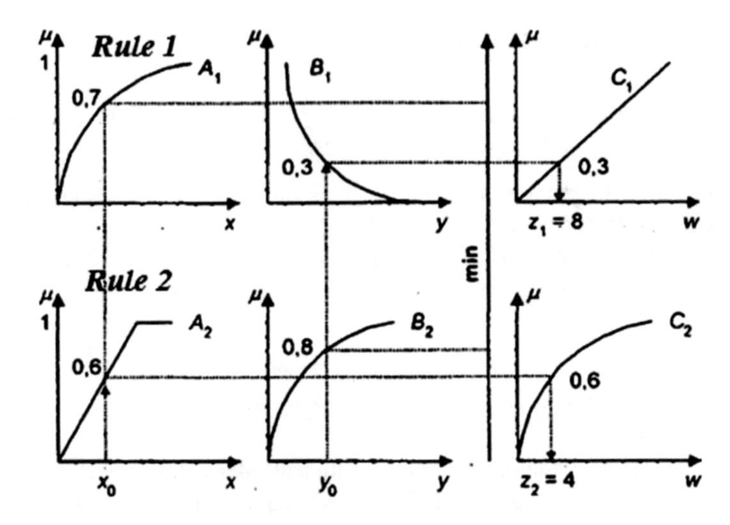

Fig. 3.2 Illustration of Tsukamoto algorithm

as a result clearly defined (non-fuzzy) values z_1 and z_2 for each of the original rules are determined.

3. Composition. Determine the clear value of the total output variable (as the weighted average of z_1 and z_2):

$$z_0 = \frac{\alpha_1 z_1 + \alpha_2 z_2}{\alpha_1 + \alpha_2}. \tag{3.2}$$

In general, the discrete version of centroid method takes the form:

$$z_0 = \frac{\sum_{i=1}^{n} \alpha_i z_i}{\sum_{i=1}^{n} \alpha_i}. \tag{3.3}$$

3.2.3 *Sugeno Fuzzy Inference*

Takagi and Sugeno have used a set of rules in the following form (as before, we give an example of two rules):

$$P_1: \text{ if } x \text{ is } A_1 \text{ and y is } B_1, \text{ then } z_1 = a_1 x + b_1 y,$$
$$P_2: \text{ if } x \text{ is } A_2 \text{ and y is } B_2, \text{ then } z_2 = a_2 x + b_2 y,$$

where a_i, b_i, are some constants.

Description of the algorithm (Fig. 3.3).

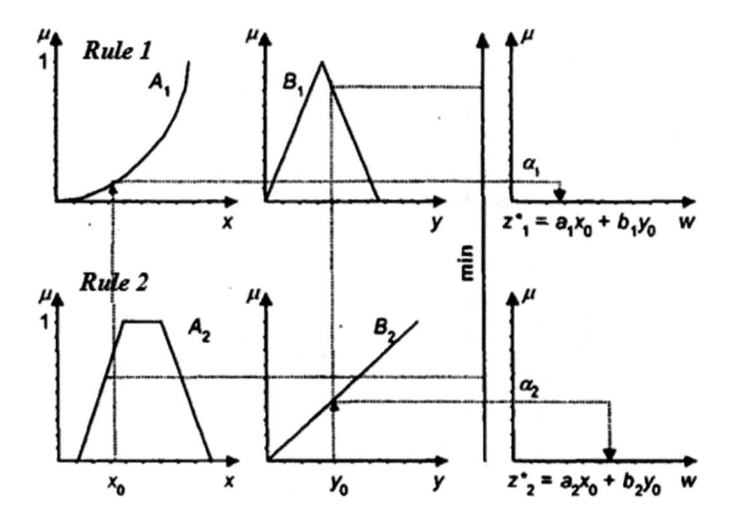

Fig. 3.3 Illustration of the algorithm Sugeno

1. Introduction of fuzziness (like Mamdani algorithm).
2. Fuzzy inference.
   The values $\alpha_1 = A_1(x_0) \wedge B_1(y_0)$, $\alpha_2 = A_2(x_0) \wedge B_2(y_0)$ are determined and individual outputs of the rules are calculated:

$$\dot{z}_1 = a_1 x_0 + b_1 y_0$$

$$\dot{z}_2 = a_2 x_0 + b_2 y_0$$

3. Composition of the output variables and total output determination

$$z_0 = \frac{\alpha_1 \dot{z}_1 + \alpha_2 \dot{z}_2}{\alpha_1 + \alpha_2} \tag{3.4}$$

3.2.4 *Larsen Fuzzy Inference*

The Larsen fuzzy inference algorithm is modeled using the multiplication operator.

Description of the algorithm (Fig. 3.4).

1. Fuzzification (as Mamdani algorithm)
2. Fuzzy inference. First, as in the algorithm of Mamdani, values:

$$\alpha_1 = A_1(x_0) \wedge B_1(y_0) = A_1(x_0) * B_1(y_0)$$
$$\alpha_2 = A_2(x_0) \wedge B_2(y_0) = A_2(x_0) * B_2(y_0).$$

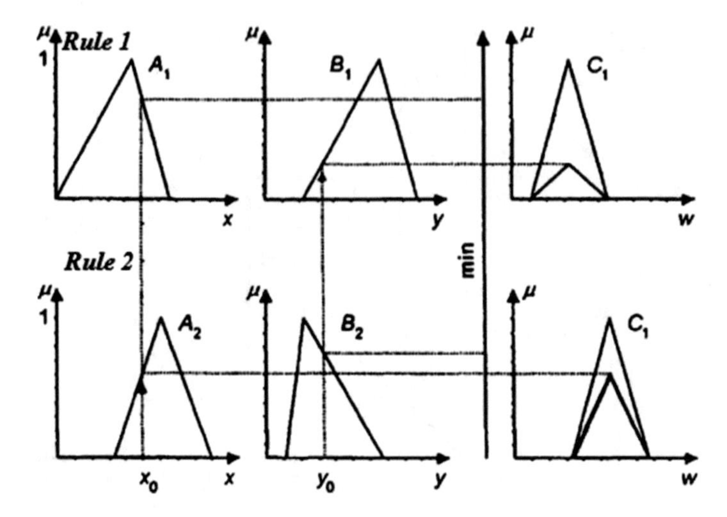

Fig. 3.4 Illustration of the algorithm Larsen

are determined. In this algorithm operation $\wedge$ is realized as product * and then output fuzzy subsets for each rule are calculated using fuzzy implication as product operation (Larsen form):

$$\alpha_1 * C_1(z),\ \alpha_2 * C_2(z).$$

3. Composition. The resulting output fuzzy subset is determined (like Mamdani):

$$\mu_\Sigma(z) = C(z) = (\alpha_1 C_1(z)) \vee (\alpha_2 C_2(z)). \tag{3.5}$$

(in general for n rules: $\mu_\Sigma(z) = C(z) = \bigvee_{i=1}^{n} (\alpha_i C_i(z))$)

4. If necessary, defuzzification (as in the previously discussed Mamdani algorithm) is performed.

3.3 Methods of Defuzzification

1. One of these methods—centroid has already been considered. Here the corresponding formula is presented again in general case, [2, 3]

$$z_0 = \frac{\int_\Omega z \cdot C(z) dz}{\int_\Omega C(z) dz}, \tag{3.6}$$

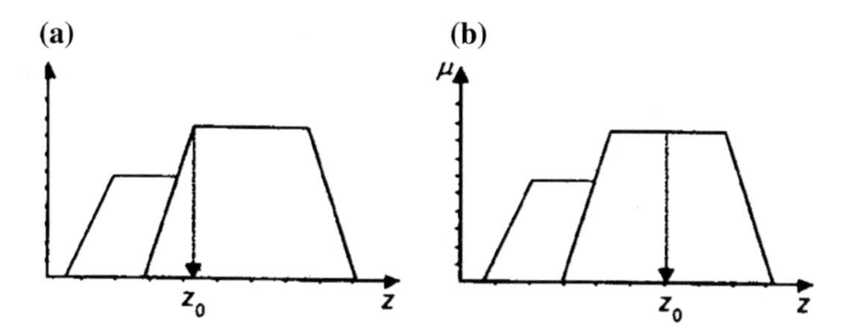

Fig. 3.5 Illustration of the method of defuzzification **a** the first maximum; **b** the average maximum

discrete version

$$z_0 = \frac{\sum_{i=1}^{n} \alpha_i z_i}{\sum_{i=1}^{n} \alpha_i}$$

2. The first maximum (First-of-Maxima-FOM). The crisp output value is the smallest value at which the maximum of the final fuzzy MF is attained (Fig. 3.5a): $z_0 = \min\left\{ z \middle| C(z) = \max_U C(U) \right\}$.
3. The average maximum (Middle-of-Maxima MOM). Crisp value is given by:

$$z_0 = \frac{\int_G z dz}{\int_G dz}, \tag{3.7}$$

where G—subset of elements maximizing C (Fig. 2.5b).
For discrete version (set G is discrete)

$$z_0 = \frac{1}{n} \sum_{i=1}^{n} z_i. \tag{3.8}$$

4. The criterion of maximum (Max-Criterion). Crisp value is selected randomly among the many elements, which maximize C

$$z_0 \in \left\{ z : C(z) = \max_u C(U) \right\}. \tag{3.9}$$

5. (Height defuzzification). Elements of the domain Ω, for which membership function values are less than a certain level α, are not taken into account, and a crisp value is calculated accordingly:

$$z_0 = \frac{\int_{C_\alpha} z \cdot C(z)dz}{\int_{C_\alpha} C(z)dz}, \tag{3.10}$$

3.4 Fuzzy Approximation Theorems

The possibility of wide applications of fuzzy logic is based on the following results [6]:

1. In 1992, Wang (Wang) showed that a fuzzy system is an universal approximator, i.e. it can approximate any continuous function on a compact U with arbitrary precision, when using a set of n ($n \to \infty$) rules [1]:

   P_1: if x is A_i and y is B_i, then z is C_i, $i = 1 \ldots n$, under the following conditions:

- Gaussian membership functions:

$$A_i(x) = \exp\left[-\frac{1}{2}\left(\frac{x - \alpha_{i1}}{\beta_{i1}}\right)^2\right]$$

$$B_i(y) = \exp\left[-\frac{1}{2}\left(\frac{y - \alpha_{i2}}{\beta_{i2}}\right)^2\right] \tag{3.11}$$

$$C_i(z) = \exp\left[-\frac{1}{2}\left(\frac{z - \alpha_{i3}}{\beta_{i3}}\right)^2\right];$$

- intersection as a product: $[A_i(x) \text{ and } B_i(y)] = A_i(x)B_i(y)$;
- implications in the form of Larsen:

$$[A_i(x) \text{ and } B_i(y)] \to C_i(z) = A_i(x)B_i(y)C_i(z);$$

- centroid method of defuzzification.

$$z_0 = \frac{\sum_{i=1}^{n} \alpha_{i3} A_i B_i}{\sum_{i=1}^{n} A_i B_i}, \quad (3.12)$$

where α_{i3}—centers C_i.

In other words, Wang proved the theorem: for every real continuous function g, given on a compact U for an arbitrary $\varepsilon > 0$ there exists a fuzzy system, which generates the function $f(x)$ such that

$$\sup_{x \in U} \|g(x) - f(x)\| \leq \varepsilon,$$

where $\|\cdot\|$—a symbol of adopted distance between the functions.

2. In 1995, Castro (Castro) showed that the Mamdani fuzzy logic controller is also universal approximator under the following conditions [7]:

- Number of rules $n \to \infty$,
- symmetrical triangular membership functions:

$$A_i(x) = \begin{cases} 1 - |a_i - x|/\alpha_i, & if\ |a_i - x| \leq \alpha_i; \\ 0, & if\ |a_i - x| > \alpha_i; \end{cases}, \quad (3.13)$$

$$B_i(y) = \begin{cases} 1 - |b_i - y|/\beta_i, & if\ |b_i - y| \leq \beta_i; \\ 0, & if\ |b_i - y| > \beta_i; \end{cases}. \quad (3.14)$$

- intersection using the operation minimum:

$$[A_i(x)\ and\ B_i(y)] = \min\{A_i(x), B_i(y)\};$$

- implication in the form of Mamdani and centroid method of defuzzification:

$$z_0 = \frac{\sum_{i=1}^{n} c_i \min\{A_i(x), B_i(y)\}}{\sum_{i=1}^{n} \min\{A_i(x), B_i(y)\}}, \quad (3.15)$$

where c_i—centers C_i.

In general, fuzzy logic systems it's reasonable to apply in the following cases [3, 4]:

- for complex processes where there is no simple mathematical model;
- if the expert knowledge about the object or the process can be formulated only in linguistic form.

Systems that are based on fuzzy logic, are applied inappropriately:

- If a desired result can be obtained by some other (standard) method;
- When for an object or process an adequate and easily researched mathematical model has already been or may be found.

As a conclusion note the main advantages of fuzzy logic systems:

1. they enable to work with fuzzy, incomplete and qualitative information; under uncertainty conditions;
2. they enable to use expert information in the form of fuzzy inference rules. Apart from great advantages fuzzy system have several drawbacks, namely:

Base of the rules formulated by an expert, may be incomplete or contradictory; membership functions of linguistic variables may be inadequate to real simulated processes.

To eliminate these shortcomings it's needed to use the training of fuzzy logic systems, i.e., make them adaptive.

3.5 Fuzzy Controller Based on Neural Networks

One of the first practical applications of fuzzy systems was the sphere of management, in which are widely used so-called fuzzy controllers (FC) [3, 8].

For the design of a fuzzy controller the linguistic rules and MF (membership function) are to be given that to represent linguistic variables. Specification of good linguistic rules depends on expert knowledge of management system. But the translation of this knowledge into fuzzy sets—this task is not formalized and need to make a choice on the basis of, for example, forms of MF. The quality of the fuzzy controller (FC) is achieved by changing the shape of MF.

Artificial neural network (ANN) is a highly parallel architecture and consist of similar elements interacting via connections which are defined by weights. Using ANN, we can not only approximate the function, but also study (investigate) control objects, applying learning and self-learning procedures. The problem lies herein training is wasting a lot of time but the result is not always guaranteed. But it is possible to implement the previously acquired knowledge in the form of rules in already trained NN to simplify learning.

Connecting NN and fuzzy logic systems (FL) allows you to combine all their advantages and avoid their weaknesses. To integrate these two technologies (FL) and ANN the structure of FL system should be presented as a corresponding

NN. This approach uses the NN for optimization of parameters of conventional neural controller (NC) or for retrieving rules from the data. Choice of MF, which represents linguistic term, is more or less arbitrary. For example, consider the linguistic term “approximately zero”. It is obvious that the corresponding fuzzy set MF is to be unimodal and reach its maximum at the point zero. The correct choice of membership functions has become the main and most important task of the NC.

NN offers the opportunity to solve this problem. The method of feed propagation (of signal) in a neural network allows to choose MF form, which depends on several parameters and can be adjusted in the learning process. As MF one can choose a symmetrical triangular shape, which depends on two parameters, one of which is the value at which AF reaches a maximum value, and the second—the length of the interval.

Training data should be divided into r cluster $R_1, \ldots, R_r$. Each cluster R_i corresponds to some decision rule R_i. Cluster elements are presented as values in the form of (X, Y), where $X = [x_1, \ldots, x_n]$—vector of input variables, Y—the output variable.

Consider the dynamic system S, in which the control is carried out by means of a variable C, and its state can be described by n variables $x_1, \ldots, x_n$.

Linguistic variables are modeled and manipulated by membership functions, and control action which leads to the desired system state is described by the fuzzy rule “if-then”. To get the desired value (i.e. control value), it is necessary to solve the problem of defuzzification, for which we use the monotonic membership function of Tsukamoto, (see Fig. 3.6), where defuzzification reduced to the application of the inverse function.

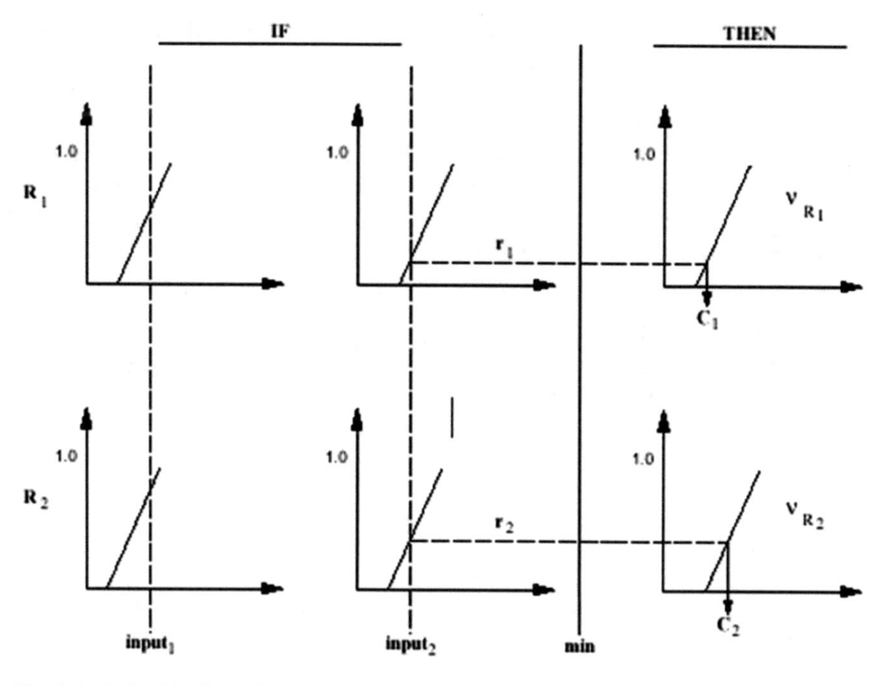

Fig. 3.6 Defuzzification using monotonous membership function of Tsukamoto

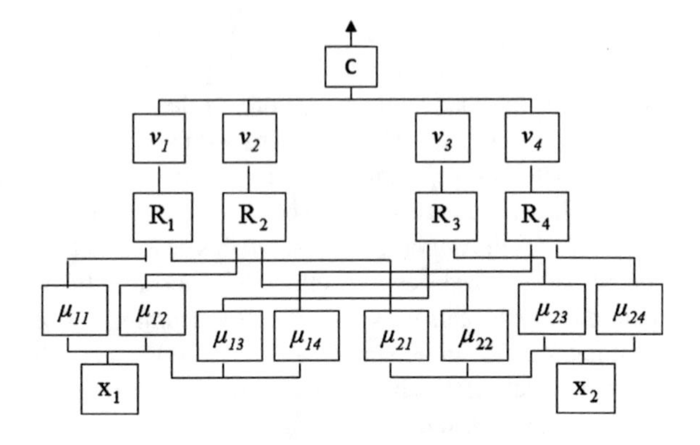

Fig. 3.7 The structure of a fuzzy neural controller

This membership function μ is characterized by two parameters a and b of the MF, and it is defined as

$$\mu(x) = \begin{cases} (-x+a)/(a-b), & \text{if } x \in [a,b] \vee (x \in [b,a] \wedge (a>b)) \\ 0, & \text{otherwise} \end{cases}. \tag{3.16}$$

Defuzzification is carried out in such a way:

$$x = \mu^{-1}(y) = -y(a-b) + a = a + y(b-a);\ y \in [0;1]$$

For our purposes, we must limit the monotonous MF to represent the linguistic value of the original variable. For the input value is usually used triangular or trapezoidal MF.

In Fig. 3.7 the structure of fuzzy neural controller is presented [4].

Modules x_1 and x_2 here are the input variables, and they send their values to their μ-modules that contain the appropriate MF. μ-modules are linked to R-modules, which represent fuzzy rules "if-then". Every μ-module transmits to all R-modules MF values $\mu_{ij}(x_i)$ of its input variable x_i. each R-module performs the intersection operation, finds $\min_i\{\mu_{ij}(x_i)\}$ and transmits this value further to v-modules which comprise MF describing output variables. v-modules using monotonous membership functions, calculate the value r_i and $v^{-1}(r_i)$ and transmit them to the C-modules which calculate the final output variable—control action C according to formula (3.17), that is, uses the centroid defuzzification algorithm

$$C = \frac{\sum_{i=1}^{n} r_i v^{-1}(r_i)}{\sum_{i=1}^{n} r_i}, \tag{3.17}$$

where n—the number of inference rules; r_i—is a degree to which the rule R_i is executed.

It is easy to see the system in the Fig. 3.7 resembles a sequential multilayer neural network, where x-, R-и C-modules play the role of neurons, μ- and ν-modules play the role of adaptive weights of network connections. The learning process is determined by the fuzzy error and works in parallel for each fuzzy rule. After C is generated by the controller and the new state of the object becomes known, the error E is calculated, which propagates in the opposite direction. Each rule analyzes its own contribution to the control output and assesses its conclusion (i.e., it has resulted in an increase or decrease of error). Thus, if the rule acted in the right direction, it should be made more sensitive and to work out such a conclusion which increases control. Conversely, if the rule acted in the wrong direction, the rule should be made less sensitive. taking into account that the MF are described by two parameters (b_i, a_i), one of them is fixed, and the second has to change. In this case, it's assumed that a_i changes.

The learning algorithm. In [8] the following training algorithm FNN was proposed. It is used to adjust membership functions of rules antecedents and consequences.

1. Each rule R_i computes the rule error

$$R_i: \ e_{R_i} = \begin{cases} -r_i E, & signC_i = signC_{opt} \\ r_i E, & signC_i \neq signC_{opt} \end{cases}, \tag{3.18}$$

where r_i—the level of activation of rule R_i;

C_i is output of ν_i-module; C_{opt}—optimal control (FNN desired output)

2. A value e_{R_i} is transmitted to ν- and μ-module;

First, a change of MF of ν-module is produced;

$$a_k^{new} = \begin{cases} a_k - \sigma e_{R_i}|a_k - b_k|, & a_k < b_k \\ a_k + \sigma e_{R_i}|a_k - b_k|, & a_k > b_k \end{cases}, \tag{3.19}$$

where σ is a training speed.

If the module ν_k is used by several R-modules, then their MF should change as many times as a number of R-modules connected with it.

3. Adjustment of antecedents (parameters of MF):

$$a_{jk}^{new} = \begin{cases} a_{jk} - \sigma e_{R_i}\left|a_{jk} - b_{jk}\right|, & a_{jk} < b_{jk} \\ a_{jk} + \sigma e_{R_i}\left|a_{jk} - b_{jk}\right|, & a_{jk} > b_{jk} \end{cases}. \tag{3.20}$$

Note: the module x_j is connected with the module R_k through module μ_{jk}. This algorithm is called the algorithm with reinforcement.

If the learning process has been successful, it means the rule base was built correctly. It is possible to arrange the learning process so that not only to adjust the rules MF, but correct the rules themselves. If the rule output corresponds to the semantics of the desired control, then the rule is stored in the rule base. If a rule acts the opposite to the desired output, it changes to the opposite consequence or generally is completely removed from rule base.

3.6 The Gradient Learning Algorithm of FNN Mamdani and Tsukamoto

Considered above in Sect. 3.5 Mamdani learning algorithm is empirical, formulas (3.18)–(3.20) for adjusting membership functions aren't theoretically grounded. This comes from the fact that Mamdani and Tsukamoto fuzzy controller uses triangular MF, and the intersection of the rules antecedents is performed in the form of min. In the result the corresponding MF are non-differentiable.

In this regard, it is advisable to construct the analytical learning algorithm, which convergence is strictly proved. For this purpose, it is reasonable to use the Gaussian MF for the antecedents and consequences [3].

So let MF of the ith μ-module associated with the rule R_k be described as follows:

$$\mu_{ik}(x_i) = \exp\left\{-\frac{1}{2} * \frac{(x_i - a_{ik})^2}{\sigma_{ik}^2}\right\}, \tag{3.21}$$

where a_{ik}, σ_{ik} are parameters to be set up in the learning process and MF of v_k—module is following,

$$\mu_k(y_i) = \exp\left\{-\frac{1}{2} * \frac{(y_i - a_k)^2}{\sigma_k^2}\right\},$$

while the antecedents intersection in rules is specified as a product;

$$\alpha_k = \prod_{i=1}^{n} \mu_{ik}(x_i) = \exp\left\{-\sum_{i-1}^{n} \frac{1}{2} \cdot \frac{(x_i - a_{ik})^2}{\sigma_{ik}^2}\right\}.$$

Assume that the centroid defuzzification method is used, whereas total output is determined as:

$$z_0 = \frac{\sum_k z_k \alpha_k}{\sum_k \alpha_k}$$

Let the rules consequences use monotonous MF *C(z)*. Then z_k is determined by solving the following equation (controller Tsukamoto).

$$C_k(z_k) = \alpha_k, \tag{3.22}$$

where

$$C_k(z_k) = \exp\left\{-\frac{1}{2} * \frac{(z_k - a_k)^2}{\sigma_k^2}\right\}, \tag{3.23}$$

Then, solving the Eq. (3.22) $\exp\left\{-\frac{1}{2} * \frac{(z_k - a_k)^2}{\sigma_k^2}\right\} = \alpha_k$, we find two roots:

$$z_k = a_k \pm \sqrt{2\ln\frac{1}{\alpha} * \sigma_k}.$$

The first root is $z_{1k} = a_k - \sqrt{2\ln\frac{1}{\alpha} * \sigma_k}$. It is located on the site of a monotonically increasing curve $z_r(a_k)$ and second z_{2k} is located on site of monotonically decreasing curve.

Let criterion be $E(z) = \frac{1}{2}(z_0 - z*)^2 \rightarrow \min$,
where z^* is desired output; z_0 is FNN output,
then we find the derivative [4]

$$\frac{\partial E}{\partial a_k} = \frac{\partial E}{\partial z_0}\frac{\partial z_0}{\partial z_k}\frac{\partial z_k}{\partial a_k} = +(z_0 - z^*)\frac{\alpha_k}{\sum_{k=1}^{K} \alpha_k}, \tag{3.24}$$

$$\frac{\partial E}{\partial \sigma_k} = \frac{\partial E_0}{\partial z_0}\frac{\partial z_0}{\partial z_k}\frac{\partial z_k}{\partial \sigma_k} = -(z_0 - z^*)\frac{\alpha_k}{\sum_{k=1}^{K} \alpha_k} \cdot \sqrt{2\ln\frac{1}{\alpha}}, \tag{3.25}$$

on the increasing monotonically part of the curve OP μ_k;

$$\frac{\partial E}{\partial \sigma_k} = +(z_0 - z*)\frac{\alpha_k}{\sum_{k=1}^{K} \alpha_k}\sqrt{2\ln\frac{1}{\alpha}}, \tag{3.26}$$

on a monotonically decreasing part of the curve.

For input μ-modules

$$\frac{\partial E}{\partial \alpha_{ik}} = \frac{\partial E_0}{\partial z_0}\frac{\partial z_0}{\partial \alpha_k}\frac{\partial \alpha_k}{\partial \alpha_{ik}} = +(z_0 - z^*)\frac{z_k \sum_k^K \alpha_k - \sum_k^K z_k \cdot \alpha_k}{\left(\sum_k \alpha_k\right)^2} \cdot \prod_{k-1}^{K} \exp\left\{-\frac{(x_i - a_{ik})^2}{2 \cdot \sigma_{ik}^2}\right\} \cdot \frac{(x_i - a_{ik})}{\sigma_{ik}^2} =$$
$$= (z_0 - z^*) \cdot \alpha_k \cdot \frac{z_k \sum_k^K \alpha_k - \sum_k^K z_k \cdot \alpha_k}{\left(\sum_k \alpha_k\right)^2} \cdot \frac{(x_i - a_{ik})}{\sigma_{ik}^2}$$

$$\frac{\partial E}{\partial \sigma_{ik}} = \frac{\partial E}{\partial z_0}\frac{\partial z_0}{\partial \alpha_k}\frac{\partial \alpha_k}{\partial \sigma_{ik}} = \\ = \left((z_0 - z^*)\alpha_k \frac{z_k \cdot \sum_{k=1}^{K} \alpha_k - \sum_k z_k \cdot \alpha_k}{\left(\sum_k \alpha_k\right)^2} \cdot \frac{(x_i - a_{ik})^2}{\sigma_{ik}^3}\right). \tag{3.27}$$

then gradient learning algorithm for FNN Mamdani-Tsukamoto is as follows [4]:

1. for the output modules

$$a_k(n+1) = a_k(n) - \gamma_n \frac{\partial E}{\partial \alpha_k} = a_k(n) - \gamma_n (z_0 - z*) \frac{a_k}{\sum_k a_k}, \tag{3.28}$$

$$\sigma_i(n+1) = \sigma_k(n) - \gamma_n' \frac{\partial E}{\partial \sigma_k} = \\ = \sigma_k(n) \pm \gamma_n (z_0 - z_\Sigma^*) \frac{\alpha_k}{\sum_k \alpha_k} * \sqrt{2 \ln \frac{1}{\alpha_k}}. \tag{3.29}$$

for input μ-modules

$$a_{ik}(n+1) = a_{ik}(n) - \gamma_{n_2} \frac{\partial E}{\partial \alpha_{ik}} = a_{ik}(n) - \gamma_n (z_0 - z^*) \cdot \\ \cdot \frac{\alpha_k \left(z_k \cdot \sum_k \alpha_k - \sum_k z_k \cdot \alpha_k\right)}{\left(\sum_k \alpha_k\right)^2} \cdot \frac{(x_i - a_{ik})}{\sigma_{ik}^2}. \tag{3.30}$$

$$\sigma_{ik}(n+1) = \sigma_{ik}(n) - \gamma_{n_3} (z_0 - z_\Sigma^*) \frac{\alpha_k \left(z_k \cdot \sum_k \alpha_k - \sum_k z_k \cdot \alpha_k\right)}{\left(\sum_k \alpha_k\right)^2} \cdot \frac{(x_i - a_{ik})^2}{\sigma_{ik}^3}. \tag{3.31}$$

here $\gamma_n, \gamma_{n_1}, \gamma_{n_2}, \gamma_{n_3}$—step sizes.

For the convergence of the method the following conditions are to be true [3, 4]:

(a) $\gamma_n \to 0,\ n \to \infty$, (б) $\sum_{n=0}^{\infty} \gamma_n = \infty,\ \sum_{n=0}^{\infty} \gamma_n^2 < \infty$.

3.7 Fuzzy Neural Network ANFIS. The Structure and Learning Algorithm

Consider an adaptive fuzzy system with the inference mechanism proposed by Sugeno on the basis of IF-THEN rules [3, 4, 9], which is called the network ANFIS (Adaptive Network Based Fuzzy Inference System). This system can be successfully used for adjusting membership functions and the rule base of fuzzy expert system. Below Sugeno fuzzy model and a block diagram of a network ANFIS (for two inputs and two rules) are presented (Fig. 3.8.).

$$f_1 = a_1 x + b_1 y + r_1$$

$$f_2 = a_2 x + b_2 y + r_2$$

$$f = \frac{w_1 f_1 + w_2 f_2}{w_1 + w_2} = \overline{w_1} f_1 + \overline{w_2} f_2$$

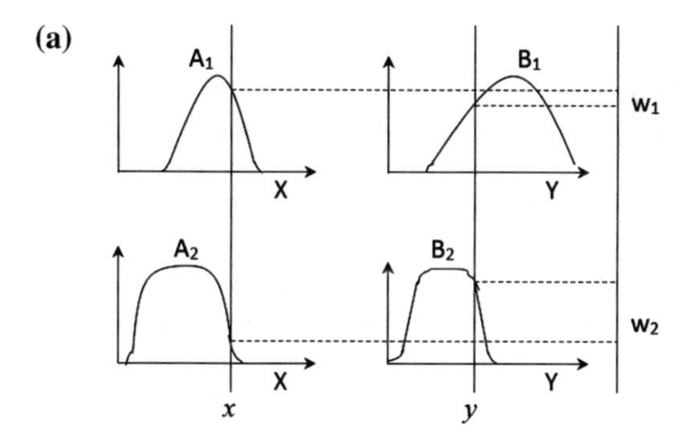

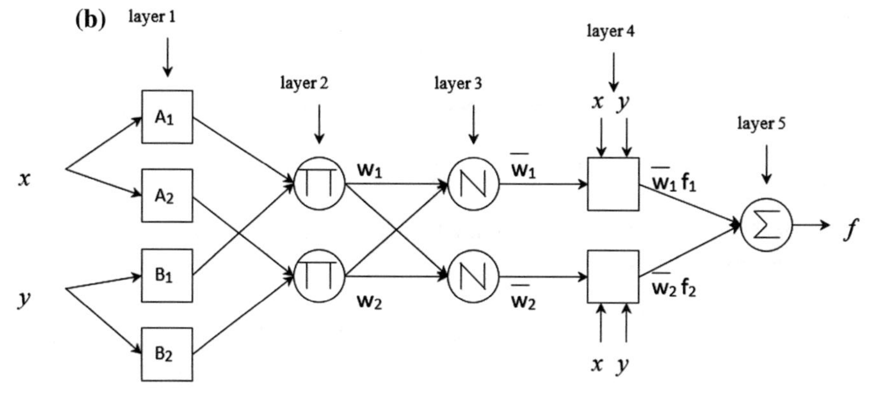

Fig. 3.8 **a** Schematic of Sugeno inference, **b** an equivalent structure of the neural network ANFIS

ANFIS system uses the following rule base:

$$\begin{cases} if\ x = A_1\ and\ y = B_1\ then\ f_1 = a_1x + b_1y + r_1 \\ if\ x = A_2\ and\ y = B_2\ then\ f_2 = a_2x + b_2y + r_2 \end{cases}$$

where A_i and B_i are linguistic variables.

a_i, b_i, r_i—some constants.

The layers of the fuzzy neural network perform the following functions:

Layer 1. Each neuron of this layer converts the input signal x or y using MF (fuzzificator). The most frequently used membership function is bell-shaped

$$\mu_{Ai}(x) = \frac{1}{1 + \left[\left(\frac{x - c_i}{a_i}\right)^2\right]}, \tag{3.32}$$

or Gaussian function

$$\mu_{Ai}(x) = \exp\left[-\left(\frac{x - c_i}{a_i}\right)^2\right]. \tag{3.33}$$

Layer 2. Each neuron in this layer, marked as Π, performs the intersection of input signals, simulating a logical AND operation, and sends to the output a value:

$$w_i = \mu_{Ai}(x) \times \mu_{Bi}(y), i = 1, 2. \tag{3.34}$$

Essentially, each neuron determines an activating rule force w_i. In fact, any operator of T-norm, which summarizes the AND operation can be used in these neurons.

Layer 3. Each neuron in this layer calculates normalized rules weight:

$$\overline{w_i} = \frac{w_i}{w_1 + w_2}, i = 1, 2. \tag{3.35}$$

Layer 4. In this layer neurons generate values of the output variables:

$$O_i^4 = \overline{w_i} f_i = \overline{w_i}(a_i x + b_i y + r_i). \tag{3.36}$$

Layer 5. In the last layer the output of the neural network is obtained and defuzzification is carried out:

$$O^5 = overall\,output = \sum_i \overline{w_i} f_i = \frac{\sum_i w_i f_i}{\sum_i w_i}. \tag{3.37}$$

The neural network ANFIS is trained by the method of steepest descent.

Rule Base Construction and Setting the Parameters of the Membership Function.

In existing systems with fuzzy neural networks one of the most important problems is the development of an optimum method of building the fuzzy rule base, based on the training sample. Basically fuzzy rules are described by experts according to their knowledge and experience of the relevant processes. But in the case of the development of fuzzy systems it is sometimes quite difficult or almost impossible to immediately get clear rules and membership functions as a result of ambiguity, incompleteness or complexity of the systems.

In such cases, the most appropriate is to consider the generation and refinement of fuzzy rules using special training algorithms. Currently widely used is back-propagation algorithm for fuzzy networks, described in Chap. 1, which allows to generate the optimum model of fuzzy systems and the rule base. This algorithm was proposed independently Ichiashi (Ichihashi), Nomura (Nomura), Wang and Mendel (Wang and Mendel) [3, 4]. Along with them, Shi Mitsumoto suggested another method that can be used for practical systems [4].

The main feature of this approach is that adjusting the parameters of fuzzy rules is carried out without modifying the rules table. Without loss of generality, consider the algorithm for a model which comprises the two input linguistic variables (x1, x2) and one output variable y. Network diagram is shown in Fig. 3.9.

Suppose we have a rule base that contains all possible combinations A_{1i} and A_{2j} ($i = 1 \ldots r$; $j = 1 \ldots k$) such that:

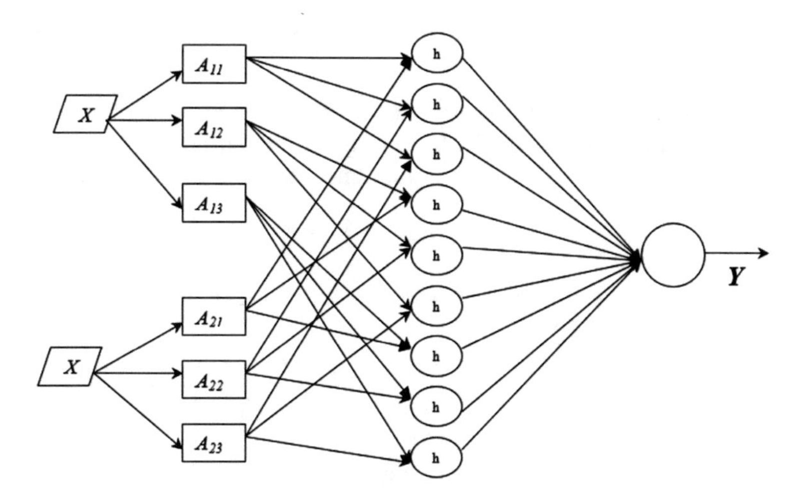

Fig. 3.9 Fuzzy neural network ANFIS structure

$$
\begin{aligned}
\text{Rule } 1: &\quad A_{11}, A_{21} \Rightarrow y_1, \\
2: &\quad A_{11}, A_{22} \Rightarrow y_2, \\
&\quad \dots \\
k: &\quad A_{11}, A_{2k} \Rightarrow y_k, \\
k+1: &\quad A_{12}, A_{21} \Rightarrow y_{k+1}, \\
2k: &\quad A_{12}, A_{2k} \Rightarrow y_{2k}, \\
\text{Rule} &\quad \dots \\
(i-1)k+j: &\quad A_{1i}, A_{2j} \Rightarrow y_{(i-1)k+j}, \\
&\quad \dots \\
\text{Rule } r \times k: &\quad A_{1r}, A_{2k} \Rightarrow y_{r\times k},
\end{aligned}
$$

where A_{1i} and A_{2j} are fuzzy sets for variables accordingly X_1 and X_2, $y_{(i-1)k+j}$ is real value of Y.

Clearly, this set of rules can be represented in tabular form:

$x_1 \backslash x_2$	A_{21}	A_{22}	…	A_{2j}	…	A_{2k}
A_{11}	y_1	y_2	…	y_j	…	y_k
A_{12}	y_{k+1}	y_{k+2}	…	y_{k+j}		y_{2k}
…	…	…	…	…	…	…
A_{1i}	…	…	…	$y_{(i-1)k+j}$	…	…
…	…	…	…	…	…	…
A_{1r}	$y_{(r-1)\ k+1}$	…	…	…	…	y_{rk}

So, if we are given a set of values (x_1, x_2), then, according to the fuzzy rule base, the output y can be obtained on the basis of fuzzy logic methods.

First of all, we denote the degree of fulfillment of the prerequisites as follows:

$$h_{(i-1)k+j} = A_{1i}(x_1)A_{2j}(x_2). \tag{3.38}$$

According to the centroid method network output is determined:

$$y = \frac{\sum_{i=1}^{r}\sum_{j=1}^{k} h_{(i-1)k+j}y_{(i-1)k+j}}{\sum_{i=1}^{r}\sum_{j=1}^{k} h_{(i-1)k+j}} = \frac{\sum_{i=1}^{r}\sum_{j=1}^{k} A_{1i}(x_1)A_{2j}(x_2)y_{(i-1)k+j}}{\sum_{i=1}^{r}\sum_{j=1}^{k} A_{1i}(x_1)A_{2j}(x_2)}. \tag{3.39}$$

Assume this learning system uses a training sample $(x_1, x_2; y^*)$, then a system error can be described as $E = (y^* - y)^2/2$.

Based on the description of fuzzy values for A_{1i} denote a_{1i}—center of membership function, b_{1i}—width for a given function, similarly for A_{2j} denote a_{2j} and b_2. According to the method of gradient descent to minimize the output error E the formulas for the calculation of coefficients a_{1i}, a_{2j}, b_{2j} и $y_{(i-1)k+j}$ (i = 1, 2, …, r; j = 1, 2, …, k) take the form [3]:

$$\begin{aligned} a_{1i}(t+1) &= a_{1i}(t) - \alpha \partial E/\partial a_{1i}(t) = \\ &= a_{1i}(t) - \alpha(\partial E/\partial y)(\partial y/\partial h_{(i-1)k+j})(\partial h_{(i-1)k+j}/\partial A_{1i})(\partial A_{1i}/\partial a_{1i}(t)) = \\ &= a_{1i}(t) + \frac{\alpha(y^* - y)[\sum_{j=1}^{k} (y_{(i-1)k+j} - y)A_{2j}](\partial A_{1i}/\partial a_{1i}(t))}{\sum_{i=1}^{r} \sum_{j=1}^{k} h_{(i-1)k+j}}, \end{aligned} \tag{3.40}$$

$$\begin{aligned} b_{1i}(t+1) &= b_{1i}(t) - \beta \partial E/\partial b_{1i}(t) = \\ &= b_{1i}(t) - \beta(\partial E/\partial y)(\partial y/\partial h_{(i-1)k+j})(\partial h_{(i-1)k+j}/\partial A_{1i})(\partial A_{1i}/\partial b_{1i}(t)) = \\ &= b_{1i}(t) + \frac{\beta(y^* - y)[\sum_{j=1}^{k} (y_{(i-1)k+j} - y)A_{2j}](\partial A_{1i}/\partial b_{1i}(t))}{\sum_{i=1}^{r} \sum_{j=1}^{k} h_{(i-1)k+j}}, \end{aligned} \tag{3.41}$$

$$\begin{aligned} a_{2j}(t+1) &= a_{2j}(t) - \alpha \partial E/\partial a_{2j}(t) = \\ &= a_{2j}(t) - \alpha(\partial E/\partial y)(\partial y/\partial h_{(i-1)k+j})(\partial h_{(i-1)k+j}/\partial A_{2j})(\partial A_{2j}/\partial a_{2j}(t)) = 0 \\ &= a_{2j}(t) + \frac{\alpha(y^* - y)[\sum_{i=1}^{r} (y_{(i-1)k+j} - y)A_{1i}](\partial A_{2j}/\partial a_{2j}(t))}{\sum_{i=1}^{r} \sum_{j=1}^{k} h_{(i-1)k+j}}, \end{aligned} \tag{3.42}$$

$$\begin{aligned} b_{2j}(t+1) &= b_{2j}(t) - \beta \partial E/\partial b_{2j}(t) = \\ &= b_{2j}(t) - \beta(\partial E/\partial y)(\partial y/\partial h_{(i-1)k+j})(\partial h_{(i-1)k+j}/\partial A_{2j})(\partial A_{2j}/\partial b_{2j}(t)) = \\ &= b_{2j}(t) + \frac{\beta(y^* - y)[\sum_{i=1}^{r} (y_{(i-1)k+j} - y)A_{1i}](\partial A_{2j}/\partial b_{2j}(t))}{\sum_{i=1}^{r} \sum_{j=1}^{k} h_{(i-1)k+j}}, \end{aligned} \tag{3.43}$$

$$\begin{aligned} y_{(i-1)k+j}(t+1) &= y_{(i-1)k+j}(t) - \gamma \partial E/\partial y_{(i-1)k+j}(t) = \\ &= y_{(i-1)k+j}(t) - \gamma(\partial E/\partial y)(\partial y/\partial y_{(i-1)k+j}(t)) = \\ &= y_{(i-1)k+j}(t) + \frac{\gamma(y^* - y)h_{(i-1)k+j}}{\sum_{i=1}^{r} \sum_{j=1}^{k} h_{(i-1)k+j}}, \end{aligned} \tag{3.44}$$

where α, β, γ is a training speed, t is iteration number in the process of learning.

3.8 Fuzzy Neural Networks TSK and Wang-Mendel

FNN TSK

A generalization of the neural network ANFIS is a fuzzy neural network TSK (Takagi, Sugeno, Kang'a). The generalized scheme of output in the model using a TSK with M rules and N variables can be written as follows [3, 4]

$$R_1: \ if\ x_1 \in A_1^{(1)},\ x_2 \in A_2^{(1)},\ \ldots,\ x_n \in A_n^{(1)},\ \text{then}\ y_1 = p_{10} + \sum_{j=1}^{N} p_{1j}x_j;$$

$$R_M: \ if\ x_1 \in A_1^{(M)}, x_2 \in A_2^{(M)},\ \ldots,\ x_n \in A_n^{(M)},\ \text{then}$$

$$y_M = p_{M0} + \sum_{j=1}^{N} p_{Mj}x_j$$

where $A_i^{(k)}$—the value of linguistic variable x_i for the rule R_k with MF (membership function)

$$\mu_A^{(k)}(x_i) = \frac{1}{1 + \left(\frac{x_i - c_i^{(k)}}{\sigma_i^{(k)}}\right)^{2b_i^{(k)}}}. \tag{3.45}$$

$$i = \overline{1, N}; k = \overline{1, M}.$$

At the intersection of the TSK network rule conditions R_k MF is defined as a product

$$\mu_A^{(k)}(x) = \prod_{j=1}^{N} \left[\frac{1}{1 + \left(\frac{x_j - c_j^{(k)}}{\sigma_j^{(k)}}\right)^{2b_j^{(k)}}} \right]. \tag{3.46}$$

With M inference rules composition results is determined by the following formula (similar to the inference of Sugeno):

$$y(x) = \frac{\sum_{k=1}^{M} w_k y_k(x)}{\sum_{k=1}^{M} w_k}, \tag{3.47}$$

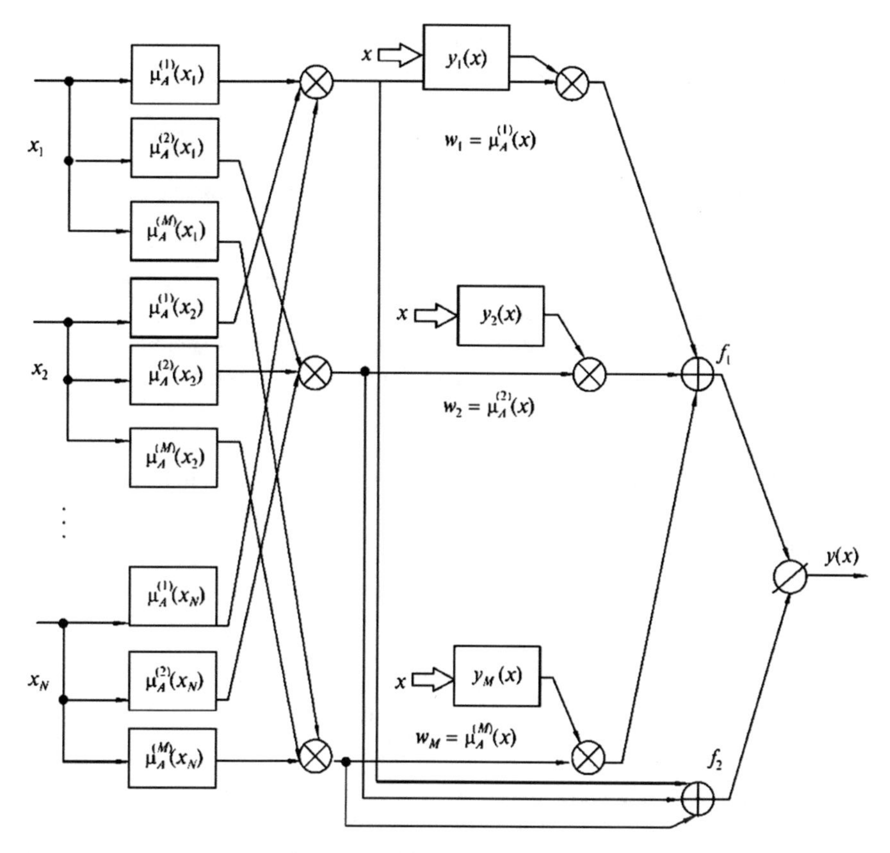

Fig. 3.10 The structure of TSK fuzzy neural network

where $y_k(x) = p_{k0} + \sum_{j=1}^{N} p_{kj}x_j$. The weights in this expression are interpreted as the degree of fulfillment of rule antecedents (conditions): $w_k = \mu_A^{(k)}(x)$, which are given by (3.46).

The fuzzy neural network TSK, which implements the output in accordance with (3.47) represents a multilayer structure network shown in Fig. 3.10. In such a network, 5 layers are present:

1. The first layer performs fuzzification separately for each variable x_i, $i = 1, 2, \ldots, N$, defining for each rule k value MF $\mu_A^{(k)}(x_i)$ in accordance with the fuzzification function, which is described, for example, by (3.45). This is a parametric layer with parameters $c_j^{(k)}$, $\sigma_j^{(k)}$, $b_j^{(k)}$, which are subject to adjustment in the learning process.

2. The second layer performs the aggregation of individual variables x_i, determining the resulting degree of membership $w_k = \mu_A^{(k)}(x)$ for the vector x. This is not a parametric layer.
3. The third layer is a function generator TSK, wherein the output values are calculated $y_k(x) = p_{k0} + \sum_{j=1}^{N} p_{kj}x_j$. At this layer also functions formed on the previous layer $y_k(x)$ and w_k are multiplied. This is a parametric layer, wherein the adaptation of linear parameters (weight) p_{k0}, p_{kj} for $j = \overline{1, N}$, $k = \overline{1, M}$, is carried out determining the rules output functions.
4. The fourth layer consists of 2 summing neurons, one of which calculates the weighted sum of the signals $y_k(x)$, and the second determines the sum of the weights $\sum_{k=1}^{M} w_k$.
5. The fifth layer is composed of a single output neuron. In it weights normalizing is performed and the output signal $y(x)$ is calculated in accordance with the expression

$$y(x) = \frac{f_1}{f_2} = \frac{\sum_{k=1}^{M} w_k y_k(x)}{\sum_{k=1}^{M} w_k}. \tag{3.48}$$

This is also non-parametric layer.

From the above description it follows that the TSK fuzzy network contains only 2 parametric layer (the first and third), the parameters of which are specified in the learning process. Parameters of the first layer $\left(c_j^{(k)}, \sigma_j^{(k)}, b_j^{(k)}\right)$ we call non-linear, and the parameters of the third layer $\{p_{kj}\}$—linear weights. The general expression for the functional dependence (3.47) for the network TSK is defined as follows:

$$y(x) = \frac{1}{\sum_{k=1}^{M} \prod_{j=1}^{M} \mu_A^{(k)}(x_j)} \sum_{k=1}^{M} \left(p_{k0} + \sum_{j=1}^{N} p_{kj}x_j\right) \prod_{j=1}^{N} \mu_A^{(k)}(x_j) \tag{3.49}$$

If we assume that at any given time the non-linear parameters are fixed, then the function $y(x)$ would be linear with respect to the variable x_j.

In the presence of N input variables each rule R_k formulates $(N+1)$ variable $p_j^{(k)}$ linear dependence $y_k(x)$. If M inference rules are present then $M(N+1)$ linear network parameters are obtained. In turn, each MF uses 3 parameters (c, σ, b), which are subject to adjustment. With M inference rules 3MN nonlinear parameters are obtained. In total this gives $M(4N+1)$ linear and nonlinear parameters that must be determined in the learning process. This is very large value. In order to reduce the number of parameters for adaptation, we operate with fewer number of MF. In particular, it can be assumed that some of the parameters of one function MF $\mu_A^{(k)}(x_j)$ are fixed, e.g. $\sigma_j^{(k)}$ and $b_j^{(k)}$.

The Structure of Wang-Mendel Network

If instead of the rules output functions $y_k(x)$ we select the output v_k, then we obtain the structure of the FNN, which is called Wang-Mendel neural network (see Fig. 3.11).

This is the four-layer structure, wherein the first layer performs the fuzzification of input variables, the second—aggregation (crossing) of the rules conditions, a third (linear)—composition of outputs of inference rules (the first neuron) and generation of a normalizing signal (second neuron), whereas a last neuron layer produces an output signal $y(x)$, which is calculated as follows:

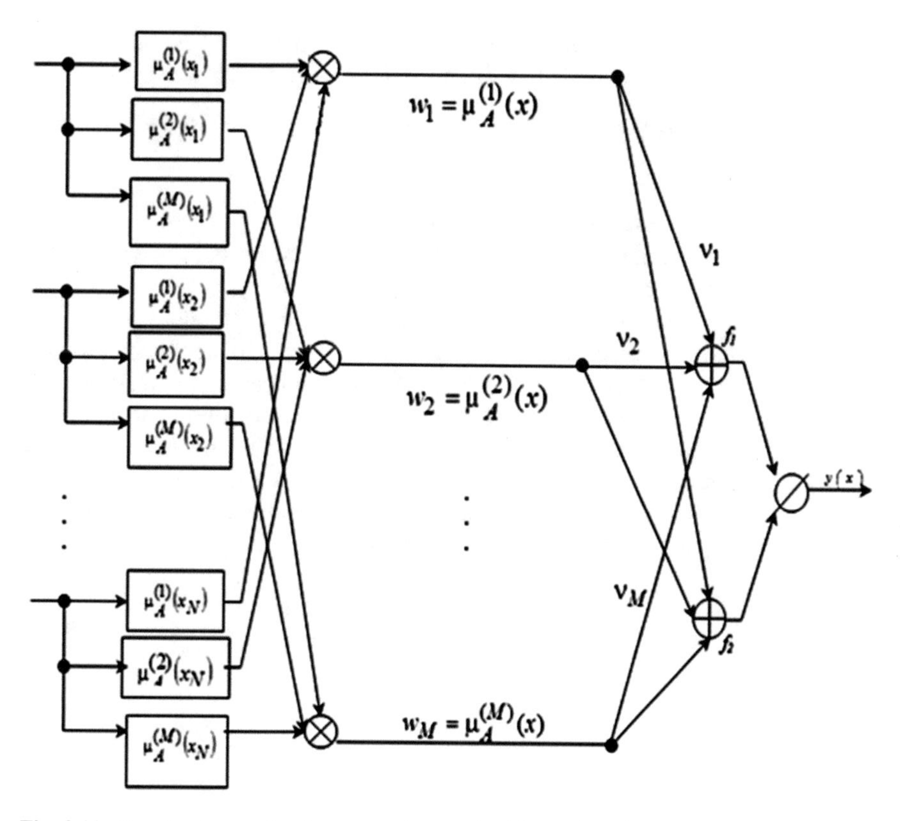

Fig. 3.11 The structure of the fuzzy neural network of Wang-Mendel

$$y(x) = \frac{\sum_{k=1}^{M} v_k \left[\prod_{j=1}^{N} \mu_A^{(k)}(x_j)\right]}{\sum_{k=1}^{M} \mu_A^{(k)}}. \tag{3.50}$$

Note the great similarity of structures of both networks. Parts defining antecedents in the first and second layer are identical (i.e., they are responsible for rule component "if…"), the differences are manifested in the representation of the consequences of the rules ("then…").

TSK output function is represented by a polynomial of the first order, and in a network of Wang-Mendel by constant $v_k = c_k$, where the value of c_k can be interpreted as the center of MF. Thus, FNN Wang-Mendel is a special case of FNN TSK.

The objective of both networks is to find such a mapping in data pairs, in which the expected value d, which corresponds to the input vector x, will be formed by network output function $y(x)$.

The training of fuzzy neural networks, as well as crisp NN may be carried out in accordance with the supervised algorithm with the teacher, which uses the objective function $E = \frac{1}{2}\sum_{l=1}^{p}\left(y(x)^{(l)} - d^{(l)}\right)^2 \rightarrow \min$, and with the unsupervised algorithm of self-organization.

Hybrid Learning Algorithm Fuzzy Neural Networks

Consider a hybrid learning algorithm FNN, which is used for both networks TSK and Wang-Mendel (whose $p_{kj} = 0$, a $p_{k0} = v_k$) [3, 10].

In the hybrid algorithm the adapted parameters can be divided into 2 groups. The first group includes linear parameters p_{kj} of the third layer, and the second group—nonlinear parameters (MF) of the first layer. Adaptation occurs in two stages.

In the first stage after fixing the individual parameters of the membership function (in the first iteration—the values that are obtained by initializing) by solving a system of linear equations, linear parameters of polynomial p_{kj} are calculated. With the known values of MF dependence input-output can be represented as a linear form with respect to the parameters p_{kj}:

$$y_k(x) = \sum_{k=1}^{M} w_k' \left(p_{k0} + \sum_{j=1}^{N} p_{kj} x_j \cdot\right), \tag{3.51}$$

where

$$w_k^{'} = \frac{\prod_{j=1}^{N} \mu_A^{(k)}(x_j)}{\sum_{r=1}^{M} \prod_{j=1}^{N} \mu_A^{(r)}(x_j)}, \ k = \overline{1, M}. \tag{3.52}$$

With the dimension L of training sample $\left(x^{(l)}, d^{(l)}\right)$, $(l = 1, 2, \ldots, L)$ and replacement of the network output by expected value $d^{(l)}$ we get a system of L linear equations of the form:

$$\begin{bmatrix} w_{11}^{'} & w_{11}^{'}x_1^{(1)} \ldots & w_{11}^{'}x_N^{(1)} \ldots & w_{1M}^{'} & w_{1M}^{'}x_1^{(1)} \ldots & w_{1M}^{'}x_N^{(1)} \\ w_{21}^{'} & w_{21}^{'}x_1^{(2)} \ldots & w_{21}^{'}x_N^{(2)} \ldots & w_{2M}^{'} & w_{2M}^{'}x_1^{(2)} \ldots & w_{2M}^{'}x_N^{(2)} \\ \ldots & \ldots & \ldots & \ldots & \ldots & \ldots \\ w_{L1}^{'} & w_{L1}^{'}x_1^{(L)} & w_{L}^{'}x_N^{(L)} & w_{LM}^{'} & w_{LM}^{'}x_1^{(L)} & w_{LM}^{'}x_N^{(L)} \end{bmatrix} \times \begin{bmatrix} p_{10} \\ p_{11} \\ \ldots \\ p_{1N} \\ \ldots \\ p_{M0} \\ p_{M1} \\ \ldots \\ p_{MN} \end{bmatrix} = \begin{bmatrix} d^{(1)} \\ d^{(2)} \\ \ldots \\ d^{(L)} \end{bmatrix} \tag{3.53}$$

where $w_{\ell i}^{'}$ means the level of activation (w) of the ith rule conditions at presentation of ℓth input vector x^{ℓ}. This expression can be written in matrix form:

$$Ap = d.$$

Matrix A dimension is equal to $L(N+1)M$. By thus a number of rows L usually is much greater than a number of columns $(N+1)M$. The solution of this equations system may be obtained by conventional methods as well as using pseudo-inverse matrix A at one step:

$$p = A^{+}d,$$

where A^{+}—pseudo-inverse matrix.

In the second stage, after fixing the values of linear parameters p_{kj} the actual output signals $y^{(\ell)}$, $\ell = 1, 2, \ldots, L$, are calculated using a linear equations system:

$$y^{(L)} = Ap. \tag{3.54}$$

After this error vector $\varepsilon = (y - d)$ and the criterion E are calculated $E = \frac{1}{2}\sum_{\ell=1}^{L}\left(y\left(x^{(\ell)}\right) - d^{(\ell)}\right)^2$.

The error signals are sent through the network in the reverse direction according to the method of Back Propagation until the first layer, at which gradient vector components of the objective function with respect to parameters $\left(c_j^{(k)}, \sigma_j^{(k)}, b_j^{(k)}\right)$ can be calculated.

After calculating the gradient vector a step of gradient descent method is made. The corresponding formulas (for the simplest method of the steepest descent) take the form:

$$c_j^{(k)}(n+1) = c_j^{(k)}(n) - \eta_c \frac{\partial E(n)}{\partial c_j^{(k)}}, \tag{3.55}$$

$$\sigma_j^{(k)}(n+1) = \sigma_j^{(k)}(n) - \eta_\sigma \frac{\partial E(n)}{\partial \sigma_j^{(k)}}, \tag{3.56}$$

$$b_j^{(k)}(n+1) = b_j^{(k)}(n) - \eta_b \frac{\partial E(n)}{\partial b_j^{(k)}}, \tag{3.57}$$

where n is a number of iteration.

After verifying the nonlinear parameters the process of adaptation of linear parameters TSK (first phase) restarts and nonlinear parameters are further adapted (second stage). This cycle continues until all the parameters will be stabilized.

Formulas (3.55)–(3.57) require the calculation of the gradient of the objective function with respect to the parameters of the MF. The final form of these formulas depends on the type of MF. For example, using the generalized bell functions:

$$\mu_A(x) = \frac{1}{1 + \left(\frac{x-c}{\sigma}\right)^{2b}} \tag{3.58}$$

the corresponding formulas for gradient of the objective function for one pair of data (x, d) take the form [3, 4, 10]:

$$\frac{\partial E}{\partial c_j^{(k)}} = (y(x) - d) \sum_{r=1}^{M} \left(p_{r0} + \sum_{j=1}^{N} p_{rj} x_j\right) \cdot \frac{\partial w_r'}{\partial c_j^{(k)}}, \tag{3.59}$$

$$\frac{\partial E}{\partial \sigma_j^{(k)}} = (y(x) - d) \sum_{r=1}^{M} \left(p_{r0} + \sum_{j=1}^{N} p_{rj} x_j\right) \cdot \frac{\partial w_r'}{\partial \sigma_j^{(k)}}.$$

$$\frac{\partial E}{\partial b_j^{(k)}} = (y(x) - d)\sum_{r=1}^{M}\left(p_{r0} + \sum_{j=1}^{N} p_{rj}x_j\right)\cdot\frac{\partial w_r'}{\partial b_j^{(k)}}.$$

Derivatives $\frac{\partial w_r'}{\partial c_j^{(k)}}, \frac{\partial w_r'}{\partial \sigma_j^{(k)}}, \frac{\partial w_r'}{\partial b_j^{(k)}}$, derived from dependency (3.52) take the following form [3]:

$$\frac{\partial w_r'}{\partial c_j^{(k)}} = \frac{\delta_{rk} m(x_j) - \ell(x_j)}{[m(x_j)]^2}\prod_{i=1, i\neq j}^{M}\frac{\left[\frac{2b_j^{(k)}}{\sigma_j^{(k)}}\left(\frac{x_j - c_j^{(k)}}{\sigma_j^{(k)}}\right)^{2b_j^{(k)}-1}\right]}{\left[1 + \left(\frac{x_j - c_j^{(k)}}{\sigma_j^{(k)}}\right)^{2b_j^{(k)}}\right]^2}\mu_A^{(k)}(x_i), \tag{3.60}$$

$$\frac{\partial w_r'}{\partial \sigma_j^{(k)}} = \frac{\delta_{rk} m(x_j) - \ell(x_j)}{[m(x_j)]^2}\prod_{i=1, i\neq j}^{M}\frac{\left[\frac{2b_j^{(k)}}{\sigma_j^{(k)}}\left(\frac{x_j - c_j^{(k)}}{\sigma_j^{(k)}}\right)^{2b_j^{(k)}}\right]}{\left[1 + \left(\frac{x_j - c_j^{(k)}}{\sigma_j^{(k)}}\right)^{2b_j^{(k)}}\right]^2}\mu_A^{(k)}(x_i), \tag{3.61}$$

$$\frac{\partial w_r'}{\partial b_j^{(k)}} = \frac{\delta_{rk} m(x_j) - \ell(x_j)}{[m(x_j)]^2}\prod_{i=1, i\neq j}^{M}\frac{\left[-2\left(\frac{x_j - c_j^{(k)}}{\sigma_j^{(k)}}\right)^{2b_j^{(k)}}\ln\left(\frac{x_j - c_j^{(k)}}{\sigma_j^{(k)}}\right)\right]}{\left[1 + \left(\frac{x_j - c_j^{(k)}}{\sigma_j^{(k)}}\right)^{2b_j^{(k)}}\right]^2}\cdot\mu_A^{(k)}(x_i), \tag{3.62}$$

for $r = 1, 2, \ldots, M$, where δ_{rk}—Kronecker delta, $\ell(x_j) = \prod_{j=1}^{N}\mu_A^{(k)}(x_j)$; $m(x_j) = \sum_{k=1}^{M}\prod_{j=1}^{N}\mu_A^{(k)}(x_j)$.

In the practice of the hybrid learning method application the dominant factor in adaptation is considered to be the first step in which weights p_{kj} are selected using pseudoinverse in one step. To balance its impact the second phase should be repeated many times in each cycle.

The described hybrid algorithm is one of the most effective ways of learning fuzzy neural networks. Its characteristic feature is the division of the process into two stages separated in time. Since the design complexity of each nonlinear optimization algorithm depends non-linear on the number of parameters subject to optimization, the reduction in the dimensions of optimization significantly reduces the total amount of calculations and increases the speed of convergence of the algorithm. Due to this hybrid algorithm is one of the most efficient in comparison with conventional gradient methods.

3.9 Adaptive Wavelet-Neuro-Fuzzy Networks

Along with the neuro-fuzzy systems, wavelet transform is becoming more and more popular as a method that gives a compact representation of signals in both the time and frequency domains.

Based on the research of neural networks (ANN), and the theory of wavelet neural networks methods for the analysis of non-stationary processes with significant non-linear trends have been developed.

The natural step was the following: to make the combination of clarity and interpretability of fuzzy neural networks, strong approximation and educational properties of artificial neural networks (ANN) with the flexibility of wavelet transforms in the context of hybrid systems of computational intelligence, which were called the adaptive wavelet-neuro-fuzzy networks (AWNFN) [11].

The key to ensuring the effectiveness of such systems, is a choice of learning algorithm, which is usually based on gradient procedures to minimize the chosen criterion. A combination of gradient optimization method and back propagation (error) significantly reduces the rate of learning hybrid CI systems.

In the case when the data processing must be done in real time, and the predicted sequence is unsteady and distorted by noise, the traditional conventional learning algorithm of gradient descent, appears to be ineffective.

In this section we consider the problem of synthesis of adaptive predictive wavelet neural network, which has a higher rate of learning in comparison with systems using conventional back-propagation algorithm of gradient type.

3.9.1 The Architecture of Adaptive Wavelet Fuzzy Neural Network

Introduce the five-layer architecture shown in Fig. 3.12, in a way similar to the previously considered FNN TSK, which is a learning system with the inference of Takagi-Sugeno-Kang (Fig. 3.10).

In this case, at the layer 4 values of signals are calculated

$$\bar{w}_j(k)(p_{j0} + \sum_{i=1}^{n} p_{ji}x(k-i)) = \bar{w}_j(k)p_j^T\bar{X}(k) \tag{3.63}$$

where $\bar{X}(k) = \left(1,\ X^T(k)\right)^T$, $p_j = \left(p_{j0},\ p_{j1},\ \ldots,\ p_{jn}\right)^T$,.

here $h(n+1)$ parameters $p_j, j = 1, 2, \ldots., h,\ i = 0, 1, 2, \ldots.,$ are to be determined.

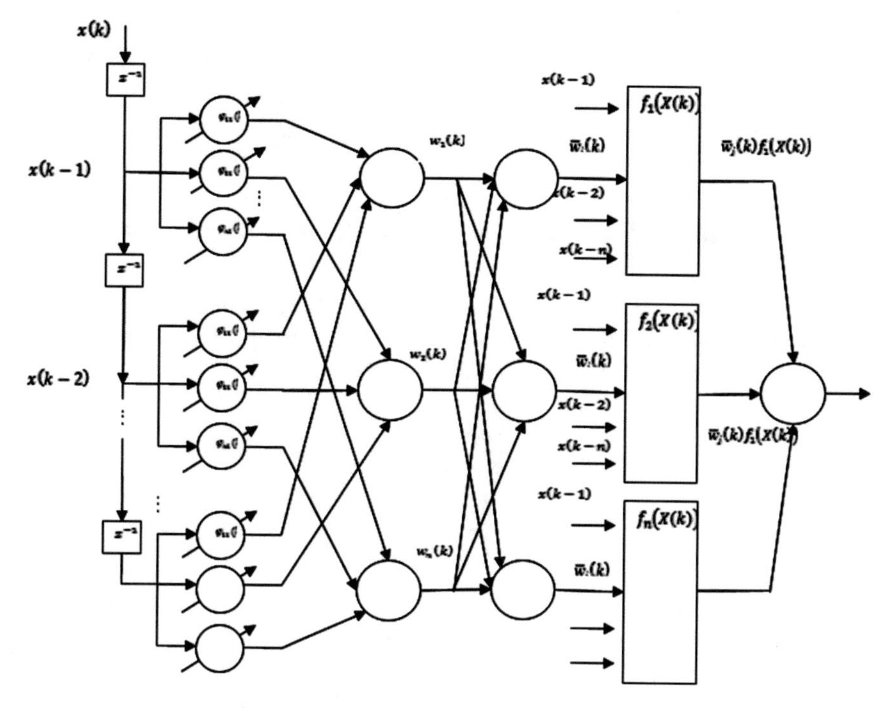

Fig. 3.12 Structure of adaptive wavelet fuzzy neural network

Finally, the output signal (prediction $x(k)$) is calculated at the 5th output layer [11]

$$\begin{aligned}\hat{x}(k) &= \sum_{j=1}^{h} \bar{w}_j(k) f_j(X(k)) = \sum_{j=1}^{h} \frac{w_j(k)}{\sum_{j=1}^{h} w_j(k)} f_j(X(k)) = \\ &= \frac{\prod_{i=1}^{n} \varphi_{ji}(x(k-i), c_{ji}, \sigma_{ji})}{\sum_{j=1}^{h} \prod_{i=1}^{n} \varphi_{ji}(x(k-i), c_{ji}, \sigma_{ji})} f_j(X(k))\end{aligned} \tag{3.64}$$

This expression vectors by introducing variables

$$f(X(k)) = \begin{pmatrix} \bar{w}_1(k), \bar{w}_1(k)x(k-1), \ldots, \bar{w}_1(k)x(k-n), \bar{w}_2(k), \\ \bar{w}_2(k)x(k-1), \ldots, \bar{w}_2(k)x(k-n), \ldots \\ \ldots, \bar{w}_n(k), \bar{w}_n(k)x(k-1), \ldots, \bar{w}_n(k)x(k-n) \end{pmatrix}^T$$

and $p = (p_{10}, p_{11}, \ldots, p_{20}, p_{21}, \ldots, p_{2n}, p_{oh}, p_{h1}, \ldots, p_{hn})^T$ dimension $h(n+1)$ can be rewritten in a compact form $\hat{x}(k) = p^T f(X(k))$.

Adjustable parameters of the network are located only in the first and fourth hidden layers. These are 2 hn wavelet parameters σ_{ji} and c_{ji} and $h(n + 1)$ linear autoregression model parameters p_{ji}.

They can be identified throughout the learning process.

3.9.2 Learning of Adaptive Neuro-Fuzzy Wavelet Network

Since configurable parameter vector is contained in a linear network, to configure it is possible to apply any of the algorithms used in the adaptive identification, primarily RLSM method. It is written in the form [11]

$$\begin{cases} p(k+1) = p(k) + \dfrac{P(k)(x(k) - p^T(k)f(X(k)))}{\alpha + f^T(X(k))P(k)f(X(k))} f(X(k)) \\ P(k+1) = \frac{1}{\alpha}\left(P(k) - \dfrac{P(k)f(x(k+1))f^T(X(k))P(k))}{\alpha + f^T(X(k+1))P(k)f(X(k+1))}\right) \end{cases} \quad (3.65)$$

where $x(k) - p^T(k)f(X(k)) = x(k) - \hat{x}(k) = e(k)$ is the prediction error, $0 \le \alpha \le 1$—forgetting factor of outdated information.

For training the optimum speed Kachmarzh algorithm can be used [11]

$$p(k+1) = p(k) + \frac{x(k) - p^T(k)f(X(k))}{f^T(X(k))p(k)f(X(k))} f(X(k)) \quad (3.66)$$

or algorithm of Gulvin-Ramadji Keynes [11–18]

$$\begin{cases} p(k+1) = p(k) + r^{=1}(k)(x(k) - p^T(k)f(X(k)))f(X(k)) \\ r(k+1) = r(k) + \|f(X(k+1))\|^2 \end{cases} \quad (3.67)$$

which is the well-known procedure of stochastic approximation.

To adjust the first hidden layer parameters in the network AWNFN can be used learning algorithm back propagation, based on chained differentiation and gradient descent

$$E(k) = \frac{1}{2}e^2(k) = \frac{1}{2}(\hat{x}(k) - x(k))^2$$

In general, the learning procedure for this layer is given by [11]

$$\begin{cases} c_{ji}(k+1) = c_{ji} - \eta_c(k)\dfrac{\partial E(k)}{\partial c_{ji}(k)} \\ \sigma_{ji}(k+1) = \sigma_{ji} - \eta_\sigma(k)\dfrac{\partial E(k)}{\partial \sigma_{ji}(k)} \end{cases} \quad (3.68)$$

Its properties are completely determined by the parameters of speed training $\eta_c(k)$, $\eta_\sigma(k)$, selected from empirical considerations. It should be noted that if the parameters of the fourth layer can be adjusted more quickly, the speed of operations is lost on the first layer.

Increasing the speed of convergence can be achieved using more complex than the gradient procedures such as the Marquardt algorithm or Hartley which can be written as follows [11]

$$F(k+1) = F(k) + \lambda(J(k)J^{T(k)} + \eta I)^{(-1)}J(k)e(k) \tag{3.69}$$

where $F(k) = (c_{11}(k), \sigma_{11}^{-1}(k), c_{21}(k), \sigma_{21}^{-1}(k), \ldots, c_{ji}(k), \sigma_{ji}^{-1}(k), \ldots\ldots$
$\ldots, c_{hn}(k), \sigma_{hn}^{-1}(k))^T$

is *2h(n + 1)*—vector of adjustable parameters (to reduce the complexity of the computational procedure not parameter $\sigma_{ji}(k)$, but its reciprocal is used) $J(k)$—is the vector gradient of the output signal $\hat{x}(k)$ of variable parameters, *I*—identity matrix *2 hn* * *2 hn*, η—scalar manipulated variable; α—positive scalar;

$$J(k) = \begin{pmatrix} \frac{\partial \hat{x}(k)}{\partial c_{11}}, \frac{\partial \hat{x}(k)}{\partial \sigma_{11}^{-1}(k)}, \frac{\partial \hat{x}(k)}{\partial c_{21}}, \frac{\partial \hat{x}(k)}{\partial \sigma_{21}^{-1}(k)}, \ldots, \frac{\partial \hat{x}(k)}{\partial c_{ji}}, \frac{\partial \hat{x}(k)}{\partial \sigma_{ji}^{-1}(k)}, \ldots \\ \ldots, \frac{\partial \hat{x}(k)}{\partial c_{hn}}, \frac{\partial \hat{x}(k)}{\partial \sigma_{hn}^{-1}(k)} \end{pmatrix}^T$$

To reduce the computational complexity of the algorithm, the following expression can be used

$$\left(JJ^T + \eta I\right)^{-1} = \eta^{-1}I - \frac{\eta^{-1}IJJ^T\eta^{-1}I}{1 + J^T\eta^{-1}IJ} \ldots \tag{3.70}$$

using which you can easily obtain the relation

$$\lambda\left(JJ^T + \eta I\right)^{-1}J = \lambda\frac{J}{\eta + \|J\|^2}$$

Substituting this relation in the algorithm (3.69) we obtain the following algorithm for learning parameters of the first layer in the form [11, 14]

$$F(k+1) = F(k) + \lambda\frac{J(k)e(k)}{\eta + \|J\|^2} \tag{3.71}$$

It is easily seen that the algorithm (3.71) is a nonlinear multiplicative modification of Kachmarzh adaptive algorithm and if $\lambda = 1$, $\eta = 0$ coincides with it in the structure.

To ensure the filtering properties of the training algorithm, introduce an additional procedure for setting regularizing parameter η in the form

$$\begin{cases} F(k+1) = F(k) + \lambda \frac{J(k)e(k)}{\eta(k)} \\ \eta(k+1) = \alpha\eta_p(k) + \|f(X(k))\|^2 \end{cases} \quad (3.72)$$

If $\alpha = 0$, then this procedure is the same as (3.66) and has a top speed of convergence, and if $\alpha = 1$, the procedure to get the properties of stochastic approximation and is a generalization of the procedure (3.71) in the nonlinear case.

3.9.3 Simulation Results

To study the effectiveness of the proposed neuro-fuzzy wavelet network and its learning algorithm AWNFN network was trained to predict chaotic time series Macki-Glass. Forecasting time series Macki-Glass is a standard test widely used to assess and compare the functioning of neural and neuro-fuzzy networks in time series forecasting problems. Time series of Macki-Glass describes the differential equations with delay [11]:

$$\dot{x}(t) = \frac{0.2x(t-\tau)}{1 + x^{10}(t-\tau)} - 0.1x(t) \quad (3.73)$$

The value of the time series (3.73) were obtained for each integer value t by the Runge-Kutt fourth order algorithm. Integration time step in this method was 0.1 the initial condition was taken $x(0) = 1.2$ and the delay $\tau = 17$ and $x(t)$ calculated for $t = 0, 1, \ldots, 51000$. The values $x(t-18)$, $x(t-17)$, $x(t-16)$, $x(t)$ were used for the prediction $x(t+6)$.

In the online mode network AWNFN was trained on the procedure (3.69) during 50000 iterations (50000 training samples for $t = 118, \ldots, 50117$.

Training algorithm parameters were $\alpha = 0.45$, $\lambda_p = 2$, $\lambda = 1$. The initial value was $\eta(0) = 1$, $\eta_p = 10000$. After *50,000* iterations the training stopped and the next *500* points for the values *50118 … 50617* were used as test data for the calculation of the forecast. As an activation function was chosen wavelet function "Mexican hat". The initial value of synaptic weights generated randomly in the interval *[−0.1, 0.1]*. As a criterion for the quality of the forecast RMSE criterion was used

$$RMSE = \frac{1}{N}\sum_{k=1}^{N}(x(k) - \hat{x}(k))^2$$

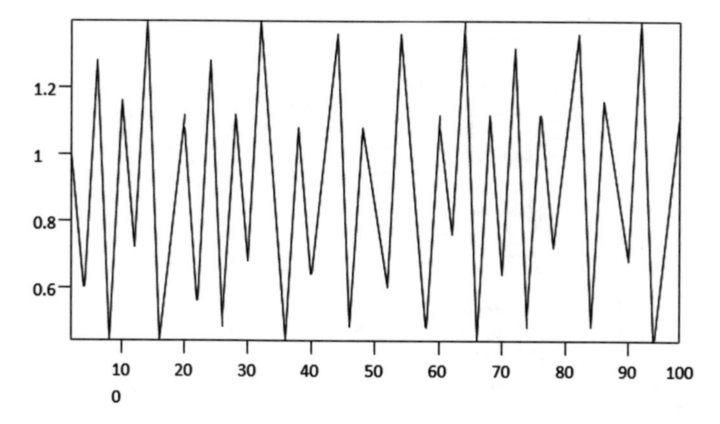

Fig. 3.13 Results of prediction time series Macki-Glass

Table 3.1 The results of the chaotic time series prediction

Neural network/learning algorithm	RMSE
The adaptive wavelet-neuro-fuzzy network/Prop. algorithm (12), (13)	0.0120
Network ANFIS algorithm backpropagation	0.2312

In Fig. 3.13 the results of prediction are presented. The two curves represent the real data (dotted line) and forecast (solid line) are almost indistinguishable.

Table 3.1 presents the results of forecasting adaptive wavelet—neuro-fuzzy network in comparison with the results of the prediction based on the standard fnn ANFIS learning algorithm and backpropagation.

As can be easily seen from the experimental results, the proposed prediction algorithm using adaptive neuro-fuzzy network with wavelet functions and learning algorithms (3.63), (3.68) provides a significantly better quality of forecasting and increased the speed of learning in comparison with classic network ANFIS.

3.10 Neo-Fuzzy-Cascade Neural Networks

Consider the neo-fuzzy-neuron with multiple inputs and a single output, which is shown in Fig. 3.14. It is described in the following expression:

$$\hat{y} = \sum_{i=1}^{n} f_i(x_i), \tag{3.74}$$

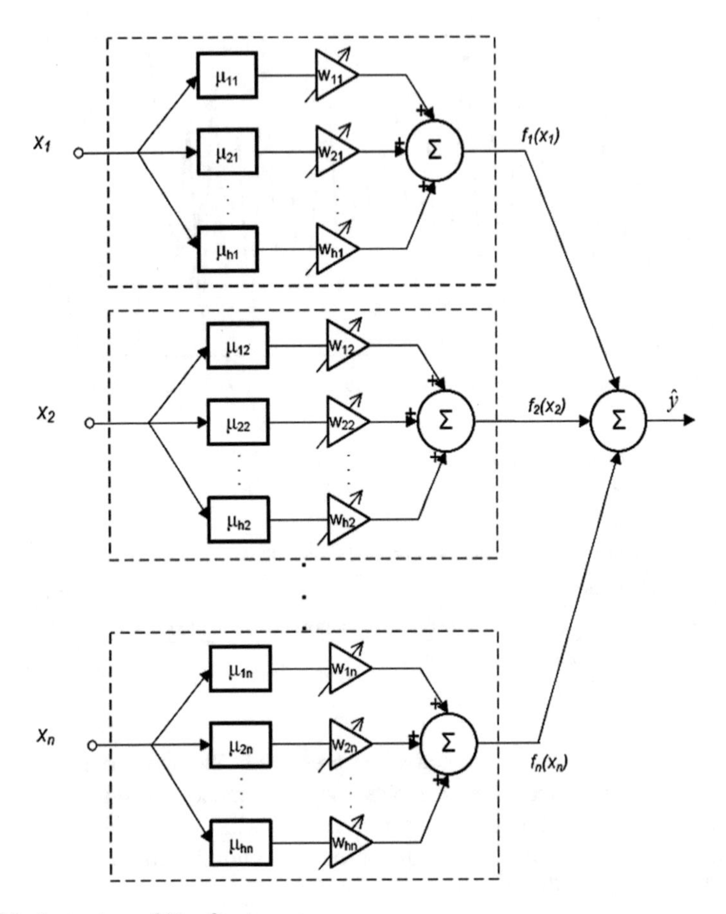

Fig. 3.14 A structure of Neo-fuzzy neuron

where x_i is the ith input $(i = 1, 2, \ldots, N)$, y is system output. Structural unit of neo-fuzzy neuron is nonlinear synapse (Nsi), which transforms the ith input signal into the form:

$$f_i(x_i) = \sum_{j=1}^{h} w_{ji}\mu_{ji}(x_i)$$

and performs fuzzy inference: If x_i is x_{ji} then output is w_{ji}, where x_{ji} is a fuzzy number, whose the membership function is μ_{ji}, w_{ji} is a synaptic weight. It is obvious that the nonlinear synapse actually implements fuzzy Takagi-Sugeno conclusion zero order.

A

When the vector signal $x(k) = (x_1(k), x_2(k), \ldots x_n(k))^T$ $(k = 1, 2, \ldots$ *discrete time*) is entered into the neo-fuzzy neuron, the output of this neuron is determined by both the membership functions $\mu_{ji}(x_i(k))$ and adjustable synaptic weights $w_{ji}(k-1)$, which were obtained in the previous training epoch [11]:

$$\hat{y}(k) = \sum_{i=1}^{n} f_i(x_i(k)) = \sum_{i=1}^{n} \sum_{j=1}^{h} w_{ji}(k-1)\mu_{ji}(x_i(k)) \tag{3.75}$$

thus neo-fuzzy neuron contains $h * n$ the synaptic weights to be determined.

Architecture of Cascade Neo-Fuzzy Neural Network

Architecture CNFNN, is shown in Fig. 3.15 and is characterized by mapping which has the following form [11, 14]:

- Neo-fuzzy neuron of the first cascade

$$\hat{y}^{[1]} = \sum\nolimits_{i=1}^{n} \sum\nolimits_{j=1}^{h} w_{ji}^{[1]} \mu_{ji}(x_i) \tag{3.76}$$

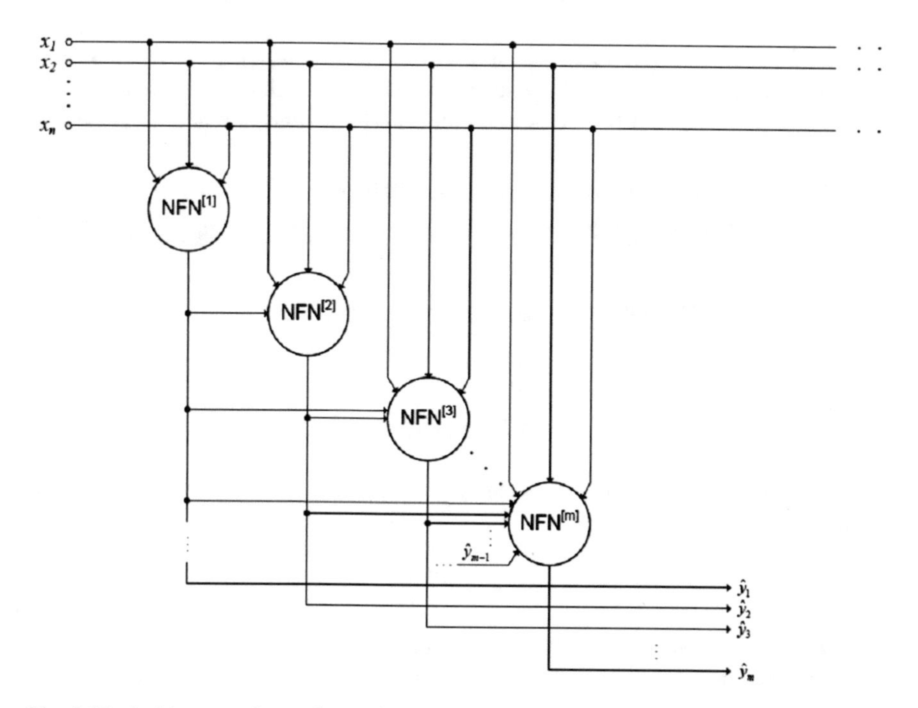

Fig. 3.15 Architecture of cascade neo-fuzzy neural network

- Neo-fuzzy neuron of the second cascade

$$\hat{y}^{[2]} = \sum_{i=1}^{n} \sum_{j=1}^{h} w_{ji}^{[2]} \mu_{ji}(x_i) + \sum_{j=1}^{h} w_{j,n+1}^{[2]} \mu_{j,n+1}\left(\hat{y}^{[1]}\right)$$

- Neo-fuzzy neuron of the 3rd cascade

$$\hat{y}^{[3]} = \sum_{i=1}^{n} \sum_{j=1}^{h} w_{ji}^{[3]} \mu_{ji}(x_i) + \sum_{j=1}^{h} w_{j,n+1}^{[3]} \mu_{j,n+1}\left(\hat{y}^{[1]}\right) + \\ + \sum_{j=1}^{h} w_{j,n+2}^{[3]} \mu_{j,n+2}\left(\hat{y}^{[2]}\right)$$

- Neo-fuzzy neuron of the mth cascade

$$\hat{y}^{[m]} = \sum_{i=1}^{n} \sum_{j=1}^{h} w_{ji}^{[m]} \mu_{ji}(x_i) + \sum_{l=n+1}^{n+m-1} \sum_{j=1}^{h} w_{jl}^{[m]} \mu_{jl}\left(\hat{y}^{[l-n]}\right) \tag{3.77}$$

Therefore, cascade neo-fuzzy neural network includes $h(n + \sum_{1}^{m-1} l)$ adjustable parameters and it is important that they are included in the linear description (3.77).

Let h(n + m − 1) × 1 be vector of membership functions mth neo-fuzzy neuron

$$\mu^{[m]} = (\mu_{11}(x_1), \ldots, \mu_{h1}(x_1), \mu_{12}(_{x2}), \ldots, \mu_{h2}(x_2), \\ \ldots, \mu_{ji}(Xi), \ldots, \mu_{hn}(Xn), \mu_{1,n+1}(\hat{y}^{[1]}), \ldots, \mu_{h,n+1}(\hat{y}^{[1]}), \ldots, \mu_{h,n+m-1}(\hat{y}^{m-1}))^T$$

And the corresponding vector of synaptic weights:

$$w^{[m]} = (w_{11}^{[m]}, w_{21}^{[m]}, \ldots, w_{h1}^{[m]}, w_{12}^{[m]}, \ldots, w_{h2}^{[m]}, \ldots\ldots, w_{ji}^{[m]} \ldots, w_{hn}^{[m]}, \\ w_{1,n+1}^{[m]}, \ldots, w_{h,n+1}^{[m]}, \ldots, w_{h,n+m-1}^{[m]})^T,$$

which has the same dimension. Then we can present the expression (3.77) in a vector form

$$y = w^{[m]T} \mu^{[m]}$$

Training of Cascade Neo-Fuzzy Neural Network

Cascade neo-fuzzy neural network training can be executed in a batch mode, or in a sequential mode (adaptive adjustment of weights).

First, let consider the situation when the training sample is determined a priori, that is, we have a sample of values *x(1), y(1); x(2), y(2); ...; x(k), y(k); ...; x(N), y (N).*

For the neo-fuzzy neuron of the first cascade NFN [1] sample values of the membership functions are $\mu^{[1]}(1), \mu^{[1]}(2), \ldots, \mu^{[1]}(k), \ldots, \mu^{[1]}(N)$ ($hn \times 1$ vector) It is defined as follows: $\mu^{[1]}(k) = (\mu_{11}(x_1(k)), \ldots, \mu_{h1}(x_1(k)), \mu_{12}(x_2(k)), \ldots, \mu_{h2}(x_2(k)), \ldots\ldots, \mu_{ji}(x_i(k)), \ldots, \mu_{hn}(x_n(k)))^T$.

Then, minimizing the learning criterion

$$E_N^{[1]} = \frac{1}{2}\sum\nolimits_{k=1}^{N}\left(e^{[1]}(k)\right)^2 = \frac{1}{2}\sum\nolimits_{k=1}^{N}\left(y(k) - \hat{y}^{[1]}(k)\right)^2,$$

vector of synaptic weights can be determined as [11]

$$\begin{aligned} w^{[1]}(N) &= \left(\sum\nolimits_{k=1}^{N}\mu^{[1]}(k)\mu^{[1]T}(k)\right)^{+}\sum\nolimits_{k=1}^{N}\mu^{[1]}(k)y(k) = \\ &= P^{[1]}(N)\sum\nolimits_{k=1}^{N}\mu^{[1]}(k)y(k) \end{aligned} \tag{3.78}$$

where $(\bullet)+$ is the Moore-Penrose pseudoinverse.

In the case of sequential data processing the recursive least squares method is used:

$$\begin{cases} w^{[1]}(k+1) = w^{[1]}(k) + \dfrac{P^{[1]}(k)\left(y(k+1) - w^{[1]T}(k)\mu^{[1]}(k+1)\right)}{1 + \mu^{[1]T}(k+1)P^{[1]}(k)\mu^{[1]}(k+1)}\mu^{[1]}(k+1) \\ P^{[1]}(k+1) = P^{[1]}(k) - \dfrac{P^{[1]}(k)\mu^{[1]}(k+1)\mu^{[1]T}(k+1)P^{[1]}(k)}{1 + \mu^{[1]T}(k+1)P^{[1]}(k)\mu^{[1]}(k+1)}, \quad P^{[1]}(0) = \beta I \end{cases} \tag{3.79}$$

where β is a large positive number, I is the unitary matrix with appropriate dimension.

It is possible to use adaptation algorithms (3.78) or (3.79), which reduce the computational complexity of the learning process. In any case, the use of procedures (3.78) (3.79) significantly reduces training time, compared with the gradient method, the underlying rules of the delta and backpropagation.

After the first cascade of the training synaptic weights of neo-fuzzy neuron NFN [1] become “frozen”, all values $y^{[1]}(1), y^{[1]}(2), \ldots Y^{[1]}(k), \ldots, Y^{[1]}(N)$ will be evaluated and a second network cascade is obtained that consists of a single neo-fuzzy neuron NFN [2]. It has one additional input for the output signal of the first cascade. Then, again the procedure (3.78) is used for adjusting the weight vector w [2] of dimension $h(n+1)$.

In the online method of training neurons are trained sequentially, based on input signals $x(k)$. Let evaluated synaptic weights of the first cascade be $w^{[1]}(k)$ and received output vector be $y^{[1]}(k)$, then using the input vector of the second cascade $(x(k), y^{[1]}(k))$ weight $w^{[2]}(k)$ and outputs $y^{[2]}(k)$ of the second cascade are calculated. For this purpose, the algorithms (3.78) and (3.79) can be used equally well.

The process of the neural network growth (increasing the number of cascades) continues until we obtain the required accuracy of the solution. To adjust the weighting coefficients of the last m-th cascade, the following expression is used [11]:

$$w^{[m]}(N) = \left(\sum_{k=1}^{N} \mu^{[m]}(k)\,\mu^{[m]T}(k)\right)^{+} \sum_{k=1}^{N} \mu^{[m]}(k)y(k) = P^{[m]}(N) \sum_{k=1}^{N} \mu^{[m]}(k)y(k)$$

In a batch mode

$$\begin{cases} w^{[m]}(k+1) = w^{[m]}(k) + \dfrac{P^{[m]}(k)(y(k+1) - w^{[m]T}(k)\,w^{[m]}(k+1))}{1 + \mu^{[m]T}(k+1)\,P^{[m]}(k)\,\mu^{[m]}(k+1)} \mu^{[m]}(k+1), \\ P^{[m]}(k+1) = P^{[m]}(k) - \dfrac{P^{[m]}(k)\,\mu^{[m]}(k+1)\,\mu^{[m]T}(k+1)\,P^{[m]}(k)}{1 + \mu^{[m]T}(k+1)\,P^{[m]}(k)\,\mu^{[m]}(k+1)}, P^{[m]}(0) = \beta I \end{cases}, \tag{3.80}$$

Or

$$\begin{cases} w^{[m]}(k+1) = w^{[m]}(k) + (r^{[m]}(k+1))^{-1}(y(k+1) - w^{[m]T}(k)\,\mu^{[m]}(k+1))\,\mu^{[m]}(k+1) \\ r^{[m]}(k+1) = \alpha\, r^{[m]}(k) + \left\|\mu^{[m]}(k+1)\right\|^{2}, 0 \le \alpha \le 1 \end{cases} \tag{3.81}$$

in a sequential mode.

Thus, the proposed CNFNN far exceeds conventional cascaded architecture by training speed and can be trained in batch mode or in sequential (adaptive) mode. Linguistic interpretation of the results significantly expands the functionality of cascade neo-fuzzy neural network.

3.11 Simulation Results

We investigated the forecasting series of numbers to evaluate the work of the proposed architecture. Cascade neo-fuzzy neural network was applied to predict chaotic process Macki-Glass defined by the equation [11]:

$$y'(t) = \frac{0.2x(t-\tau)}{1 + y^{10}(t-\tau)} - 0.1y(t) \tag{3.82}$$

The signal y(t), defined by (3.82), was digitized with step 0.1. A fragment containing 500 measurements was taken for the training sample. It is necessary to predict the value of the time series at six steps forward using its values at time k, (k −6), (k−12) and (k−18). The test sample contains 9500 measurements, i.e. the

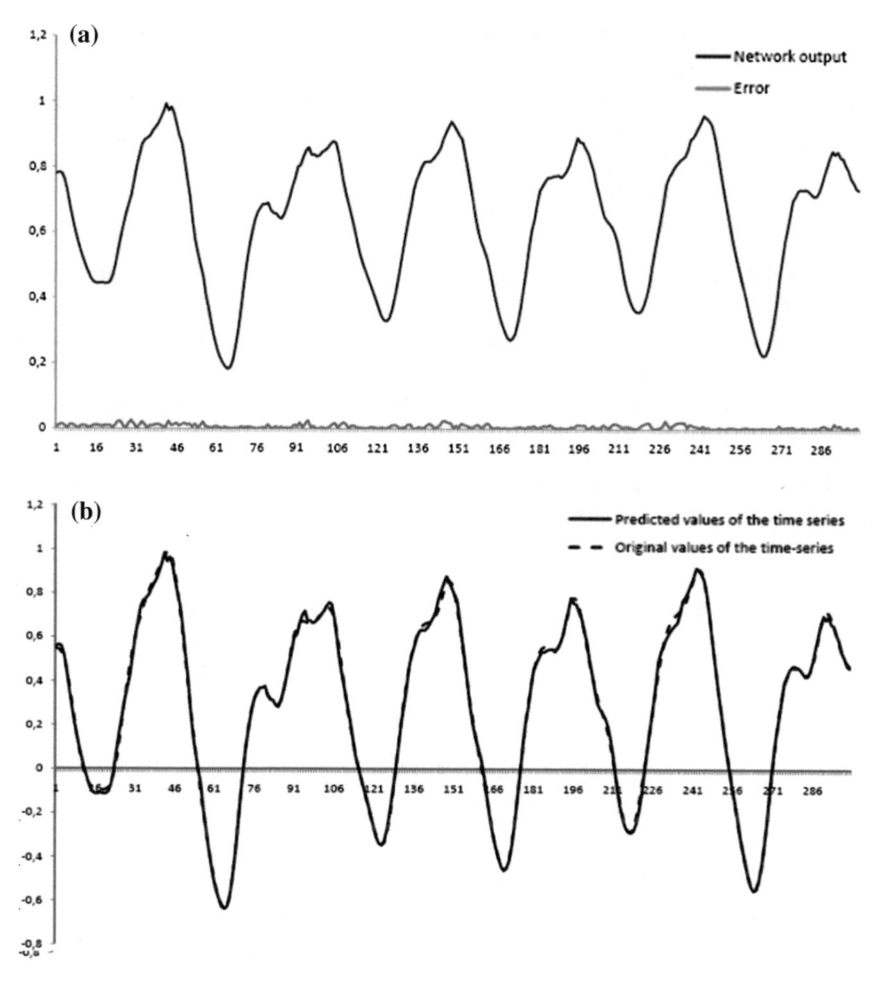

Fig. 3.16 The results of CNFNN: **a** the network outputs and error; **b** forecasted and actual time series

value of the time series in the time range from 501 to 1000. In order to evaluate the results Normalized root mean square error (NRMSE) was used.

The modeling was performed using a network CNFNN with 5 cascades. The first cascade comprises 4 nonlinear synapses for each input value of the time series and each synapse contains 10 activation functions (i.e. membership functions). Input signals were presented in the interval [*0, 1*]. The experimental results of forecasting time series and their errors are shown in Fig. 3.16a, b. After training, the CNFNN network error on test sample was 4×10^4.

Application of GMDH for Optimal Synthesis of Architecture of Neo-Cascade Fuzzy Neural Network

An important problem that must be solved using neural cascade neo-fuzzi-networks is the problem of selecting the number of cascades and the synthesis of the structure of such a network. Such a task is reasonable to solve using heuristic self-organizing method GMDH considered in the Chap. 6 [15, 16].

Consider this method and give the corresponding experimental results.

GMDH Algorithm for Synthesis of a Cascade Neo-Fuzzy Network Structure

If to use a neo-fuzzy neurons with two inputs, it is possible to apply the method GMDH for the synthesis of the optimal structure of the cascade neo-fuzzy neural network. The basic idea of the algorithm GMDH is that there is a sequential buildup of layers of the neural network until an external criterion starts to increase.

Description of the algorithm.

1. Form a pair of outputs of neurons of this layer (in the first iteration, we use a number of input signals). Each pair is applied to the inputs of the neo-fuzzy neuron.
2. Using the training sample, determine a weight for each connection of a neo-fuzzy neuron.
3. Using the test sample, we calculate the value of external regularity criterion for every neo-fuzzy neuron

$$\varepsilon_p^{[s]} = \frac{1}{N_{giv}} \sum_{i=1}^{N_{giv}} \left(y(i) - \hat{y}_p^{[s]}(i) \right)^2 \tag{3.83}$$

where N is the size of the test sample, s is a layer number,

$\hat{y}_p^{[s]}(i)$ is the output of neuron p at s-th layer at the i-th input signal.

4. Find the minimum value of the external criterion for all neurons in the current layer

$$\varepsilon^{[s]} = \min_p \varepsilon_p^{[s]}$$

Testing conditions for stop is: $\varepsilon^{[s]} > \varepsilon^{[s-1]'}$, where $\varepsilon^{[s]}, \varepsilon^{[s-1]}$—criteria value of the best neuron of the s-th and (s−1)-th layers, respectively.

If the condition is satisfied, then go back to the previous layer, find the best neuron with a minimum criterion and go to step 5, otherwise choose the best F neurons in accordance with the criterion (3.83) and go to step 1 for the construction of the next layer of neurons.

5. Determine the final structure of the network. To do this, move in the opposite direction from the best neuron along input connections and sequentially move through all layers of neurons, save in the final structure of the network only those neo-fuzzy neurons, which are used in the next layer.
6. After the end of the algorithm GMDH optimal network structure will be synthesized. It is easy to notice that we get not only the optimal structure of the network, but also a trained neural network that is ready to handle the new data. One of the major advantages of using GMDH for the synthesis cascade network architecture is its ability to use very fast learning procedures for the adjustment of neo-fuzzy neurons because learning occurs in layers sequentially (layer by layer).

The Results of Experimental Studies

Experimental studies of neo-fuzzy neural network were carried out in the prediction problem. The aim of the forecast was predicting RTS index based on the current share prices of leading Russian companies [18].

Inputs: daily share price RTS index over the period from 5 February to May 5, 2009. Output variable—RTS index next day. The size of the training sample—100 sample values.

Forecast criteria: MSE;

– The mean absolute percentage error (MAPE).

Types of Experiments with Neo-Fuzzy Neural Network

1. Variation ratio training–testing sample in the range: 25:75, 50:50, 75:25;
2. the variation of the layers number: 1-3-5;
3. the variation of the number of iterations: 1000, 10000, 100000;
4. The variation of the number of forecasted points: 1-3-5;

Change the maximum error-stop condition: from 0.01 to 0.09.

Some experimental results are given below

Experiment 1. Variation Ratio Learning/Verification Sample.

Experiment (A) ratio of 75:25. Charts of real and predicted values are shown in Fig. 3.17.

Experiment (B) ratio of 50:50 MSE = 0.053562

Experiment (C) ratio of training/test—25:75. Results are shown in Fig. 3.18 MSE = 0.068489

Experiment Type 2. Comparison of the algorithm by varying the number of layers: 1-3-5-7. The forecast for one point at a ratio of training—test sample 75:25.

Experiment (A) the number of layers—1 MSE = 0.04662. Charts of the forecasting results are presented in Fig. 3.19.

Experiment (B) the number of layers—3, MSE = 0.255

Experiment (C) layers number 5, MSE = 0.0446

Experiment (D) layers number 7, MSE = 0.0544

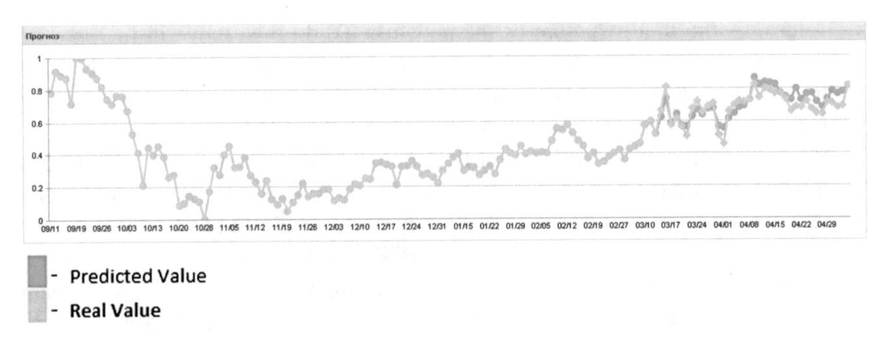

Fig. 3.17 Forecasting results for learning /test samples ratio 75:25

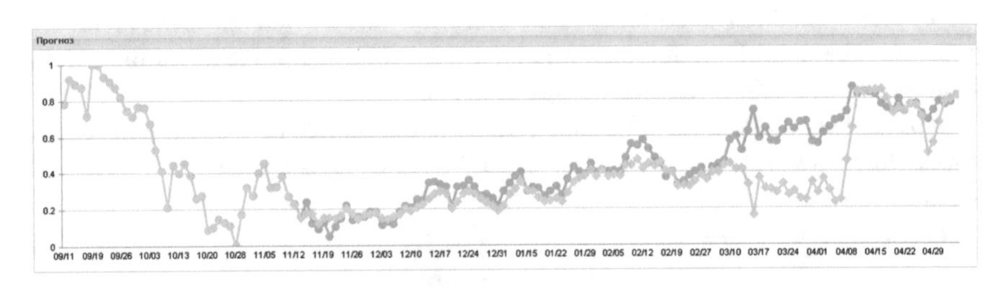

Fig. 3.18 Forecasting results for learning/test ratio 25:75

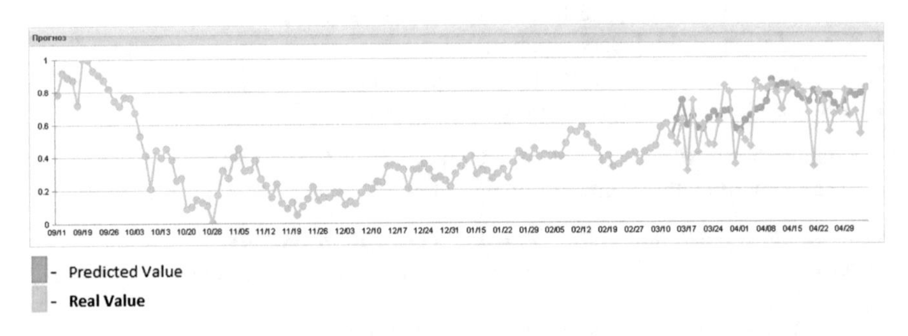

Fig. 3.19 Forecasting results for network with one layer

Experiment Type №3 Variation of iterations-: 1000, 10000, 100000

Experiment (A) The number of iterations—1000. MSE = 0.0588

Experiment (B) the number of iterations—10000, MSE = 0.0575

Experiment (C) the number of iterations—100000, MSE = 0.0525

Experiment Type №4 Comparison of the accuracy of prediction algorithm by varying the number of forecasted points 1–3–5, using the ratio of training/testing sample 75:25

Experiment (A) the number of forecasted points—1

Results are shown in Fig. 3.20.

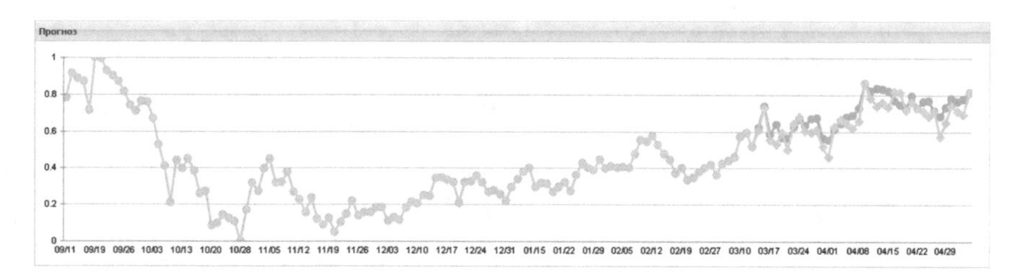

Fig. 3.20 Forecasting results for one point

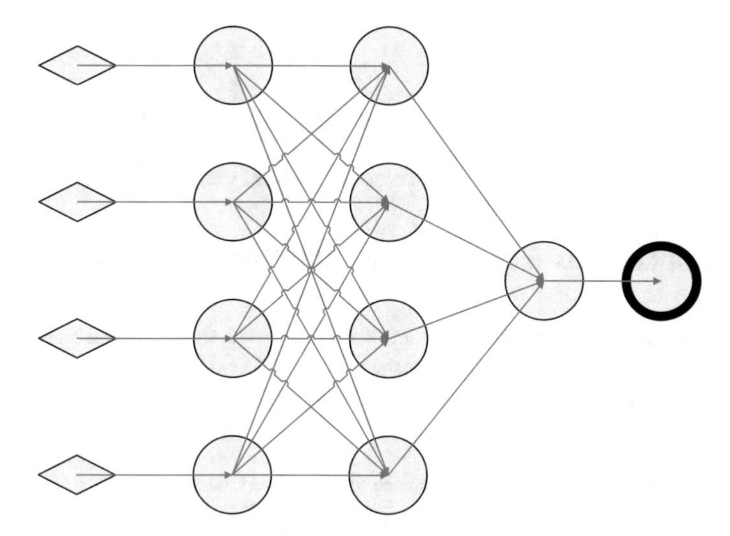

Fig. 3.21 The structure of the full neo-fuzzy neural network

MSE = 0.0495

Experiment (B) the number of forecasted points—3

MSE = 0.4469

Experiment (C) the number of forecasted points—5

MSE = 1.0418

Full structure of the neo-fuzzy neural network is shown in Fig. 3.21, and optimal network structure, synthesized by GMDH-in the Fig. 3.22

- 4 inputs
- 1 output (1 point predicted);
- 1 hidden layer
- 4 inputs; 1 output (1 point forecast); 1 hidden layer

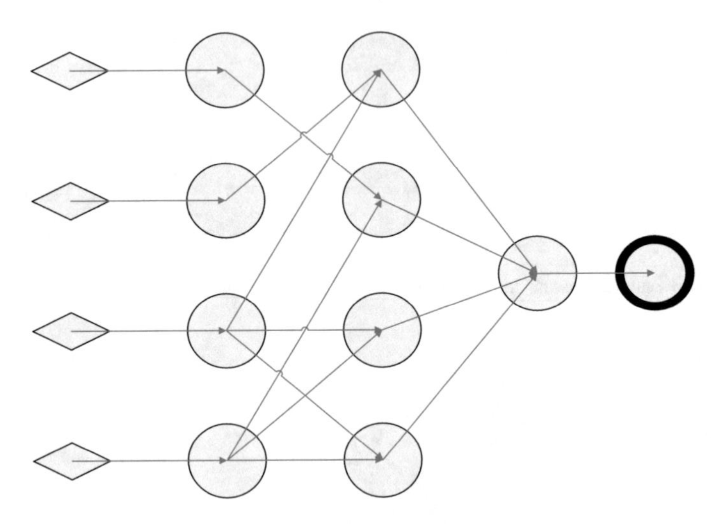

Fig. 3.22 The optimum structure of neo-fuzzy neural network, which was synthesized by GMDH

3.12 Conclusions from Experimental Studies

After performing a series of experiments with the neo-fuzzy neural network with full structure and optimal structure synthesized using GMDH following final results were obtained presented in the Table 3.2 [18].

Table 3.2 Experimental results for synthesized and standard CNFN networks

	Comparison CNFNN with full structure and CNFNN synthesized by GMDH		
Type of the experiment	Experiment parameters	CNFNN synthesized by GMDH	Full structure CNFNN
Variation of ratio training/testing sample	75 %: 25 %	0.0484	0.0501
	50 %: 50 %	0.0532	0.0536
	25 %: 75 %	0.0608	0.0684
Number of layers variation	1	0.0628	0.0626
	3	0.0381	0.0544
	5	0.0434	0.0652
Iterations number	1000	0.0588	0.0674
	10000	0.0479	0.0485
	100000	0.0459	0.0482
Number of forecasted points	1	0.0495	0.0587
	3	0.4469	1.0844
	5	1.0418	1.3901

It is easy to notice the neo-fuzzy neural network synthesized using GMDH gives much better results than conventional full network structure. This can be explained by impact of the network self-organization mechanism used by GMDH. However, the optimum network structure synthesized by GMDH shows slower convergence.

For convenience of the experimental results analysis number of figures are presented below displaying the found dependencies. Figure 3.23 shows the dependence of the value of MAPE error on the number of predicted points.

As one can see when predicting only one point a very high accuracy is reached: the average relative error is less than 15 %. By increasing the number of predicted points accuracy falls, the error is in the range of 15–45 %.

By analyzing the results presented in Fig. 3.24, one can conclude that the best results are achieved when the number of layers is equal to 3. The average relative error is less than 10 %. In one hidden layer error is low, but is not uniformly distributed and may exceed 30 %.

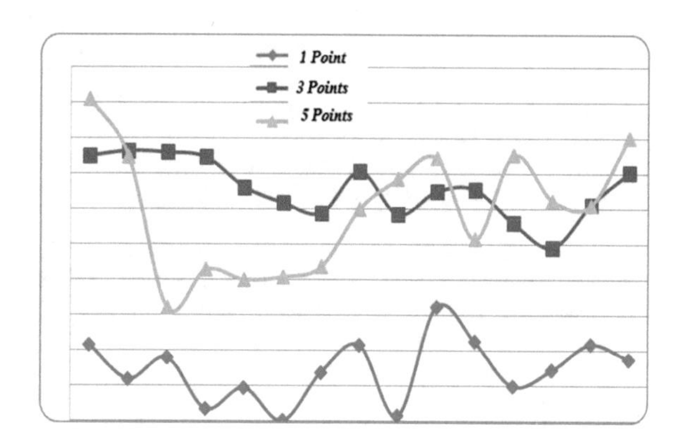

Fig. 3.23 Curves MAPE error in the forecast of 1.3 or 5 points

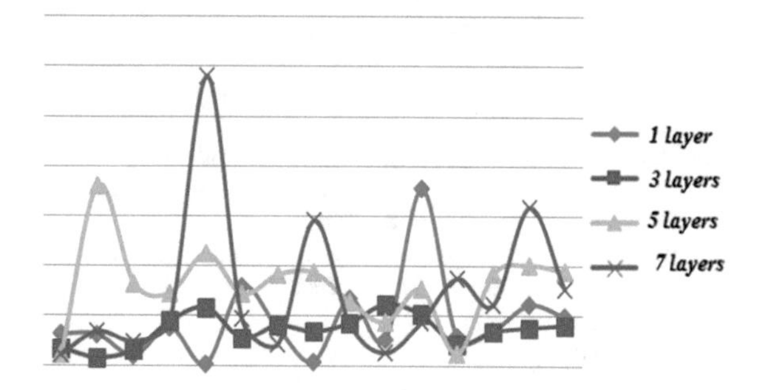

Fig. 3.24 Prediction error (MAPE) depending on the number of layers (1, 3, 5, 7)

In addition, during experimental studies the optimal parameters of algorithm for the full neo-fuzzy neural network were determined [18]:

- the best ratio of "training/testing sample" −75 to −25 %.;
- the best possible number of layers—3;
- the best results were obtained when the number of training iterations—100000.

References

1. Wang, F.: Neural networks genetic algorithms, and fuzzy logic for forecasting. In: Proceedings, International Conference on Advanced Trading Technologies, New York, July 1992, pp. 504–532
2. Zadeh, L.A.: The Concept of a Linguistic variable and its application to approximate reasoning, Part 1 and 2, Information Sciences, vol. 8, pp. 199–249, 301–357 (1975)
3. Zaychenko, Y.P.: Fuzzy Models and Methods in Intellectual Systems. Publisher House "Slovo", Kiev, 354 p. (2008). Zaychenko, Y.: Fuzzy group method of data handling under fuzzy input data. In: System Research and Information Technologies, № 3, pp. 100–112 (2007) (in Russian)
4. Zaychenko, Y.P.: Fundamentals of Intellectual Systems Design. Publisher House "Slovo", Kiev, 352 p. (2004) (in Russian)
5. Yarushkina, N.G. Fuzzy neural networks. News Artif. Intell. № 36, 47–51 (2001) (in Russian)
6. Kosko, B.: Fuzzy systems as universal approximators. IEEE Trans. Comput. (11), 1329–1333 (1994)
7. Castro, J.L., Delgado, M.: Fuzzy systems with defuzzification are universal approximators. IEEE Trans. Syst. Man Cybern. 1996 (1), 361–367 (1998)
8. Nauck, D: Building neural fuzzy controllers with NEFCON-I. In: Kruse, R., Gebhardt, J., Palm, R. (eds.) Fuzzy Systems in Computer Science. Artificial Intelligence, pp. 141–151. Vieweg, Wiesbaden (1994)
9. Rutkovska, D. Pilinsky, M., Rutkovsky, L.: Neural networks, genetic algorithms and fuzzy systems (Transl. from pol. Rudinsky, I.D.), 452c. Hot Line Telecom (2006) (in Russian)
10. Osovsky, S.: Neural networks for information processing (Transl. from pol.), 344 p. Publisher House Finance and Statistics (2002) (in Russian)
11. Bodyanskiy, Y., Viktorov, Y., Pliss, I.: The cascade NFNN and its learning algorithm. Вісник Ужгородського національного університету. Серія «Математика і інформатика», Вип. 17, с. 48–58 (2008)
12. Aleksandr, I., Morton, H.: An Introduction to Neural Computing. Chapman & Hall, London (1990)
13. Dorigo, M., Stutzle, T.: An experimental study of the simple ant colony optimization algorithm. In: Proceedings of the WSES International Conference on Evolutionary Computation, pp. 253–258 (2001)
14. Bodyanskiy, Y., Gorshkov, Y., Kokshenev, I., Kolodyazhniy, V.: Robust recursive fuzzy clustering algorithms. In: Proceedings of East West Fuzzy Colloquim 2005. HS, Goerlitz, Zittau, pp. 301–308 (2005)
15. Zaychenko, Y.P., Zayets, I.O. Ю. П. Synthesis and adaptation of fuzzy forecasting models on the basis of self-organization method. Scientific Papers of NTUU "KPI", № 3, pp. 34–41 (2001) (in Russian)
16. Zaychenko, Y.P., Kebkal, A.G., Krachkovsky, V.F.: Fuzzy group method of data handling and its application for macro-economic indicators forecasting. Scientific Papers of NTUU "KPI", № 2, pp. 18–26 (2000) (in Russian)

17. Zaychenko, Y.P., Petrosyuk, I.M., Jaroshenko, M.S.: The investigations of fuzzy neural networks in the problems of electro-optical images recognition. System Research and Information Technologies, № 4, pp. 61–76 (2009) (in Russian)
18. Bodyanskiy, Y., Gorshkov, Y., Kokshenev, I., Kolodyazhniy, V.: Outlier resistant recursive fuzzy clustering algorithm. In: Reusch, B. (ed.) Computational Intelligence: Theory and Applications. Advances in Soft Computing, vol. 38, pp. 647–652. Springer, Berlin (2006)

Chapter 4
Application of Fuzzy Logic Systems and Fuzzy Neural Networks in Forecasting Problems in Macroeconomy and Finance

4.1 Introduction

This chapter is devoted to numerous applications of fuzzy neural networks in economy and financial sphere. In the Sect. 4.2 the problem of forecasting macroeconomic indicators of Ukraine with application of FNN is considered. The goal of this investigation was to estimate the efficiency of different fuzzy inference algorithms. Fuzzy algorithms of Mamdani, Tsukamoto and Sugeno were compared in forecasting Consumer Price Index (CPI) and GDP of Ukraine. As result of this investigation the best forecasting algorithm is detected.

In the Sect. 4.3 the problem of forecasting in the financial markets with application of FNN is considered. The forecasting variable are share prices of "Lukoil" at the stock exchange RTS. In these experiments the same FNN are used as in the Sect. 4.2 and the results of these experiments confirmed the conclusions of Sect. 4.2.

In the Sect. 4.4 the problem of forecasting corporations bankruptcy risk under uncertainty is considered and investigated. At first classical method by prof. Altman is presented, its properties and drawbacks are analyzed. As an alternative matrix method based on fuzzy sets theory is described and discussed. The application of fuzzy networks with inference algorithms of Mamdani and Tsukamoto for corporations bankruptcy risk forecasting is considered and the experimental comparative results of application all the considered methods for solution of this problem for Ukrainian enterprises are presented and analyzed which confirmed the preference of FNN.

In the Sect. 4.5 the problem of banks financial state analysis and bankruptcy forecasting under uncertainty is considered. For its solution the application of FNN TSK and ANFIS is suggested. The experimental results of the efficiency of considered FNN for forecasting are presented and analyzed and the best class of FNN with the least MSE and MAPE for Ukrainian banks bankruptcy forecasting is determined. In the Sect. 4.6 the similar problem of banks financial state analysis

M.Z. Zgurovsky and Y.P. Zaychenko, *The Fundamentals of Computational Intelligence: System Approach*, Studies in Computational Intelligence 652,
DOI 10.1007/978-3-319-35162-9_4

and bankruptcy forecasting for leading European banks is considered. The application of FNN for its solution is investigated and the experimental results are presented and analyzed.

4.2 Forecasting in Macroeconomics and Financial Sector Using Fuzzy Neural Networks

Problem Statement

The goal of investigation is to assess the effectiveness of the fuzzy neural networks (FNN) with different inference algorithms in the forecasting problems in macroeconomics and financial sphere.

As the initial data macroeconomic indicators Ukraine's economy were selected, presented in the form of statistical time series (see Table 4.1) [1].

The task was forecasting of the following macroeconomic indicators: CPI, NGDP and Volume of industrial production (VIP) by known macroeconomic indicators.

For the construction of rule base significant variables and their lags were identified. As the degree of the relationship between the input variables $x_1, x_2, \ldots, x_n$ and the output variable Y were used correlation coefficients, by the value of which were selected essential variables.

4.2.1 A Comparative Analysis of Forecasting Results in Macroeconomics Obtained with Different FNN

Experimental investigations were conducted comparing the prediction efficiency obtained using the following methods [1, 2]:

- Tsukamoto neuro-fuzzy controller (NFC) with linear membership functions;
- Tsukamoto neuro-fuzzy controller with monotonous membership functions;
- Mamdani neuro-fuzzy controller with Gaussian membership functions;
- Fuzzy neural network ANFIS.

Table 4.2 and Fig. 4.1 show the comparative results of forecasting the CPI index, obtained by different methods of fuzzy inference.

As it can be easily seen from the presented results, all three fuzzy controllers coped well with the task. We also see that the network ANFIS provides also acceptable result, but worse than the neural fuzzy controllers Mamdani, Tsukamoto and Sugeno. This reflects both the effectiveness of controllers learning algorithms and preference of application fuzzy logic systems to predict macroeconomic indices [2, 3].

Table 4.1 Macroeconomic indicators in Ukraine

Data	NGDP	VIP	IRIP	CPI	WPI	RIP	M2	MB	TL
1	2358.3	102.2	100.3	103.1	99.6	113.1	1502	840.1	73.7
2	2308.5	102	99.7	101.2	100.3	111	1522.9	846.1	84.6
3	2267.7	103.7	99.9	101.1	99.2	108.1	1562.4	863.5	96.5
4	2428.5	104.3	102.2	101.2	100.7	118	1621.3	917.7	98.2
5	2535.6	102.8	102.5	101.7	102.2	107.8	1686	977.7	118.2
6	2522.8	104.4	103.1	100.5	104.4	105.8	1751.1	1020.7	138.6
7	2956.4	107.8	102.6	100.7	105.4	112.3	1776.1	1019.8	142.6
8	3025.9	103.4	101.7	100.1	105	109	1812.5	1065.6	157.8
9	3074.5	10.55	101.2	100.4	105.3	106.6	1846.6	1067.9	165.5
10	2854.3	103.9	102.1	101.1	105.3	108.5	1884.6	1078.6	158.9
11	2812.5	100.8	101.6	101.6	100.2	107.8	1930	1128.9	163.4
12	2998.4	103.2	99.8	101.5	105.7	106.9	2119.6	1232.6	262.5
13	2725.6	104.9	100.4	102.4	100.5	114.4	2026.5	1140.1	93.8
14	2853.4	1065	101.4	101.6	101.2	116.8	2108	1240.7	110.3
15	2893.1	106.7	101.3	101.1	103.3	115.4	2208.5	1284.5	125.9
16	3014.2	107.1	101.4	101	103.6	109.1	2311.2	1386.8	130.1
17	3102.6	108.5	99.8	100.8	103.9	119.7	2432.4	1505.7	158.8
18	3110.7	107	100.7	100.8	103.9	113.8	2604.5	1534	158.8
19	3192.4	107.1	102.2	100.7	104.9	112.7	2625.4	1510.8	181.9
20	3304.7	105.5	101.4	99.6	105.9	109.8	2683.2	1500.8	185
21	3205.8	108	101.4	100.3	106.9	112.6	2732.1	1484.5	205.8

NGDP—nominal GDP
VIP—volume of industrial production, % of the corresponding period of the previous year
IRIP—the index of real industrial production, % of the corresponding period of the previous year
CPI—consumer price index, % to the corresponding period of the previous year
WPI—the wholesale price index in % to the corresponding period of the previous year
RIP—the real incomes of the population
M2—aggregate M2
MB—monetary base
TL—Total loans, including loans in foreign currency

4.3 FNN Application for Forecasting in the Financial Sector

To verify these conclusions further experimental studies of fuzzy neural networks application in forecasting problems in the financial markets were carried out. As an example share prices of "Lukoil" have been chosen for prediction, admitted to trading at the "Stock Exchange Russian Trading System" (RTS).

For training a sample of 267 daily values of stock prices "Lukoil" was used at the period since 30.12.2005 to 04.01.2006 [1, 4].

Table 4.2 Comparative results of forecasting CPI index, obtained by different methods

Real values real	Network ANFIS		NFC Tsukamoto with linear MF		NFC Tsukamoto with monotonous MF		NFC Mamdani with Gaussian MF	
	Forecasting	Error	Forecasting	Error	Forecasting	Error	Forecasting	Error
101.5	101.28	0.22	101.32	0.18	101.32	0.18	101.34	0.16
102.4	101.69	0.71	102.15	0.25	102.16	0.24	102.34	0.06
101.6	101.43	0.17	101.28	0.32	101.30	0.30	101.48	0.12
101.1	101.54	0.44	100.86	0.24	100.89	0.21	101.07	0.03
101.0	100.92	0.08	100.70	0.30	100.71	0.29	100.94	0.06
100.8	100.73	0.07	100.65	0.15	100.65	0.15	100.70	0.10
100.8	99.83	0.97	100.13	0.67	100.24	0.56	100.73	0.07
100.7	99.88	0.82	100.22	0.48	100.22	0.44	100.65	0.05
99.6	98.86	0.74	99.09	0.51	99.13	0.47	99.44	0.16
100.3	99.54	0.76	99.78	0.52	99.80	0.50	100.18	0.12
	MSE: 0.3524		MSE: 0.1577		MSE: 0.1309		MSE: 0.0930	

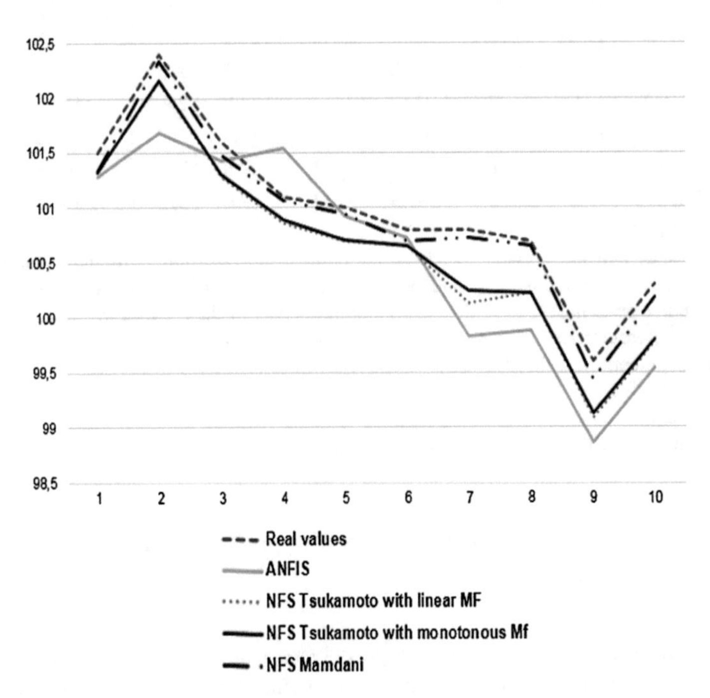

Fig. 4.1 Forecasting results by different FNN, row 1-real data, row 2-ANFIS, row 3-NFC Tsukamoto with linear MF, row 4-NFC Tsukamoto with monotonous MF, row 5-NFC Mamdani

During testing it was found experimentally that the best was to use from three to five terms and the rules, since just with these parameters minimum MSE was obtained. Gradient method was used for training MF with a step of 0.04.

Table 4.3 Forecasting results for NFC Mamdani with Gaussian MF

Data	Real values	The predicted value	Deviation	The square deviation
01.12.2005	58.1	58.23	0.13	0.0169
02.12.2005	58.7	58.54	0.16	0.0256
05.12.2005	59.4	59.14	0.26	0.0676
06.12.2005	59	59.11	0.11	0.0121
07.12.2005	59.85	59.97	0.12	0.0144
08.12.2005	59.6	59.416	0.184	0.033856
09.12.2005	59.9	60.12	0.22	0.0484
12.12.2005	60.65	60.5	0.15	0.0225
13.12.2005	60.65	60.54	0.11	0.0121
14.12.2005	61.15	61.32	0.17	0.0289
15.12.2005	60.25	60.1	0.15	0.0225
16.12.2005	61	61.2	0.2	0.04
19.12.2005	61.01	61.24	0.23	0.0529
20.12.2005	60.7	60.54	0.16	0.0256
				MSE = 0.030239714

1. The use of NFC (neuro-fuzzy controller) Mamdani in predicting stock prices. Using NFC Mamdani with triangular and Gaussian MF the following results were obtained while predicting the stock price of "Lukoil". They are presented in the Table 4.3.
   After that prediction using NFC Mamdani for triangular MF was performed. As it was shown by the experiments, the best proved to be Mamdani controller with Gaussian MF (MSE on the test sample of 14 points is only 0.03024, the average relative error is 3.02 %).
2. Further experiments were conducted to predict with NFC Tsukamoto triangular and Gaussian MF. Forecasting results for NFC Tsukamoto with Gaussian MF are presented in Table 4.4.
   Comparison of forecast errors for NFC Tsukamoto with triangular and Gaussian MF are shown in Fig. 4.2.
   As the second experiment has shown the best is controller Tsukamoto with Gaussian MF (MSE on the test sample of 14 points is only 0.0667, and the average relative error of the forecast is 6.67 %).
3. In the next experiment investigations of NFC with Sugeno inference algorithm were performed. Prediction results using the NFC Sugeno for Gaussian MF are presented in Table 4.5, and for triangular MF in a Table 4.6.
   The third experiment has shown that the best is Sugeno controller with Gaussian MF as in the two previous cases (MSE on the test sample of 14 points 0.0967).
4. Comparative analysis of stock prices forecasting by different methods.

Table 4.4 Forecasting results for NFC Tsukamoto with Gaussian MF

Data	Real values	The predicted value	Deviation	The square deviation
01.12.2005	58.1	58.37	0.27	0.0729
02.12.2005	58.7	58.47	0.23	0.0529
05.12.2005	59.4	59.1	0.3	0.09
06.12.2005	59	59.25	0.25	0.0625
07.12.2005	59.85	60.19	0.34	0.1156
08.12.2005	59.6	59.37	0.23	0.0529
09.12.2005	59.9	60.27	0.37	0.1369
12.12.2005	60.65	60.48	0.17	0.0289
13.12.2005	60.65	60.42	0.23	0.0529
14.12.2005	61.15	61.4	0.25	0.0625
15.12.2005	60.25	60.06	0.19	0.0361
16.12.2005	61	61.22	0.22	0.0484
19.12.2005	61.01	61.28	0.27	0.0729
20.12.2005	60.7	60.48	0.22	0.0484
			0.252857143	MSE = 0.0667

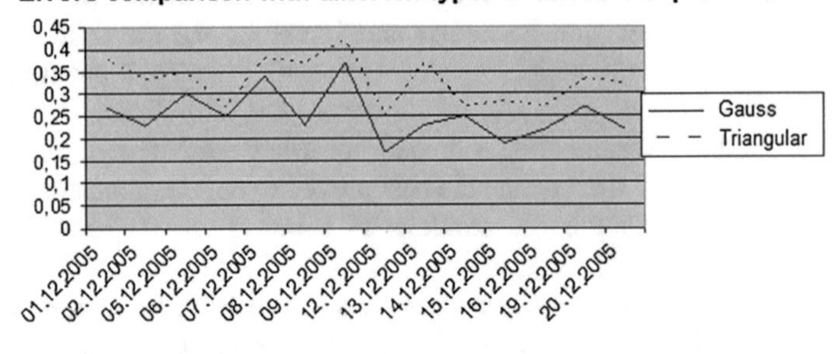

Fig. 4.2 Forecasting results for NFC Tsukamoto

The results of forecasting by FNC Mamdani, Sugeno and Tsukamoto with Gaussian MF and the NFC ANFIS are presented in the Fig. 4.3, and with the triangular MF- in the Fig. 4.4 correspondingly.

As these results demonstrate [4], the best one proved to be Mamdani controller with Gaussian MF. Its standard deviation is only 0.028236. Next by the quality of the forecasting is Tsukamoto controller with Gaussian MF. It shows a slightly better result than NFC Tsukamoto with triangular MF. But in general, their predictions are very close (MSE = 0.0728 and MSE = 0.0817, respectively). This enables to assume that the selection of more adequate type of membership functions may allow to further improve the forecast results. Finally, at the last position are results obtained using the NFC ANFIS (MSE = 0.34312). Its drawback lies in that this FNN only MF of input rules are trained but not the rules outputs.

Table 4.5 Forecasting results for NFC Sugeno with Gaussian MF

Data	Real values	The predicted value	Deviation	The square deviation
01.12.2005	58.1	58.42	0.32	0.1024
02.12.2005	58.7	58.34	0.36	0.1296
05.12.2005	59.4	59.02	0.38	0.1444
06.12.2005	59	59.29	0.29	0.0841
07.12.2005	59.85	60.24	0.39	0.1521
08.12.2005	59.6	59.29	0.31	0.0961
09.12.2005	59.9	60.3	0.4	0.16
12.12.2005	60.65	60.41	0.24	0.0576
13.12.2005	60.65	60.4	0.25	0.0625
14.12.2005	61.15	61.43	0.28	0.0784
15.12.2005	60.25	60.04	0.21	0.0441
16.12.2005	61	61.25	0.25	0.0625
19.12.2005	61.01	61.31	0.3	0.09
20.12.2005	60.7	60.4	0.3	0.09
			0.3057142	MSE = 0.0967

Table 4.6 Forecasting results for NFC Sugeno with trianmgular MF

Data	Real values	The predicted value	Deviation	The square deviation
01.12.2005	58.1	58.5	0.4	0.16
02.12.2005	58.7	58.31	0.39	0.1521
05.12.2005	59.4	59.01	0.39	0.1521
06.12.2005	59	59.33	0.33	0.1089
07.12.2005	59.85	60.3	0.45	0.2025
08.12.2005	59.6	59.16	0.44	0.1936
09.12.2005	59.9	60.39	0.49	0.2401
12.12.2005	60.65	60.31	0.34	0.1156
13.12.2005	60.65	60.38	0.27	0.0729
14.12.2005	61.15	61.49	0.34	0.1156
15.12.2005	60.25	59.94	0.31	0.0961
16.12.2005	61	61.3	0.3	0.09
19.12.2005	61.01	61.37	0.36	0.1296
20.12.2005	60.7	60.32	0.38	0.1444
			0.370714286	MSE = 0.14096

Experimental studies of the prediction in macroeconomics and financial sphere with different classes of fuzzy neural networks enable to make the following **conclusions** [1]:

1. Fuzzy Mamdani and Tsukamoto controllers coped with the task of forecasting the test sample the best. This demonstrates the efficiency of the controllers

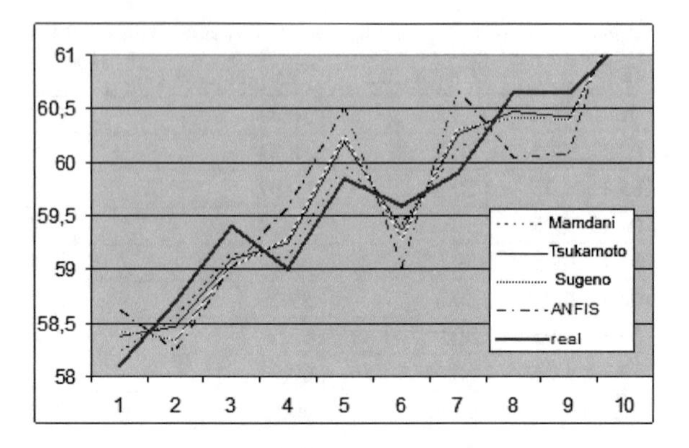

Fig. 4.3 Forecasting results by FNC Mamdani, Sugeno and Tsukamoto with Gaussian MF

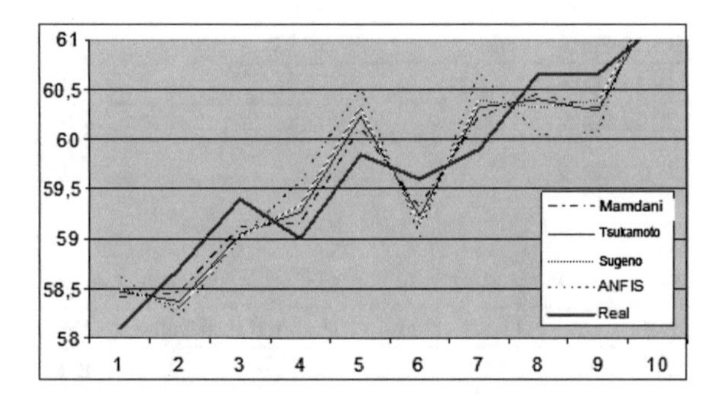

Fig. 4.4 Forecasting results by FNC Mamdani, Sugeno and Tsukamoto with triangular MF

training algorithm, as well as the high representativeness of the sample. As it was shown by the first test, the best proved to be Mamdani controller with Gaussian membership functions and fuzzy intersection as a product (MSE on the test sample is 0.09).

2. Analysis of the impact of the inference rules has revealed that the best results are achieved with comprehensive number of rules, increasing the number of rules forecasting error (MSE) first falls (50 % of the total number of rules), attains minimum and then begins to rise slightly, indicating existence of optimal number of rules for the specific forecasting task.
3. A full set of rules should be used for not large number of variables (n = 4−8). For more variables, it is advisable to use an incomplete rule base, which greatly simplifies the structure of the NFC and gives acceptable results in the accuracy of forecasting.
4. In a whole experiments showed the great potential of NFC and confirmed their effectiveness in the tasks of macroeconomic and financial forecasting.
5. Comparative analysis of the accuracy of prediction using NFC Mamdani, Tsukamoto, Sugeno and ANFIS shown that FNN ANFIS has significantly poorer performance as compared with FNN Mamdani and Tsukamoto.

4.4 Predicting the Corporations Bankruptcy Risk Under Uncertainty

Introduction

One of the actual problems related to strategic management is the analysis of enterprise financial state and assessment of the bankruptcy risk. Timely detection of signs of possible bankruptcy allows managers to take urgent measures to improve the financial state and cut the risk of possible bankruptcy.

For many years, classical statistical methods were widely used to predict the risk of bankruptcy. These methods also have the name of the one-dimensional ('single-period') classification techniques. They include the classification procedure, which classifies a company to a group of potential bankrupts, or to a group of companies with a favorable financial position with a certain degree of accuracy. Using these models, there may be two types of errors.

The first type of error occurs when a future bankrupt company is classified as a company with a favorable financial situation. The error of the second type occurs when an enterprise with a normal financial condition is classified as a potential bankrupt. Both mistakes can lead to serious consequences and losses. For example, if a credit institution refuses to companies with "healthy" financial situation to provide credit (error of type 2), it can lead to the loss of future profits of this institution. This error is often called a "commercial risk". Conversely, if the credit institution makes a decision on granting a loan to a company that is a potential bankrupt (error type 1), it can lead to the loss of interest on the loan, a significant part of the loan funds, opportunity cost, and other losses. Therefore, this error is called "credit risk".

Up to date there are several generally accepted methods and techniques of bankruptcy risk assessment. The most famous and widely used is professor Altman method (so-called Z-model) [5].

Z- model of Altman is a statistical model that is based on an assessment of indicators of financial state and solvency of the company that allows to estimate the risk of bankruptcy and divide business corporations on potential bankrupts and non-bankrupts. However, Altman's model has a number of shortcomings, and its application to the economy of Ukraine is associated with certain difficulties. Therefore, in recent years, alternative approaches and methods were developed which take into account the specificity of the analysis and decision- making under uncertainty. These include the fuzzy sets theory and fuzzy neural networks technique.

The purpose of this section is to review and analyze the efficiency of different methods, both classical and computational intelligence methods for bankruptcy risk assessment in relation to the economy of Ukraine.

4.4.1 Altman Model for Bankruptcy Risk Estimates

The most well-known bankruptcy risk assessment model is professor E. Altman's model [5]. Altman's model is constructed using a multi-variable discriminant analysis (MDA), which allows to choose such indicators, that the variance between the groups would be the maximum and within the group be minimum. In this case, the classification is carried out for two groups of companies some of which later went bankrupt, while others were able to survive and strengthen their financial positions.

As a result, a model by Altman (Z-score)was developed, which has the following form:

$$Z = 1.2K_1 + 1.4K_2 + 3.3K_3 + 0.6K_4 + 1.0K_5 \tag{4.3.1}$$

where

K1 = own working capital/sum of assets;
K2 = retained earnings/sum of assets;
K3 = earnings before interest/sum of assets;
K4 = market value of equity/debt capital cost;
K5 = sales/sum of assets.

As a result of calculating the Z—score for a particular company, the following inference is made:

if Z < 1,81—a very high probability of bankruptcy;
if 1,81 $\leq$ Z $\leq$ 2,7—high probability of bankruptcy;
if 2,7 $\leq$ Z $\leq$ 2,99—perhaps bankruptcy;
if Z $\geq$ 3,0—probability of bankruptcy is very low.

Altman's model gives a fairly accurate prediction of the probability of bankruptcy within a time interval of 1–2 years.

As a result of discriminant analysis application to a group of companies that have declared bankruptcy later, using financial indicators taken a year before the possible default this fact was truly confirmed in 31 cases of 33 (94.5 %), and in 2 cases was a mistake (6 %). In the second group of companies—non—bankrupts, the model predicted a bankruptcy in only 1 case (error 3 %), while in the remaining 32 cases (97 %) was estimated very low probability of bankruptcy, that was confirmed in practice [5].

The corresponding results are given in Table 4.7.

Similar calculations were carried out based on the financial indicators for the two years prior to the possible bankruptcy. As it can be seen in Table 4.8, the results were not good, especially for a group of companies that have declared the default, whereas in group 2, the accuracy of the calculations remained approximately at the same level. The overall accuracy of the classification by Altman's model was 95 % for one year and 82 % for two years before the bankruptcy.

Table 4.7 Forecasting results of Altman' model one year before possible bankruptcy

Group	Number of companies	Forecasting: belonging to group 1	Forecasting belonging to group 2
Group 1 (the bankrupt companies)	33	31 (94.0 %)	2 (6.0 %)
Group 2 (the non—bankrupt companies)	33	1 (3.0 %)	32 (97.0 %)

Table 4.8 The forecasting results of Altman's model (for two years before possible bankruptcy)

Group	Number of companies	Prediction: belonging to group 1	Prediction: belonging to group 2
Group 1 (the bankrupt company)	33	23 (72.0 %)	9 (28.0 %)
Group 2 (the non—bankrupt companies)	33	2 (6.0 %)	31 (94.0 %)

The above considered Z-score is applicable only for large companies, whose shares are traded at the stock exchange. In 1985 E. Altman proposed a new model that allows to correct this shortcoming. The formula for determining the probability of bankruptcy prediction for companies whose shares are not represented on the stock exchange is [5, 6]:

$$Z = 0.717K_1 + 0.847K_2 + 3.107K_3 + 0.42K_4 + 0.995K_5 \tag{4.3.2}$$

where K_1—the value of own equity in relation to the borrowed capital.

If $z < 1.23$ then the risk of bankruptcy is very large.

Altman's approach was repeatedly applied and extended by his followers in many countries (UK, France, Brazil, China, Russia etc.).

An example of application of this model are the research results obtained by Altman himself for 86 companies-bankrupts in the period 1969–1975 years, 110 companies-bankrupts in the period 1976–1995 years and 120 companies-bankrupts in the period 1997–1999. Using a threshold value of 2,675, the accuracy of the method proved to be in the range from 82 to 96 %.

In the last years the investigations were performed on application of Altman's model for CIS countries with transient economy. One of the most successful MDA models was model developed by Davidova and Belikov for Russian economy.

Davidova—Belikov's model takes the form [7]

$$R = 8.38K_1 + K_2 + 0.054K_3 + 0.63K_4 \tag{4.3.3}$$

where

K_1—ratio own working capital/sum of assets;
K_2—ratio return/own equity;

K_3—ratio total sales (cost of sales)/sum of assets; (turn-over coefficient);
K_4—ration net return/production costs.

If R < 0- the probability of bankruptcy is maximal (90–100 %); 0 < R < 0, 18 —the probability of bankruptcy is large (60–80 %); 0,18 < R < 0, 32—the probability of bankruptcy is average (35–50 %); 0, 32 < R < 0, 42—the probability of bankruptcy is low (15–20 %); R < 0, 42 the probability of bankruptcy is minimal.

Application of MDA model is based on several assumptions [6].

The first assumption is that the input data are dichotomous, i.e. groups are disjoint sets. The other assumptions are the following:

- The independent variables included in the model are normally distributed;
- Variance-covariance matrices of a group of successful companies and bankrupt group are equal;
- The cost of misclassification and the a priori probability of failure are known.

In practice, all three of the above mentioned assumptions are rarely met, therefore the use of MDA often happens to be inadequate and correctness of the results obtained by its use, is questionable.

Weakness of Altman's model lies is therein this model is purely empirical, fitted on a sample, and has no theoretical basis. In addition, the model coefficients should be determined for various industries, and will naturally vary.

In Ukrainian economy Altman's model was not widely used for the following reasons:

1. It requires the calculation of the relevant coefficients at the terms K_i, $i = 1, 5$, which, of course, differ from their values for foreign countries;
2. information about the financial condition of the analyzed companies is generally not trustworthy as managers of many companies often "correct" economic indices in their financial accounts, making it impossible to find reliable estimates of the coefficients in Z- model.

Therefore, the problem of estimating the probability of the bankruptcy risk should be solved in the conditions of uncertainty and incompleteness of initial information, and for its solution should be proposed to use an adequate approach: methods of computational intelligence—fuzzy sets and fuzzy neural networks (FNN).

4.4.2 Forecasting the Corporation Bankruptcy Risk on the Basis of the Fuzzy Sets Theory

One of the first methods of the corporations bankruptcy risk forecasting under uncertainty based on fuzzy sets was suggested by O.A Nedosekin known as matrix method [8, 9]. Consider its main ideas

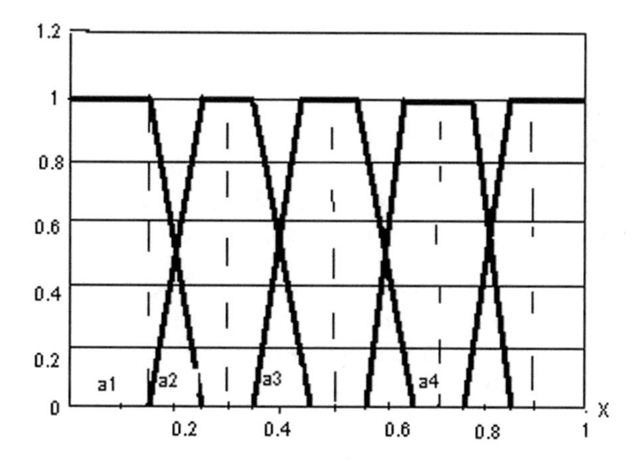

Fig. 4.5 Trapezoidal membership function

1. Expert builds linguistic variable with its term-set values. For example, "The level of management" can have the following term set of values " very low, low, medium, high, very high."
2. In order to constructively describe linguistic variable, the expert selects the appropriate quantitative feature—for example, a specially designed indicator of the level of management, which takes values from zero to one.
3. Further to each value of linguistic variable which is a fuzzy subset of on the interval [0,1], the expert assigns a corresponding membership function of a fuzzy set. As a rule, it's a trapezoidal or triangular membership function. The upper base of the trapezoid corresponds to the full assurance of the correctness of the expert classification, and the lower- assurance that no other values of the interval [0,1] belong to the selected fuzzy set (see Fig. 4.5).

The **matrix method** comprises the following steps

Step 1 (Linguistic variables and fuzzy sets)

a. Linguistic variable **E "State enterprise"** has five values
$\mathbf{E_1}$—fuzzy subset of states "Limit level of trouble";
$\mathbf{E_2}$—fuzzy subset of states " high level of trouble";
$\mathbf{E_3}$—fuzzy subset of states "average level";
$\mathbf{E_4}$—fuzzy subset of states "relative prosperity";
$\mathbf{E_5}$—fuzzy subset of states, "the ultimate well-being."
b. The corresponding to the variable E linguistic variable **G "Bankruptcy risk"** also has 5 values:
$\mathbf{G_1}$—fuzzy subset of states, "the ultimate risk of bankruptcy"
$\mathbf{G_2}$—fuzzy subset of states "high risk of bankruptcy"
$\mathbf{G_3}$—fuzzy subset of states "average risk of bankruptcy"
$\mathbf{G_4}$—fuzzy subset of states "low risk of bankruptcy"
$\mathbf{G_5}$—fuzzy subset of states "negligible risk of bankruptcy."
Carrier of fuzzy set G—indicator of bankruptcy risk g—takes values from zero to one by definition.

c. For certain financial and management indicator X_i assign the linguistic variable $\mathbf{B_i}$ "level of indicator X_i" in the following term set of values:
$\mathbf{B_{i1}}$—a subset of the "very low level of indicator X_i"
$\mathbf{B_{i2}}$—subset of "low level of indicator X_i"
$\mathbf{B_{i3}}$—a subset of the "average level of indicator X_i"
$\mathbf{B_{i4}}$—a subset of "high level of indicator X_i"
$\mathbf{B_{i5}}$—subset of "very high level of indicator X_i".

Step 2 (Indicators). We construct a set of individual indicators $X = \{X_i\}$ total number N, which according to the expert, on the one hand affect the assessment of the bankruptcy risk, and on the other hand, reflect the different parts of the enterprise business and financial activity. For example, in the matrix method such indicators are used [8]:

- X_1—autonomy coefficient (the ratio of equity to the balance sheet);
- X_2—ratio of current assets to ensure their own needs (the ratio of net working capital to current assets);
- X_3—intermediate liquidity ratio (the ratio of cash and receivables to short-term liabilities);
- X_4—absolute liquidity ratio (the ratio of cash to short-term liabilities);
- X_5—the turnover of all assets for the year (the ratio of revenue from sales to the average the value of assets in the year period;
- X_6—Return on total capital (the ratio of net profit to the average assets value for the period of year).

Step 3 (Significant indicators). For each indicator X_i its significance r_i is determined. To evaluate this level, it is necessary to range all indicators in order of decrease their significance according to the relation:

$$r_1 \geq r_2 \geq \ldots \geq r_N.$$

If the system indicators are ranked in decreasing order of importance, the weight of the ith indicator should be determined by the Fishburn's rule:

$$r_i = \frac{2(N - i + 1)}{(N + 1)N}. \tag{4.3.4}$$

If all indicators have the same weight, then $r_i = 1/N$.

Step 4 (Classification of risk). A classification of the current values of the g indicator of risk as a criterion for the partition of this set into fuzzy subsets (Table 4.9).

Step 5 (Classification of indicator values). We construct a classification of current values of parameters X as the split criteria to the full set of values into the fuzzy subset of B_i. One example of such a classification is given below in Table 4.10. The table cells are trapezoidal fuzzy numbers that characterize the respective membership functions.

Table 4.9 The classification of risk

The range of values g	The classification level of the parameter	The degree of confidence (membership function value)
$0 \leq g \leq 0.15$	G_5	1
$0.15 < g < 0.25$	G_5	$\mu_5 = 10 \times (0.25\text{-}g)$
	G_4	$1 - \mu_5 = \mu_4$
$0.25 \leq g \leq 0.35$	G_4	1
$0.35 < g < 0.45$	G_4	$\mu_4 = 10 \times (0.45\text{-}g)$
	G_3	$1 - \mu_4 = \mu_3$
$0.45 \leq g \leq 0.55$	G_3	1
$0.55 < g < 0.65$	G_3	$\mu_3 = 10 \times (0.65\text{-}g)$
	G_2	$1 - \mu_3 = \mu_2$
$0.65 \leq g \leq 0.75$	G_2	1
$0.75 < g < 0.85$	G_2	$\mu_2 = 10 \times (0.85\text{-}g)$
	G_1	$1 - \mu_2 = \mu_1$
$0.85 \leq g \leq 1.0$	G_1	1

Step 6 (Assessment of indicators level). We will assess the current level of indicator and put the results in the Table 4.11.

Step 7 (Classification level indicators). We carry out a classification of the current values of indicator x using the criterion in table constructed in step 5 (Table 4.10). The result of the classification is a table of values $\lambda_{ij} - (i,j)$—levels of membership of x_i to fuzzy subsets v_j.

Step 8 (Risk Assessment). We perform computing operations to assess the degree of bankruptcy risk g:

$$g = \sum_{j=1}^{5} g_j \sum_{i=1}^{N} r_i \lambda_{ij}, \tag{4.3.5}$$

$$\text{where } g_j = 0.9 - 0.2 * (j - 1), \tag{4.3.6}$$

The sense of formulas (4.3.5) and (4.3.6) is as follows. First, we estimate the weight of a subset of **B** in the assessment of the corporation state **E** and risk assessment **G**. These weights participate in the external sum to determine the average value of the indicator g, where g_j is nothing but the average estimate g of the appropriate range of the Table 4.9 at Step 4.

Step 9 (linguistic recognition). The resulting risk value is classified on the database of Table 4.9. The results of the classification are the linguistic description of the bankruptcy risk and the expert level of confidence in the correctness of his classification (Table 4.10).

Table 4.10 Classification of certain financial indicators

Index	T-number{γ} for the value of linguistic variable "The parameter":				
	"very low"	"low"	"average"	"high"	"very high"
X_1	(0,0,0.1,0.2)	(0.1,0.2,0.25,0.3)	(0.25,0.3,0.45,0.5)	(0.45,0.5,0.6,0.7)	(0.6,0.7,1,1)
X_2	(−1,−1,−0.005,0)	(−0.005,0,0.09,0.11)	(0.09,0.11,0.3,0.35)	(0.3,0.35,0.45,0.5)	(0.45,0.5,1,1)
X_3	(0,0,0.5,0.6)	(0.5,0.6,0.7,0.8)	(0.7,0.8,0.9,1)	(0.9,1,1.3,1.5)	(1.3,1.5,∞, ∞)
X_4	(0,0,0.02,0.03)	(0.02,0.03,0.08,0.1)	(0.08,0.1,0.3,0.35)	(0.3,0.35,0.5,0.6)	(0.5,0.6,∞, ∞)
X_5	(0,0,0.12,0.14)	(0.12,0.14,0.18,0.2)	(0.18,0.2,0.3,0.4)	(0.3,0.4,0.5,0.8)	(0.5,0.8,∞, ∞)
X_6	(-∞, -∞,0,0)	(0,0,0.006,0.01)	(0.006,0.01,0.06, 0.1)	(0.06,0.1,0.225, 0.4)	(0.225,0.4,∞,)

Table 4.11 The current level of performance

Index	Present value
X_I	x_I
–	–
X_i	x_i
–	–
X_N	x_N

Table 4.12 The results of the forecast by Altman model for "enterprises—bankrupts"

Group	Number of companies	Forecast bankruptcy risk		
		High	Middle	Low
Two years before bankruptcy	26	35 % (9)	27 % (7)	38 % (10)
The year before bankruptcy	26	42 % (11)	35 % (9)	23 % (6)
Average for two years	26	38 %	31 %	31 %

The main advantages of fuzzy matrix method are as follows:

1. the possibility of using not only the quantitative but qualitative factors;
2. taking into account inaccurate fuzzy information on the approximate values of the factors.

4.4.3 *The Application of Fuzzy Neural Network in Bankruptcy Risk Prediction Problem*

Now consider the application of computational intelligence technologies -FNN of Mamdani and Tsukamoto for the same problem of predicting the risk of bankruptcy. The method comprises the following steps [5, 6, 10].

Stage 1 (Linguistic variables and fuzzy subsets).

Like fuzzy-matrix method define the fuzzy sets E, G, B.

Step 2 (Performance). Building a set of individual indicators $X = \{X_i\}$ (the total number N), which at the discretion of the expert-analyst affect the assessment of the bankruptcy risk of enterprises and evaluate different aspects of the business and financial activity. We choose a system of six indicators, the same ones as in fuzzy-set matrix method

Step 3 (Building the rule base of fuzzy inference system).

An expert builds rules base on the subject area as a set of fuzzy rules of the predicate type:

$$\Pi_1 : \mathit{if}\ x \in A_1 \text{ and } y \in B_1 \text{ then } z \in C_1$$

$$\Pi_2 : \mathit{if}\ x \in A_2 \text{ and } y \in B_2 \text{ then } z \in C_2$$

Introduce the following linguistic variables for the implementation of fuzzy inference algorithms of Mamdani and Tsukamoto.

- **X1**: (Very Low (VL), Low (L), Medium (M), High (H), Very High (VH));
- **X2**: (Very Low, Low, Medium, High, Very High);
- **X3**: (Very Low, Low, Medium, High, Very High);
- **X4**: (Very Low, Low, Medium, High, Very High);
- **X5**: (Very Low, Low, Medium, High, Very High);
- **X6**: (Very Low, Low, Medium, High, Very High);

Set the following levels of bankruptcy: (Very Low, Low, Medium, High, Very High).

For simplicity, introduce the following abbreviations:

Very Low—(VL), Low—(L), Medium—(M), High—(H), very high—(VH).

Then we can write the following rules, taking into account all the possible combinations:

- If X_1 is "VL" and X_2 is "VL" and X_3 is "VL" and X_4 is "VL" and X_5 is "VL" and X_6 is "VL" then bankruptcy risk is " VH";
- If X_1—"M" and X_2—"L" and X_3—"VL" and X_4—"VL" and X_5—"VL" and X_6—"VL" then bankruptcy risk—"VH";
- If X_1 "M" and X_2 "M" and X_3 "L" and X_4 "VL" and X_5 "VL" and X_6 "VL" then bankruptcy risk "VH";
- If X_1 "M" and X_2 "M" and X_3 "M" and X_4 "L" and X_5 "VL" and X_6 "VL" then bankruptcy risk "H";
- If X_1 "H" and X_2 "M" and X_3 "L" and X_4 "VL" and X_5 "L" and X_6 "VL" then bankruptcy risk "L";
- If X_1 "H" and X_2 "H" and X_3 "H" and X_4 "M" and X_5 "M" and X_6 "H" then bankruptcy risk "M";
- If X_1 "VH" and X_2 "VH" and X_3 "H" and X_4 "VH" and X_5 "H" and X_6 "VH" then bankruptcy risk "L";
- If X_1 "VH" and X_2 "VH" and X_3 "VH" and X_4 "VH" and X_5 "VH" and X_6 "VH", then bankruptcy risk "VL";

Since the total number of rules is very high, if we consider all possible options permutations values, to facilitate perception and brevity of the rules base introduce scores for linguistic values.

$$\text{VL} = 5; \quad \text{L} = 4; \quad \text{M} = 3; \quad \text{H} = 2; \quad \text{VH} = 1.$$

We calculate the marginal performance level of bankruptcy, using the following boundary rules:

1. If X_1 "VL" and X_2 "VL" and X_3 "VL" and X_4 "VL" and X_5 "VL" and X_6 "VL" then the SCORE = 30;
2. If X_1 "L" and X_2 "L" and X_3 "L" and X_4 "L" and X_5 "L" and X_6 "L" then the SCORE = 24;
3. If X_1 "M" and X_2 "M" and X_3 "M" and X_4 "M" and X_5 "M" and X_6 "M" then the SCORE = 18;
4. If X_1 "H" and X_2 "H" and X_3 "H" and X_4 "H" and X_5 "H" and X_6 "H" then the SCORE = 12;
5. If X_1 "VH" and X_2 "VH" and X_3 "VH" and X_4 "VH" and X_5 "VH" and X_6 "VH" then the SCORE = 6;

Then the new rules to assess the risk of bankruptcy can be written as follows:

If SCORE > 24, the bankruptcy level is VH;
If SCORE $\leq$ 24 and > 18, the bankruptcy level is H;
If SCORE $\leq$ 18 and BAL R > 12, the bankruptcy level is M;
If SCORE $\leq$ 12 and > 6, the bankruptcy level is L;
If the SCORE = 6, the bankruptcy level is VL.

This approach allows us to cover the whole set of rules.

Step 4 (fuzzification of input parameters). We perform fuzzification of the input parameters, or a description of each term set (linguistic variables) using the membership functions, and find the degree of truth for each value in the rules: *A1 (x0), A2 (x0), B1 (y0), B2 (y0).*

As the membership functions use the triangular functions. For greater clarity of the membership functions represent them graphically in Fig. 4.6 and assign to them appropriate actual values.

Step 5 (Inference). Find levels of "cut-off" for each of the rule antecedents using operation min.

$$\alpha_1 = A_{11}(x_{10}) \wedge A_{21}(x_{20}) \wedge A_{31}(x_{30}) \wedge A_{41}(x_{40}) \wedge A_{51}(x_{50}) \wedge A_{61}(x_{60});$$
$$\alpha_i = A_{1i}(x_{10}) \wedge A_{2i}(x_{20}) \wedge A_{3i}(x_{30}) \wedge A_{4i}(x_{40}) \wedge A_{5i}(x_{50}) \wedge A_{6i}(x_{60}).$$

And determine "truncated" output membership functions:

$$C_1' = (\alpha_1 \wedge C_1(z)); \; C_i' = (\alpha_i \wedge C_i(z)).$$

Step 6 (Composition). Perform composition of found output membership functions using operation max, resulting in a fuzzy subset of the final output variable with the membership function μ (z).

Step 7 (Reduction to the crisp value (defuzzification). Perform the fuzzification using centroid method

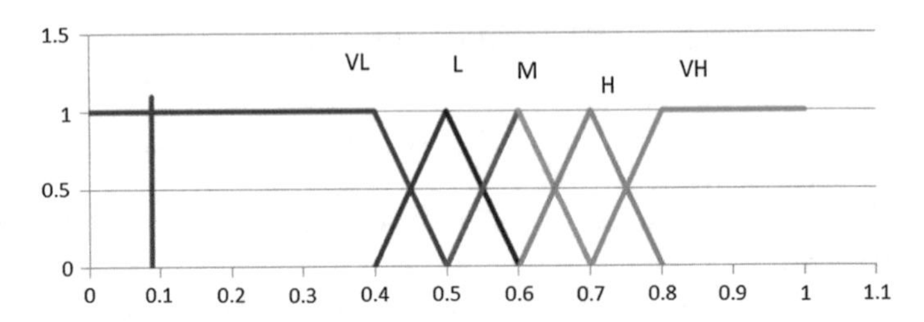

Fig. 4.6 Membership functions for the variable x_1

$$w_0 = \frac{\int_\Omega w \cdot \mu_\Sigma(w)dw}{\int_\Omega \mu_\Sigma(w)dw}$$

Experimental Studies

In order to analyze the different methods of forecasting the bankruptcy risk a software package was developed, which implements the classical methods of discriminant analysis by Altman, Davydova—Belikov, Nedosekin's matrix method and fuzzy neural networks of Mamdani and Tsukamoto [6, 10].

Using the developed software the experimental investigations were conducted for predicting bankruptcy of fifty-two enterprises in Ukraine, 26 of which were declared bankrupt by the arbitral court in 2011.

The input data for the forecasting were financial indicators, which were calculated based on data from accounting statements (balance sheet and income statement) in 2009 and 2010 years. Forecasting was conducted using models of Altman, the matrix method of Nedosekin, Mamdani and Tsukamoto FNNs. Analysis was based only on quantitative indicators the same as in the matrix method of Nedosekin.

Define the accuracy level of predicting the bankruptcy risk during the considered period. So Altman model has an accuracy of correct prediction 0.69 for companies that are potential bankrupts (see Fig. 4.7).

For the second group of companies, which we call "workable", the following results were obtained. During the two reporting years prior to the current status, 22 companies are recognized as the company with the "low" level of bankruptcy, 2 companies—"medium" level and 2 companies received the status of a "high" level of bankruptcy. During one fiscal year prior to the present state 18 enterprises received the status the "low" level of bankruptcy, 5 enterprises—"medium" level of bankruptcy, three companies—"high" level of bankruptcy (Table 4.13).

Define the accuracy level of risk prediction for this group of enterprises during the considered period. So the average prediction accuracy for enterprises—not bankrupts equals 0.77 (Fig. 4.8).

We generalized these results, and found the average level of forecasting accuracy for each fiscal year, and in the two years together (Table 4.14). After analyzing the results, one can say that Altman approach is not always able to determine correctly

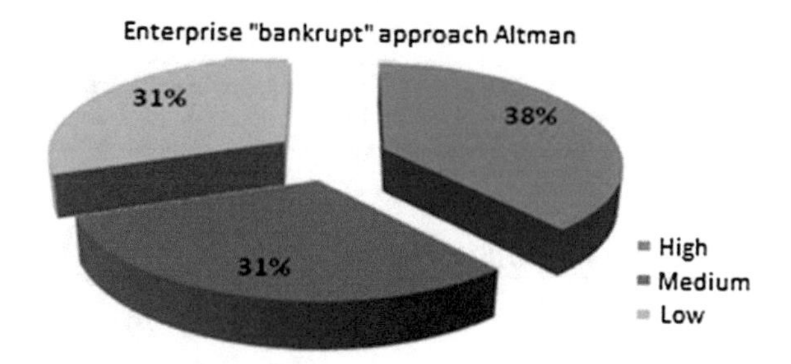

Fig. 4.7 The level of accuracy of forecasting bankrupt enterprises

Table 4.13 The results of the forecast by Altman model for "enterprise—not bankrupts"

Group	Number of companies	Forecast bankruptcy (levels)		
		High	Middle	Low
In 2009 fiscal year	26	7,5 % (2)	7.5 % (2)	85 % (22)
In 2010 fiscal year	26	12 % (3)	19 % (5)	69 % (18)
Average for two years	26	10 %	13 %	77 %

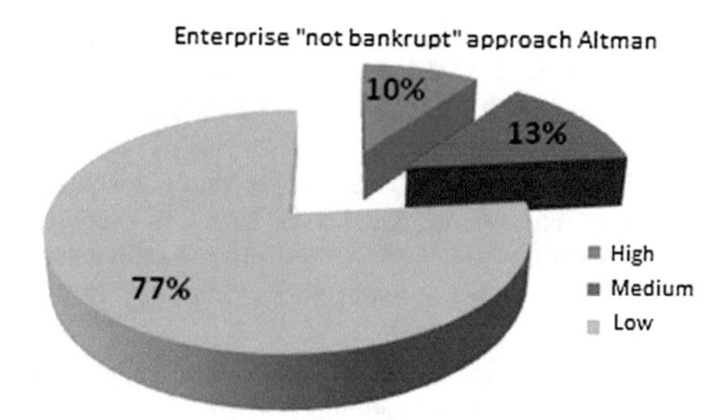

Fig. 4.8 The level of accuracy of forecasting the state of the enterprises-non-bankruptcs

Table 4.14 Generalized forecast accuracy of enterprise bankruptcy by Altman's model

Group	Number of companies	Forecast (%)	
		True	Error
Two years before bankruptcy	52	77	23
The year before bankruptcy	52	73	27
Average	52	75	25

financial state, especially if you need to detect bankruptcy at an early stage of financial instability. Also, the approach Altman does not allow to trace the dynamics of enterprise development, i.e. to conduct a financial analysis of a company that to take early measures to prevent bankruptcy.

Table 4.15 The results of the forecast by Nedosekin model for "enterprise-bankrupt"

Group	Number of companies	Forecast	
		Bankrupt	Non-bankrupt
Two years before bankruptcy	26	77 % (20)	23 % (6)
The year before bankruptcy	26	85 % (22)	15 % (4)
Average for two years	26	81 %	19 %

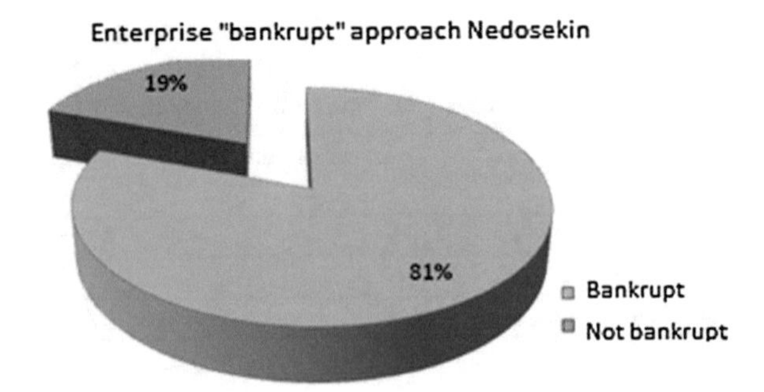

Fig. 4.9 The level of accuracy of forecasting state of enterprises-bankrupt)

Further we analyzed the financial state of the same two groups of companies that were analyzed by Altman's model using fuzzy-set Nedosekin's method. In result of 26 bankrupt companies in the two years before the default 20 enterprises were identified "very high", "high" or "medium" level of bankruptcy risk, 6 enterprises have the status of companies with "low" and "very low" the level of bankruptcy. In the year before the default 22 enterprises—with a "very high", "high" or "medium" level of bankruptcy, 4 enterprises—"low" or "very low" level of bankruptcy [10] (Table 4.15).

Define the accuracy level of predicting the risk of bankruptcy during the studied period. So Nedosekin's approach gives a correct prediction for companies that are potential bankrupts with an average accuracy of 0.81 (see Fig. 4.9).

For the second group of enterprises—"workable", the following results using Nedosekin's method were obtained [10]. During the two reporting years prior to the current status, 21 companies were recognized as the company with a "very low", "low" or "medium" level of bankruptcy, five companies obtained the status of "high" or "very high" level of bankruptcy. During one fiscal year before the present date 20 enterprises were recognized with "very low", "low" or "medium" level of bankruptcy, six companies—"high" or "very high" level of bankruptcy (Table 4.16).

Define the accuracy level of predicting the bankruptcy risk for working enterprises during the survey period. As one can see, the Nedosekin's model with an average accuracy of 0.79 gives a correct prediction for enterprises not potential bankrupts (See Fig. 4.10).

Summarizing the results, find the average value of the prediction accuracy for each year and for the two years together (Table 4.17).

Table 4.16 The results of the forecast by Nedosekin model for "enterprises-not bankrupt"

Group	Number of companies	Forecast	
		Bankrupt	Not-bankrupt
In 2009 fiscal year	26	19 % (5)	81 % (21)
In 2010 fiscal year	26	23 % (6)	77 % (20)
Average for two years	26	21 %	79 %

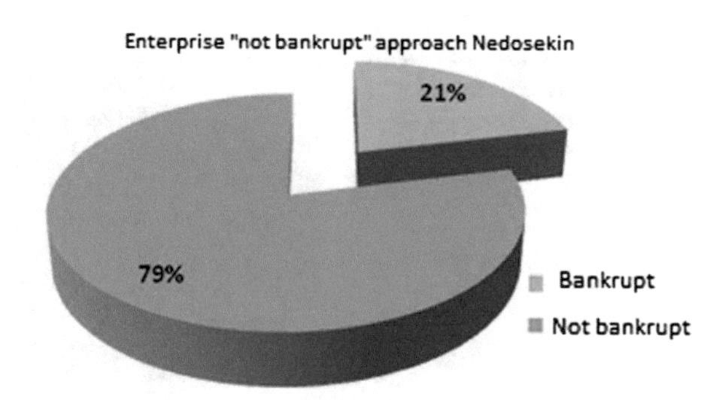

Fig. 4.10 The accuracy level of forecasting for the workable enterprises by Nedosekin model

Table 4.17 The overall accuracy of enterprise bankruptcy risk prediction by Nedosekin's model

Group	Number of companies	Forecast (%)	
		True	Error
During the two reporting years	52	79	21
During one fiscal year	52	81	19
Average	52	80	20

Analyzing these results, we can say that Nedosekin's approach allows not only to determine the financial state of enterprises, but to define it more precisely thanks to the linguistic scale consisting of five grade levels of bankruptcy, "VL, L, M, H, VH". In addition, Nedosekin's approach makes it possible to trace the dynamics development of the enterprise, i.e. allows to conduct a financial analysis of a company for the previous and current reporting periods and determine the level of bankruptcy at the "initial" stage, allowing early to take measures to prevent possible bankruptcy.

We analyzed the financial state of the same two groups of companies by Mamdani approach [10]. After analysis of the companies 'bankrupt' in two years before the default among the 26 companies 22 companies were identified with a "very high", "high" or "medium" level of risk of bankruptcy, four companies obtained the status of companies with "low" and "very low" level of bankruptcy. In one year before the default 25 enterprises were identified with a "very high", "high" or "medium" level of bankruptcy, one company—"low" or "very low" level of bankruptcy (Table 4.18).

Table 4.18 The results of the forecast by Mamdani's model for "enterprises—bankrupt"

Group	Number of companies	Forecast	
		Bankrupt	Not bankrupt
Two years before bankruptcy	26	85 % (22)	15 % (4)
The year before bankruptcy	26	96 % (25)	4 % (1)
Average for two years	26	90 %	10 %

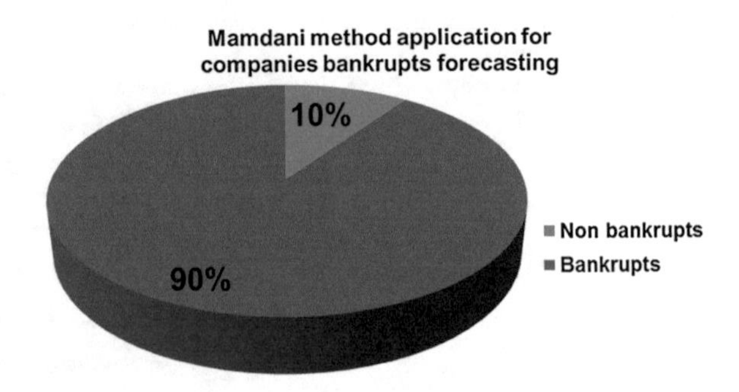

Fig. 4.11 The level of accuracy of forecasting bankrupt enterprises

Determine the average prediction accuracy level of bankruptcy risk using Mamdani approach (Fig. 4.11).

For the second group of enterprises so-called "workable", the following results using Mamdani model were obtained. In 2009 fiscal year, 23 companies were recognized as the company with a "very low", "low" or "medium" level of bankruptcy, 3 companies received the status of "high" or "very high" level of bankruptcy. In 2010, the state of some enterprises has deteriorated, they went a step below, but the overall picture remains the same (Table 4.19).

The average prediction accuracy of bankruptcy risk level for working enterprizes using Mamdani approach is presented in the Fig. 4.12 [10].

After analyzing the obtained results, the average prediction accuracy for all enterprises is presented in Table 4.20. As one can readily see Mamdani approach is similar to Nedosekin approach. It allows not only to determine the financial state of enterprises, but also to define it more precisely thanks to the linguistic scale consisting of five grade levels of bankruptcy: "VL, L, M, H, VH".

Mamdani's approach enables to monitor the development dynamics of the enterprise, i.e. allows to conduct a financial analysis of a company for the previous and current reporting periods and to determine the level of bankruptcy risk at the

Table 4.19 The results of the forecast by Mamdani model for "enterprises—not bankrupts"

Group	Number of companies	Forecast	
		Bankrupt	Do not bankrupt
Two years before bankruptcy	26	12 % (3)	88 % (23)
The year before bankruptcy	26	12 % (3)	88 % (23)
Average for two years	26	12 %	88 %

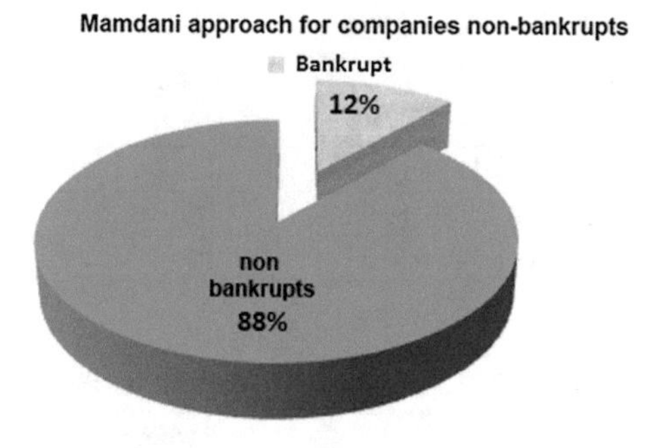

Fig. 4.12 The level of forecasting accuracy for the working enterprise by Mamdani model

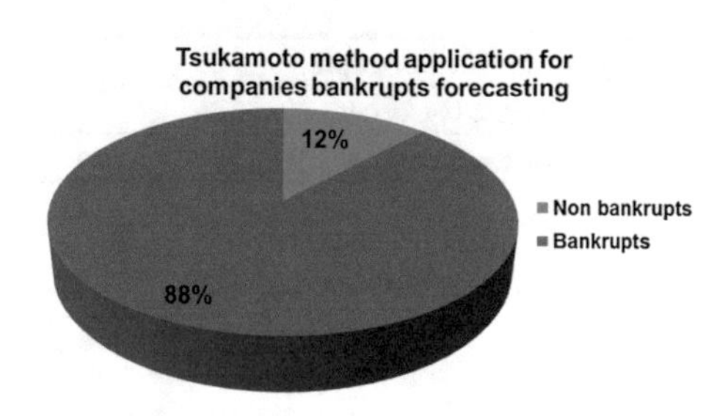

Fig. 4.13 The level of accuracy of forecasting the state of the enterprise bankrupt by Tsukamoto's model

Table 4.20 The average accuracy of the forecast of enterprise financial state by Mamdani model

Group	Number of companies	Forecast (%)	
		True	Error
During the two reporting years	52	89.42	10.58
During one fiscal year	52	92.31	7.69
Average	52	87.50	12.50

"initial" stage. Therefore Mamdani method allows to make more comprehensive analysis of the enterprises than the previously considered two methods.

After analyzing the financial state of the same companies by Tsukamoto model, the following results were received [10] (Tables 4.21, 4.22, 4.23).

As can be seen from the tables Tsukamoto approach also enables to analyze the financial state of a company at any stage and with a fairly high percentage of probability to detect its critical state at an early stage.

We generalized the results for companies 'bankrupt', using approaches by Altman Nedosekin, Mamdani, Tsukamoto, and present them below in Table 4.24 and in Fig. 4.14.

Summarizing results obtained for all the companies, using approaches Altman Nedosekin, Mamdani, Tsukamoto are presented in the Table 4.25 and in Fig. 4.15 [10].

As one can see, the method of Altman, in presented investigations, correctly predicted the state of enterprises by an average of 69 %, the method of Nedosekin

Table 4.21 The results of the forecast by Tsukamoto model for "enterprises—bankrupts"

Group	Number of companies	Forecast	
		Bankrupt	Do not bankrupt
Two years before bankruptcy	26	22 (85 %)	4 (15 %)
One year before bankruptcy	26	24 (92 %)	2 (8 %)
Average for two years	26	88 %	12 %

Table 4.22 The results of the forecast by Tsukamoto model for "enterprises—not bankrupts"

Group	Number of companies	Forecast	
		Bankrupt	Do not bankrupt
Two years before bankruptcy	26	12 % (3)	88 % (23)
The year before bankruptcy	26	15 % (4)	85 % (22)
Average for two years	26	13 %	87 %

Table 4.23 Average forecasting accuracy of bankruptcy risk by Tsukamoto model

Group	Number of companies	Forecast (%)	
		True	Error
During the two reporting years	52	87	13
During one fiscal year	52	88	12
Average	52	87	13

Table 4.24 Forecasting results for enterprises-bankrupts by Altman, Nedosekin, Mamdani, Tsukamoto models

Method	Number of companies	Forecast	
		Right (%)	Not right (%)
Altman	26	69.2	30.8
Nedosekin		80.8	19.2
FNN Mamdani		90.4	9.6
FNN Tsukamoto		88.5	11.5

correctly predicted by an average of 81 %, and Tsukamoto Mamdani FNN gave similar results, the forecast was carried out at 90 % correctly. Errors in forecasting state enterprises may be caused by several reasons. Firstly, we do not take into account certain social interest of the company or certain individuals in the elimination of the working enterprise. Second, we do not consider the existence of an interest in bankrupt enterprises or false bankruptcy. But we can surely say that Nedosekin fuzzy-set method and fuzzy neural networks of Mamdani and Tsukamoto allow to detect trends in the development of enterprises and identify the threat of possible bankruptcy in the early stages.

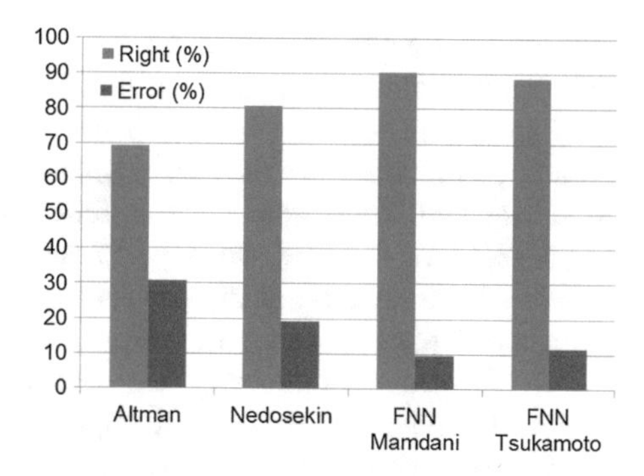

Fig. 4.14 Accuracy of bankruptcy risk prediction of bankrupt enterprises by different methods

Table 4.25 Forecasting results by Altman, Nedosekin. Mamdani, Tsukamoto models for all enterprises

Method	Number of companies	Forecast	
		Right (%)	Right (%)
Altman	52	73	27
Nedosekin		80	20
Mamdani		89.4	10.6
Tsukamoto		88	13

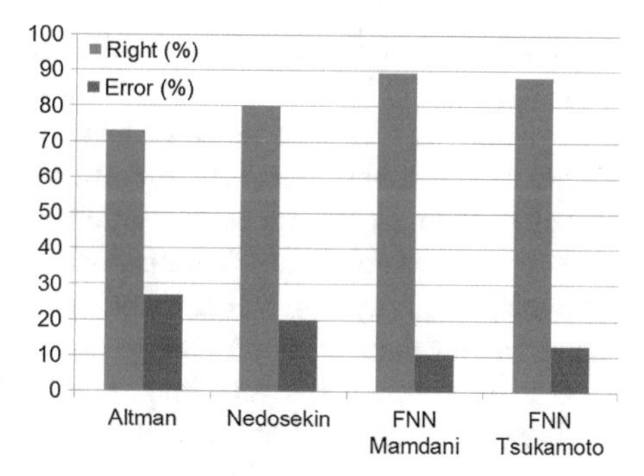

Fig. 4.15 Forecasting results of the enterprises bankruptcy risk by different methods

Conclusions

In this section methods of financial state analysis and forecasting bankruptcy risk are considered and analized: the classical method of Altman, fuzzy-set method of Nedosekin and methods based on the application of fuzzy neural networks with the inference of Mamdani and Tsukamoto.

Using the developed software package the experimental investigations were carried out predicting bankruptcy risk for 52 enterprises in Ukraine, among which 26 enterprises were potential bankrupts, and other 26 companies were solvent, (i.e. the level of bankruptcy is "low", "very low", "medium"). Among the 26 companies —potential bankrupts 24 companies were declared bankrupt on 01.02.2011 by a court decision. Among the 26 enterprises classified as having "average" level of bankruptcy 4 enterprises were recognized bankrupts, and 6 companies were reorganized in additional liability and limited liability companies.

Note that in result of the comparative analysis the highest accuracy of bankruptcy prediction have shown computational intelligence methods -FNN of Mamdani and Tsukamoto (90 %), followed by fuzzy-set method of Nedosekin (80 %) and finally, the worst performance prediction accuracy has shown the classic Altman method of discriminant analysis.

In addition, fuzzy neural networks of Mamdani, Tsukamoto enable to detect the risk of bankruptcy in the early stages, especially Mamdani approach, as well as to analyze the development of the company, taking the into account reports for few quarters or years. Besides, to the contrary to Nedosekin method FNN allow to utilize expert knowledge in the form of fuzzy inference rules (knowledge base).

4.5 Banks Financial State Analysis and Bankruptcy Forecasting

Introduction

In a modern economy problem of banks financial state analysis and possible bankruptcy forecasting plays the significant role. The timely detection of features of coming bankruptcy enables top bank managers to adopt urgent measures for stabilizing financial bank state and prevent the bankruptcy. Up to the date there are a lot of methods and techniques of banks state analysis and determination of bank rating- WEBMoney, CAMEL, Moody's S&P etc. But their common drawback is that all of them work with complete and reliable data and cannot give correct results in case of incomplete and unreliable input data. This is especially important for Ukrainian bank system where bank managers often provide the incorrect information about their financial indices that to obtain new credits and loans.

Due to aforesaid it's very important to develop new approaches and technique for banks bankruptcy risk forecasting under uncertainty. The main goal of present investigation is to develop and estimate novel methods of bank financial state analysis and bankruptcy forecasting under uncertainty and compare with classical methods.

As it is well known 2008 year was the crucial year for bank system of Ukraine. If the first three quarters were periods of fast grow and expansion the last quarter became the period of collapse in the financial sphere. A lot of Ukrainian banks faced the danger of coming default.

For this research the quarterly accountancy bank reports were used obtained from National bank of Ukraine site. For analysis the financial indices of 170 Ukrainian banks were taken up to the date 01.01. 2008 and 01.07.2009, that is about two years before crises and just before the start of crises.

The important problem which arouse before start of the investigations is which financial indices are to be used for better forecasting of possible bankruptcy. Thus, another goal of this exploration was to determine the most adequate financial indices for attaining maximal accuracy of forecasting.

For analysis the following indices of banks accountancy were taken:

- Assets; Capital; financial means and their equivalents; Physical persons entities;
- juridical persons entities; liabilities; Net incomes (losses).

The collected indices were utilized for application by fuzzy neural networks as well as classic crisp methods—method by Kromonov and method developed by association of Byelorussian banks As output data of models for Ukrainian banks were two values:

- 1, if the significant worsening of bank financial state is not expected in the nearest future;
- −1, if the bank bankruptcy or introduction of temporary administration in bank is expected in the nearest future.

4.5.1 The Application of Fuzzy Neural Networks for Financial State Forecasting

For forecasting of banks bankruptcy risk it was suggested the application of fuzzy neural networks (FNN) ANFIS and TSK considered in the Chap. 3. The application of FNN is determined by following advantages:

1. the possibility to work with incomplete and unreliable information under uncertainty;
2. the possibility to use expert information in the form of fuzzy inference rules.

A special software kit was developed for FNN application in forecasting problems enabling to perform Ukrainian banks financial state analysis using ANFIS and TSK.

As input data the financial indices of Ukrainian banks in financial accountant reports were used in the period of 2008–2009 years [11]. As the output values were +1, for bank-non-bankrupt and −1—for bank-bankrupt. In the investigations different financial indices were used, different number of rules for FNN and the analysis of data collection period influence on forecasting accuracy was performed.

Comparative Analysis of FNN ANFIS and TSK

The results of experimental investigations of FNN application for banks financial state analysis are presented below. In the first series of experiments input data at the period of January 2008 were used (that is for two years before possible bankruptcy) and possible banks bankruptcy was analyzed at the beginning of 2010 year.

Experiment №1

- Training sample—120 Ukrainian banks,
- Test sample—50 banks;
- Number of rules = 5.
- Input data—financial indices (taken from bank accountant reports):
  - Assets;
  - Capital;
  - Cash (liquid assets);
  - Households deposits;
  - Liabilities.

The results of application of FNN TSK are presented in Table 4.26.

Experiment №2

- Training sample—120 Ukrainian banks;
- test sample—50 banks; number of rules = 5.
- Input data—financial indices the same as in the experiment 1.

The results of application FNN ANFIS are presented in the Table 4.27.

As one can see comparing results of Table 4.26 and 4.27, neural network TSK gives more accurate results than FNN ANFIS.

Experiment №3 The next experiment was aimed at the detecting of influence of rules number on the forecasting results.

- Training sample—120 Ukrainian banks;
- test sample—50 banks;
- number of rules = 10.
- Input data—the same financial indices as in experiment 1.

The results of application of FNN TSK are presented in the Table 4.28.

Table 4.26 Results of FNN TSK forecasting

Results	
Total amount of errors	5
%% of errors	10 %
First type of errors	0
Second type of errors	5

Table 4.27 Results of FNN ANFIS forecasting

Results	
Total amount of errors	6
%% of errors	12 %
First type of errors	0
Second type of errors	6

Table 4.28 Results of FNN TSK forecasting

Results:	
Total amount of errors	6
%% of errors	12 %
First type of errors	1
Second type of errors	5

Table 4.29 Results of FNN ANFIS forecasting

Results:	
Total amount of errors	7
%% of errors	14 %
First type of errors	1
Second type of errors	6

The similar experiments were carried out with FNN ANFIS

Experiment №4 The next experiment was aimed at the detecting of influence of rules number on the forecasting results.

- Training sample—120 Ukrainian banks,
- Test sample—50 banks.
- Number of rules = 10.
- Input data—financial indices the same as in experiment 1.

The results of application of FNN ANFIS are presented in the Table 4.29.

The comparative analysis of forecasting results versus the number of rules is presented in the Table 4.30.

The next experiments were carried out aimed at investigation of influence of training and test samples size on accuracy of forecasting.

Experiment №5

- Training sample—120 Ukrainian banks;
- Test sample—50 banks;
- Number of rules = 10.
- Input data—financial indices:
  - Assets; Entity;
  - Cash (liquid assets);
  - Households deposits;
  - Liabilities.

Table 4.30 Comparative analysis of FNN ANFIS and TSK in dependence on rules number

Network/number of rules	Total amount of errors	% of errors (%)	Amount of 1 type errors	Amount of 2 type errors
Anfis 5	6	12	0	6
Anfis 10	7	14	1	6
TSK 5	5	10	0	5
TSK 10	6	12	1	5

Table 4.31 Results of FNN TSK forecasting

Results	
Total amount of errors	7
%% of errors	10 %
First type of errors	1
Second type of errors	6

Table 4.32 results of FNN ANFIS forecasting

Results	
Total amount of errors	7
%% of errors	10 %
First type of errors	0
Second type of errors	7

The results of FNN TSK are presented in the Table 4.31.

The similar experiment was carried out with FNN ANFIS

Experiment №6

- Training sample—120 Ukrainian banks;
- Test sample—50 banks;
- Number of rules = 10.
- Input data—financial indices are the same as in the experiment 5.

The results obtained with FNN ANFIS are presented in the Table 4.32.

After analysis of the presented experimental results the following conclusions were made:

1. FNN TSK gives more accurate forecasting than FNN ANFIS.
2. The variation of the number of rules in the training and test samples doesn't influence on the accuracy of forecasting.

The next series of experiments was aimed at determining optimal input data (financial indices) for forecasting. The period of input data was January 2008.

Experiment №7

- Training sample—120 Ukrainian banks,
- Test sample—50 banks;
- Number of rules = 10.
- Input data—financial indices (banks financial accountant reports):
  - profit of current year; net percentage income; net commission income;
  - net expense on reserves;
  - net bank profit/losses.

Table 4.33 Results of FNN TSK forecasting

Results:	
Total amount of errors	13
%% of errors	19 %
First type of errors	6
Second type of errors	7

The results of FNN TSK application are presented in the Table 4.33.

Experiment №8

- Training sample-—120 Ukrainian banks;
- Test sample—50 banks;
- Number of rules = 10.
- Input data—financial indices (banks financial accountant reports):
  - General reliability factor (Own capital/Assets);
  - Instant Liquidity Factor (Liquid assets/Liabilites);
  - Cross-coefficient (Total liabilities/working assets);
  - General liquidity coefficient (Liquid assets + defended capital + capitals in reserve fund/total liabilities);
  - Coefficient of profit fund capitalization (Own capital/Charter fund)

The results of application of FNN TSK arte presented in the Table 4.34.

Note that these financial indices are used as input data in Kromonov's method of banks bankruptcy [12], results of its application are considered below.

Experiment №9

- Training sample—120 Ukrainian banks;
- Test sample—70 banks;
- Number of rules = 5.
- Input data—financial indices:
- ROE—Return on Entity (financial results/entity);
- ROA—Return on Assets (financial results/assets);
- CIN—incomes- expenses ratio (income/expense);
- NIM—net interest margin;
- NI—net income.

The results of application of FNN TSK for forecasting with these input indices are presented in the Table 4.35.

Table 4.34 Results of FNN TSK forecasting

Results	
Total amount of errors	7
%% of errors	10 %
First type of errors	1
Second type of errors	6

Table 4.35 Results of FNN TSK forecasting

Results:	
Total amount of errors	12
%% of errors	17 %
First type of errors	5
Second type of errors	7

It should be noted that these indices are used as input in method of EuroMoney [13].

Experiment №10

- Training sample—120 Ukrainian banks;
- Test sample—70 banks;
- Number of rules = 5.
- Input data—financial indices (banks financial accountant reports):
- General reliability factor (Own capital/Assets);
- Instant Liquidity Factor (Liquid assets/Liabilities);
- Cross-coefficient (Total liabilities/working assets);
- General liquidity coefficient (Liquid assets + defended capital + capitals in reserve fund/total liabilities);
- Coefficient of profit fund capitalization (Own capital/Charter fund);
- Coefficient entity security (Secured entity/own entity).

The results of FNN TSK application with these financial indices presented in the Table 4.36.

It should be noted these indices are also used in Kromonov's method.

The comparative analysis of forecasting results using different sets of financial indices are presented in the Table 4.37.

Further the series of experiments were carried out aimed on determination of influence of data collection period on the forecasting results. It was suggested to consider two periods: January of 2008 (about 2 years before the crisis) and July of 2009 (just before the start of crisis).

Experiment №11

- Training sample—120 Ukrainian banks;
- Test sample—70 banks;
- Number of rules = 10.
- Input data—financial indices the same as in the experiment 10.

Table 4.36 Results of FNN TSK forecasting

Results	
Total amount of errors	8
%% of errors	13 %
First type of errors	1
Second type of errors	7

Table 4.37 The dependence of forecasting accuracy on sets of input financial indices

Experiment	Total number of errors	1 type of errors	2 type of errors	total % of errors
Experiment № 5	7	1	6	10
Experiment № 8	7	0	7	10
Experiment № 9	12	5	7	17
Experiment № 1	8	1	7	13

Table 4.38 Accuracy of forecasting in dependence on data collection period

Experiment/number of rules	Total amount of errors	1 type of errors	2 type of errors	Total % of errors (%)
01.01.2008 5 rules	7	0	7	10
01.07.2009 5 rules	5	0	5	7
01.07.2009 10 rules	7	3	4	10

In the Table 4.38 the comparative results of forecasting in dependence on period of input data are presented.

4.5.2 The Application of Fuzzy GMDH for Financial State Forecasting

In the process of investigations fuzzy Group Method of Data Handling (FGMDH) was also applied for financial state of Ukrainian banks forecasting [12]. As input data the same indices were used as in the experiments with FNN TSK In the Table 4.39 the corresponding results are presented in dependence on input data period.

If to compare the results of FGMDH with the results of FNN TSK one can see that FNN TSK gives better results if we use the input data for one year before

Table 4.39 Comparative results of forecasting using method FGMDH in dependence on period of input data collection

Input data period	Total error number	%% of errors (%)	1 type of errors	2 type of errors
2004	10	14	3	7
2005	9	13	3	6
2006	8	11.4	3	5
2007	7	10	2	5
2008	6	8.5	1	5
2009	6	8.5	2	4

possible bankruptcy while FGMDH gives better results using older input data and has advantages in long-term forecasting (2 or more years).

4.5.3 The Application of Conventional Methods for Financial State Forecasting

In order to compare the results of application of fuzzy methods for banks financial state forecasting the conventional crisp methods were implemented. As the crisp methods of banks state analysis the so called Kromonov's method and aggregated multilevel index of bank state method developed by the Byelorussian banks association were used [11].

In the first experiment the application of Kromonov's method was performed using quarterly data of Ukrainian banks at the beginning of 2008 year and checking the forecast on the middle of 2009 year using real data of bank bankruptcy. The goal of the experiments was the exploration of influence of input data period on the forecasting quality. The results of analysis are presented in the Table 4.40.

As one may see the Kromonov's method gives false forecast for 34 banks of total number 170, that is 20 % of error when using the data on 01.01.2008r. while false forecast for 24 banks of 170 (15 % of errors) using more fresh data on 01.07.2009. So the considerable improvement of forecasting accuracy using more fresh data is obtained. The types of errors were distributed uniformly in both cases.

The second experiment in this series was the application of method developed by Byelorussian bank association (BBA) for bankruptcy risk forecasting using the Ukrainian banks quarterly data on the beginning of 2008 year and checking of the forecasting accuracy in the middle of 2009. Experimental results are presented in the Table 4.41.

As one may readily see in the Table 4.41 Byelorussian bank association (BBA) method gives false forecast for 27 banks of 170, that is 16 % of errors. This result for the data on the beginning of 2008 year is more preferable than Kromonov's method which had 20 % of errors on this data. Using data of the middle of 2009 year BBA method gives the false forecast for 24 banks (15 %) that coincides with results of Kromonov' s method.

Table 4.40 Kromonov's method results in dependence on data collection period

	01.01.2008	01.07.2009
Total number of errors	34	24
%% of errors	20 %	15 %
1 type errors number	18	12
2 type errors number	16	12
Test sample size	170	170

Table 4.41 Byelorussian bank association method forecasting results

	01.01.2008	01.07.2009
Total number of errors	27	24
%% of errors	16 %	15 %
1 type errors number	12	4
2 type errors number	15	20
Test sample size	170	170

4.5.4 The Generalized Analysis of Crisp and Fuzzy Forecasting Methods

In the concluding experiments the comparative analysis of application of all the considered methods was carried out. The following methods were considered [14]:

- fuzzy neural network ANFIS;
- fuzzy neural network TSK;
- Kromonov's method
- Byelorussian bank association method.

As input data the financial indices of Ukrainian banks on July 2007 year were used. The results of application of all methods for bankruptcy risk analysis are presented in the Table 4.42.

Conclusions

Various methods for banks financial state forecasting were considered and investigated. The following methods were considered [11, 12]:

- fuzzy neural network ANFIS; • fuzzy neural network TSK;
- Kromonov's method; • Byelorussian bank association method.

As the input data the financial indices of Ukrainian banks were considered.

1. While experiments the adequate input financial indices were detected with which the best forecasting results for Ukrainian banks were obtained:
   - General reliability factor (Own capital/Assets);
   - Instant Liquidity Factor (Liquid assets/Liabilities);
   - Cross-coefficient (Total liabilities/working assets);
   - General liquidity coefficient (Liquid assets + defended capital + capitals in reserve fund/total liabilities);
   - Coefficient of profit fund capitalization.

Table 4.42 Comparative results analysis of various forecasting methods

Method/period)	Total amount of errors	%% of errors (%)	1 type errors	2 type errors
ANFIS	7	10	1	6
TSK	5	7	0	5
Kromonov's method	10	15	5	5
BBA method	10	15	2	8

2. It was established that FNN TSK gives much more accurate results than FNN ANFIS The variation of rules number in training sample doesn't influence on forecasting results. The fuzzy GMDH gives better results using older data that is, more preferable for long-term forecasting (two or more years).
3. In general, the comparative analysis had shown that **fuzzy forecasting methods** and techniques **give better results than conventional** crisp methods for forecasting bankruptcy risk. But at the same time the crisp methods are more simple in implementation and demand less time for their adjustment.

4.6 Comparative Analysis of Methods of Bankruptcy Risk Forecasting for European Banks Under Uncertainty

Introduction

The results of successful application of fuzzy methods for Bankruptcy risk forecasting of Ukrainian banks under uncertainty stimulated the further investigations of these methods application for financial state analysis of European leading banks

The main goal of this exploration was to investigate novel methods of European banks bankruptcy risk forecasting which may work under uncertainty with incomplete and unreliable data.

Besides, the other goal of this investigation was to determine which factors (indices) are to be used in forecasting models that to obtain results close to real data. Therefore, we used a set of financial indicators (factors) of European banks according to the International accountant standard IFRS [14]. The annual financial indicators of about 300 European banks were collected in 2004–2008 years, preceding the start of crisis of bank system in Europe in 2009 year. The data source is the information system Bloomberg[]. The resulting sample included the reports only from the largest European banks as system Bloomberg contains the financial reports only from such banks. For correct utilization input data were normalized in interval [0,1].

Application of fuzzy neural networks for European banks bankruptcy risk forecasting

The period for which the data were collected was 2004–2008 years. The possible bankruptcy was analyzed in 2009 year. The indicators of 165 banks were considered among which more than 20 banks displayed the worsening of the financial state in that year. Fuzzy neural networks and Fuzzy Group Method of Data Handling (FGMDH) were used for bank financial state forecasting.

In accordance with the above stated goal the investigations were carried out for detecting the most informative indicators (factors) for financial state analysis and bankruptcy forecasting. Taking into account incompleteness and unreliability of input data FNN ANFIS and TSK were suggested for bankruptcy risk forecasting [11].

After performing a number of experiments the data set of financial indicators was found using which FNN made the best forecast. These indicators are the following:

- Debt/Assets = (Short-term debt + Long-term debt)/Total Assets
- Loans to Deposits ratio
- Net Interest Margin (NIM) = Net Interest income/Earning Assets
- Return on Equity (ROE) = Net Income/Stockholder Equity
- Return on Assets (ROA) = Net Income/Assets Equity
- Cost/Income = Operating expenses/Operating Income
- Equity/Assets = Total Equity/Total Assets

The series of experiments were carried out for determining the influence of the number of rules and period of data collection on forecasting results.

In the first series of experiments FNN TSK was used for forecasting.

Experiment №1

- Training sample = 115 banks of Europe;
- Testing sample = 50 banks; number of rules = 5
- Input data period = 2004 year.

The results of FNN TSK application are presented in Table 4.43.

Experiment №2

- Training sample = 115 banks of Europe;
- Testing sample = 50 banks; number of rules = 5
- Input data period = 2005

The results of FNN TSK application are presented in Table 4.44

Table 4.43 Forecasting results of FNN TSK

Results	
Total number of errors	8
%% errors	16 %
first type of errors	0
Second type of errors	8

Table 4.44 Forecasting results of TSK using data of 2005 year

Results	
Total number of errors	7
%% errors	14 %
first type of errors	0
Second type of errors	7

Table 4.45 Forecasting results of TSK using data of 2006 year

Results:	
Total number of errors	5
%% errors	10 %
first type of errors	0
Second type of errors	5

Table 4.46 Forecasting results of TSK using data of 2007 year

Results:	
Total number of errors	1
%% errors	2 %
first type of errors	0
Second type of errors	1

Experiment №3

- Input data the same as in the experiment 1.
- Input data period = 2006

The results of FNN TSK application are presented in Table 4.45.

Experiment №4

- Input data the same as in the experiment 1.
- Input data period = 2007

The results of FNN TSK application are presented in Table 4.46.

Further, the similar experiments were performed with FNN ANFIS while the period of data collection varied since 2004 to 2007 year.

The experiments for detection of rules number influence on forecasting results were also carried out. The corresponding results are presented in Table 4.47 for

Table 4.47 Forecasting results for FNN TSK versus number of rules and data period

Experiment/number of rules	Total errors number	%% of errors (%)	Number of the 1-st type errors	Number of the 2-nd type errors
2004–5	8	16	0	8
2005–5	7	14	0	7
2006–5	5	10	0	5
2007–5	1	2	0	1
2004–10	8	16	0	8
2005–10	8	16	1	7
2006–10	11	22	7	4
2007–10	4		0	4

Table 4.48 Forecasting results for FNN ANFIS versus data period

Experiment/number of rules	Total errors number	%% of errors (%)	Number of the 1-st type errors	Number of the 2-nd type errors
2004–5	8	16	0	8
2005–5	8	16	1	7
2006–5	8	16	4	4
2007–5	4	8	0	4

FNN TSK. The results for FNN ANFIS are presented in Table 4.48 which display the influence of data collection period on forecasting accuracy.

After analysis of these results the *following conclusions* were made

1. FNN TSK gives better results than ANFIS while forecasting the bankruptcy risk for European banks.
2. The best input variables (indicators) for European banks bankruptcy risk forecasting are the following:

   – Debt/Assets = (Short-term debt + Long-term debt)/Total Assets
   – Loans to Deposits
   – Net Interest Margin (NIM) = Net Interest income/Earning Assets
   – Return on Equity (ROE) = Net Income/Stockholder Equity
   – Return on Assets (ROA) = Net Income/Assets Equity
   – Cost/Income = Operating expenses/Operating Income
   – Equity/Assets = Total Equity/Total Assets

Input data collection period (forecasting interval) makes influence on forecasting results.

4.6.1 *The Application of Fuzzy GMDH for Bank Financial State Forecasting*

In next experiments Fuzzy Group Method of Data Handling (FGMDH) was applied for European banks financial state forecasting. Fuzzy GMDH enables to construct forecasting models using experimental data automatically without expert

Table 4.49 Comparative analysis of forecasting results for FGMDH

Input data period	Total number of errors	%% of errors (%)	Number of the first type errors	Number of the second types of errors
2004	7	14	0	7
2005	6	12	1	5
2006	4	8	1	3
2007	2	4	0	2

Table 4.50 Forecasting results of different fuzzy methods

Method (period)	Total number of errors	%% of errors (%)	Number of the first type errors	Number of the second type of errors
ANFIS (1 year)	4	8	0	4
TSK (1 year)	1	2	0	1
FGMDH (1 year)	2	4	0	2
ANFIS (2 years)	8	16	4	4
TSK (2 years)	5	10	0	5
FGMDH (2 years)	4	8	1	3

[14]. The additional advantage of FGMDH is possibility to work with fuzzy information.

As the input data in these experiments were used the same indicators as in experiments with FNN TSK. In Table 4.49 forecasting results are presented in dependence on input data period collection for FGMDH.

If to compare the results of FGMDH application with the results of FNN TSK one can see that neural network gives better results on the short forecasting interval (one year) while fuzzy GMDH gives better results on greater intervals (2 or more years) This conclusion coincides with similar conclusion for Ukrainian banks.

In Table 4.50 the comparative results of application of different methods for bankruptcy risk forecasting are presented.

4.6.2 Application of Linear Regression and Probabilistic Models

Regression models For analysis of fuzzy methods efficiency at the problem of bankruptcy risk forecasting the comparison with crisp methods: the regression analysis of linear models was performed. As input data the same indicators were used which were optimal for FNN. Additionally, the index Net Financial Result, was also included in the input set. This index makes great impact on forecasting results. Thus, input data in this experiments were 8 financial indicators of 256 European banks according to their reports based on International standards:

- Debt/Assets—X1
- Loans/Deposits—X2
- Net Interest Margin—X3

Table 4.51 Comparative analysis of ARMA models

Input data	Testing sample	I type errors	II type errors	Total number of errors	% of errors (%)
All variables (8)	50	5	4	9	18
6 variables	50	5	4	9	18
4 variables	50	5	4	9	18

- ROE (Return on Equity)—X4
- ROA (Return on Assets)—X5
- Cost/Income—X6
- Equity/Assets—X7
- Net Financial Result—X8

The input data were normalized before the application. The experiments were carried out with full regression ARMA model, which used 8 variables and shortened models with 6 and 4 variables.

Each obtained model was checked on testing sample consisting of 50 banks. The comparative forecasting results for all ARMA models are presented in Table 4.51.

As one may see in Table 4.51, the application of all types of linear regression models gives the same error 18 %, that is much worse than application of fuzzy neural networks.

Logit-models Further the experiments were performed using logit-models for bankruptcy forecasting [14]. The training sample consisted of 165 banks and the testing sample—of 50 banks.

The first one was constructed linear logit-model using all the input variables. It has the following form (estimating and forecasting equations):

$$
\begin{aligned}
I_Y = {} & C(1) + C(2) * X_1 + C(3) * X_2 + C(4) * X_3 + C(5) * X_4 + \\
& + C(6) * X_5 + C(7) * X_6 + C(8) * X_7 + C(9) * X_8 \\
Y = {} & 1 - @CLOGISTIC(-(C(1) + C(2) * X_1 + C(3) * X_2 + \\
& + C(4) * X_3 + C(5) * X_4 + C(6) * X_5 + C(7) * X_6 + C(8) * X_7 + \\
& + C(9) * X_8))
\end{aligned}
$$

The next constructed model was a linear probabilistic logit-model with 6 independent variables. The final table including the forecasting results of all the logit-models is presented below (Table 4.52).

Probit-models The next experiments were carried out with probit- models [14]. The first constructed model was the linear prodbit –model based on 206 banks using all the input variables. It had the following form

Table 4.52 Comparative analysis of logit-models

Input data	Testing sample	I type errors	II type errors	Total number of errors	% of errors (%)
All variables (8)	50	6	2	8	16
6 variables	50	6	2	8	16

Table 4.53 Forecasting results of probit-models

Input data	Testing sample	I type errors	II type errors	Total number of errors	% of errors (%)
All variables (8)	50	5	2	7	14
6 variables	50	5	2	7	14
4 variables	50	6	3	9	18

$$\begin{aligned}I_Y &= C(1) + C(2) * X_1 + C(3) * X_2 + C(4) * X_3 + C(5) * X_4 + \\ &\quad + C(6) * X_5 + C(7) * X_6 + C(8) * X_7 + C(9) * X_8 \\ Y &= 1 - @CLOGISTIC(-(C(1) + C(2) * X_1 + C(3) * X_2 + \\ &\quad + C(4) * X_3 + C(5) * X_4 + C(6) * X_5 + C(7) * X_6 + C(8) * X_7 + \\ &\quad + C(9) * X_8))\end{aligned}$$

As the experiments had shown the inputs Net Interest Margin (X_3) and Net Financial Result (X_8) very weakly influence on the results and they were excluded in the next experiments. The next constructed probit-model included 6 variables.

Further in this model were excluded insignificant variables Debt/Assets (X_1) and Loans/Deposits (X_2) and as result linear probit-model with 4 variables was obtained.

Each of the constructed probit- models was checked on the test sample of 50 banks. The results of application of all probit- models are presented in the Table 4.53.

As one may see from the Table 4.53, the application of all the probit-models gives relative error 14–18 %, that is much worse than results obtained by fuzzy neural networks It is worth to mention the decrease of model forecasting quality

Table 4.54 Comparative analysis of methods for banks bankruptcy forecasting

Method (period)	Total number of errors	%% of errors (%)	I type errors	II type errors
ANFIS	4	8	0	4
TSK	1	2	0	1
FGMDH	2	4	0	2
ARMA	9	18	4	5
LOGIT	8	16	2	6
PROBIT	7	14	2	5

after exclusion of insignificant variables. In particularly, after exclusion of variables Debt/Assets and Loans/Deposits error value has increased error from 14 % to 18 %.

4.6.3 Concluding Experiments

In the final series of experiments investigations and detailed analysis of various methods for forecasting bankruptcy risk were performed. The following methods were investigated [14]: FNN ANFIS FNN TSK FGMDH; Regression models Logit-models; •Probit-models.

Period of input data was—2007 year (for 1 year before possible bankruptcy).

Comparative analysis of all the forecasting methods is presented in the Table 4.54.

As one may see from this table fuzzy methods and models show much better results than crisp methods: ARMA, Logit-models and probit-models When forecasting by one year prior to current date fuzzy neural network TSK shows better results than FGMDH. But when forecasting for longer intervals(several years) FGMDH is the best method.

In a whole the conclusions of experiments with European banks completely confirmed the conclusions of experiments with Ukrainian banks.

References

1. Zaychenko, Y.P.: Fuzzy models and methods in intellectual systems. Kiev—Publishing House Slovo, p. 354 (2008). Zaychenko Yu. Fuzzy Group Method of Data Handling under fuzzy input data. Syst. Res. Inf. Technol. №3, 100–112 (2007) (rus)
2. Zaychenko, Y.P.: Sevae, F., Titarenko, K.M., Titarenko N.V.: Investigations of fuzzy neural networks in macroeconomic forecasting. Syst. Res. Inf. Technol. №2, 70–86 (2004) (rus)
3. Zaychenko, Y.P., Sevaee F., Matsak A.V.: Fuzzy neural networks for economic data classification. Vestnik of National Technical University of Ukraine KPI, Section Informatic, Control and Computer Engineering. vol. 42, pp. 121–133 (2004) (rus)
4. Zaychenko Yu.P. Fundamentals of intellectual systems design. Kiev—Publishing House Slovo, p. 352 (2004) (rus)
5. Zgurovsky, M.Z., Zaychenko, Y.P.: Complex analysis of corporation bankruptcy risk uinder uncertainty. Part 1. Syst. Res. Inf. Technol. №1, 113–128 (2012) (rus)
6. Zgurovsky, M.Z., Zaychenko, Y.P.: Models and methods of decision-making under uncertainty. Kiev.-Publ. House Naukova Dumnka, p. 275 c (2011) (rus)
7. Zaychenko, Y.P., Rogoza S.V., Stolbunov, V.A.: Comparative analysis of enterprises bankruptcy risk. Syst. Res. Inf. Technol. №3, 7–20 (2009) (rus)
8. Nedosekin, A.O.: System of portfolio optimization of Siemens services. Banking technologies, № 5 (2003) (rus)—See also: http://www.finansy.ru/publ/fin/004.htm
9. Nedosekin, A.O.: The applications of fuzzy sets theory to problems of finance control. Section Audit and Financial Analysis, №2 (2000) (rus)—See also: http://www.cfin.ru/press/afa/2000-2/08-3.shtml

10. Zgurovsky, M.Z., Zaychenko Yu.P. Complex analysis of corporation bankruptcy risk uinder uncertainty. Part 2 Syst. Res. Inf. Technol. №2, 111–124 (2012) (rus)
11. Aghaei, O.N., Ghamish, A., Yuriy, Z., Voitenko, O.: Analysis of financial state and banks bankruptcy risk forecasting. Syst. Res. Inf. Technol. №2 (2015) (rus)
12. Aghaei, O.N., Ghamish, A., (Iran), Zaychenko, Y., Voitenko, O.: Banks financial state analysis and bankruptcy forecasting. Int. J. Eng. Innov. Technol. (IJEIT) **4**(6), 62–67 (2014)
13. Averkin, A.N.: Fuzzy semiotic control systems. Intellectual Control: New Intellectual Technologies in Control Tasks. pp. 141–145. Moscow.-Nauka. Fizmathlit (1999) (rus)
14. Aghaei, O.N., Ghamish, A., (Iran), Zaychenko, Y.: Comparative analysis of methods of banks bankruptcy risk forecasting under uncertainty. Int. J. Eng. Innovative Technol. (IJEIT) **4**(5) (2015)

Chapter 5
Fuzzy Neural Networks in Classification Problems

5.1 Introduction

The purpose of this chapter is consideration and analysis of fuzzy neural networks in classification problems, which have a wide use in industry, economy, sociology, medicine etc.

In the Sect. 5.2 a basic fuzzy neural network for classification—NEFClass is considered, the learning algorithms of rule base and MF of fuzzy sets are presented and investigated. Advantages and lacks of the system NEFClass are analyzed and its modification FNN NEFClass M, free of lacks of the system NEFClass is described in the Sect. 5.3.

The results of numerous comparative experimental researches of the basic and modified system NEFClass are described in Sect. 5.4. The important in a practical sense task of recognition of objects on electro-optical images (EOI) is considered and its solution with application of FNN NEFClass is presented in the Sect. 5.5. The comparative analysis of different learning algorithms of FNN NEFClass at the task of recognition of EOI objects in the presence of noise is carried out. Problem of hand-written mathematical expressions recognition is considered in the Sect. 5.6 and its solution with application of FNN NEFClass is presented.

5.2 FNN NEFClass. Architecture, Properties, the Algorithms of Learning of Base Rules and Membership Functions

A classification problem is one of the most actual spheres of application of the computational intelligence systems. For its decision different approaches and methods were suggested, among which popular solutions were offered, combining neural networks and fuzzy inference systems. One of such decisions is the system

M.Z. Zgurovsky and Y.P. Zaychenko, *The Fundamentals of Computational Intelligence: System Approach*, Studies in Computational Intelligence 652,
DOI 10.1007/978-3-319-35162-9_5

NEFClass (NEuro-Fuzzy CLASSifier), based on the generalized architecture of fuzzy perceptron and suggested by Nauck and Kruse in [1–3].

Both original and modified model of NEFClass are derivative from the general model of fuzzy perceptron [4]. A model purpose is a development of fuzzy rules from a set of data which can be divided into the several non-overlapping classes. The fuzziness arises up due to the imperfect or incomplete measurings of properties of objects, subject to classification.

Fuzzy rules, describing expert information, have the following form:

if is μ_{1i} and x_2 is μ_{2i} and … and x_n is μ_{ni},
then pattern $(x_1, x_2, \ldots x_n)$ belongs to the class of i,
where $\mu_{1i}, \ldots \mu_{ni}$, are MF of fuzzy sets.

The goal of NEFClass is to define these rules, as well as parameters of membership functions for fuzzy sets. It was assumed here, that intersection of two different sets is empty.

The system NEFClass has 3-layer successive architecture (see Fig. 5.1). The first layer U_1 contains inputs neurons which inputs patterns are fed in. Activating of these neurons does not change usually input values. The hidden layer U_2 contains fuzzy rules, and the third layer U_3 consists of output neurons (classifiers).

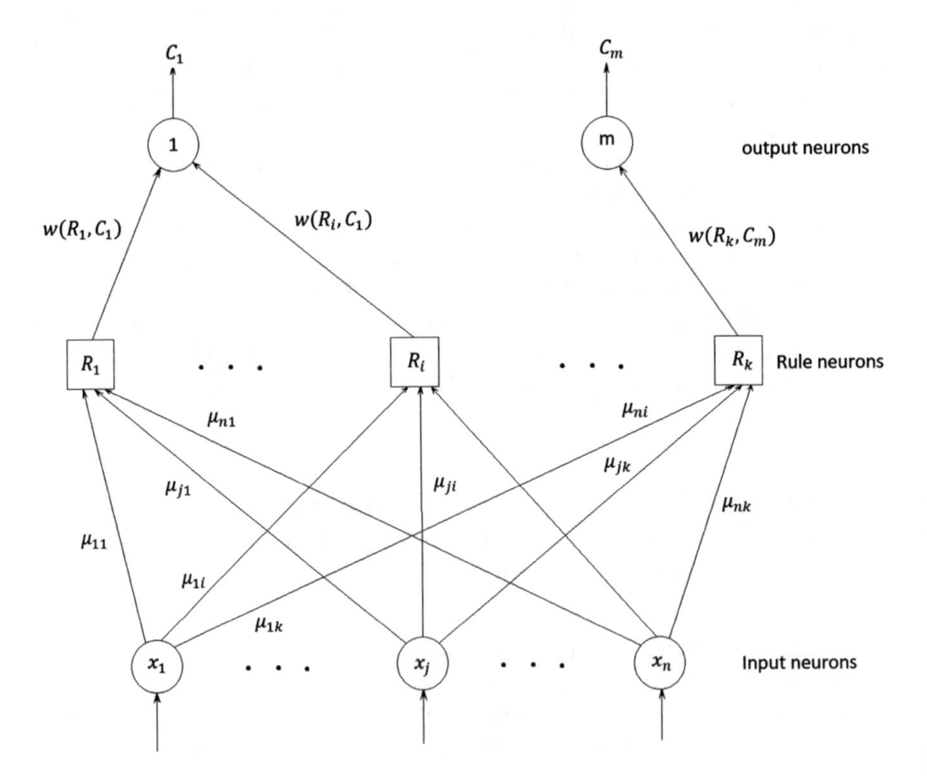

Fig. 5.1 Structure of FNN NEFCLASS

Activations of rule neurons and neurons of output layer with the pattern p are calculated so:

$$a_R^{(p)} = \min_{x \in U_1}\left\{W(x,R)(a_x^{(p)})\right\}, \tag{5.1}$$

$$a_C^{(p)} = \sum_{R \in U_2} W(c,R) \cdot a_R^{(p)}, \tag{5.2}$$

or alternatively

$$a_C^{(p)} = \max_{R \in U_2}\left\{a_R^{(p)}\right\}, \tag{5.3}$$

where $W(x,r)$ is a fuzzy weight of connection of input neuron x with a rule neuron R, and $W(R,c)$—fuzzy weight of connection of a rule neuron R with the neuron c of output layer. Instead of application of operations of maximum and minimum it is possible to use other functions of so-called "t-norm" and "t-conorm" accordingly [1].

A rule base is approximation of unknown function and describes a classification task $\varphi(x)$, such, that $c_i = 1, c_j = 0\,(j = 1\ldots m, \forall j \neq i)$, if x belongs to the class of C_i.

Every fuzzy set is marked a linguistic term, such as "large", "small", "middle" etc. Fuzzy sets and linguistic rules present approximation of classifying function and determine the result of the system NEFClass. They are obtained from a sample by learning. It's necessary, that for every linguistic value (for example, "x_1 is positive and large") there should be only one presentation of fuzzy set.

Learning in the System NEFClass

Learning of Rules Base

The system NEFClass can be built on partial knowledge about patterns. An user must define the amount of initial fuzzy sets for each of object features (number of terms) and set the value k_{max} that is a maximal number of rule nodes, which can be created in the hidden layer. For learning triangular MF are used.

Consider the system of NEFClass with n input neurons $x_1, \ldots x_n$, k $(k \leq k_{max})$ rule neurons and m output neurons $c_1, \ldots c_m$,. The learning sample of patterns is also given: $L = \{(p_1, t_1), \ldots (p_s, t_s)\}$, each of which consists of input pattern $p \in R^n$ and desired pattern $t \in \{0,1\}^m$.

A learning algorithm consists of two stages.

Stage 1. Generation of rule base.

The first stage whose purpose is to create rule neurons of the system NEFClass consists of the followings steps [1–3]:

1. Choose a next pattern (p, t) from sample L.
2. For every input neuron $x_i \in U_1$ find such membership $\mu_{J_i}^i$ that

$$\mu_{Ji}^{(p)} = \max_{j \in 1,..q_1} \{\mu_{ji}^{(p)}(p_i)\}, \tag{5.4}$$

where $x_i = p_i$.

3. If a number of rule nodes k is less than $k_{\max}$ and there is no rule node R such, that

$$W(x_1, R) = \mu_{J1}, \ldots, W(x_n, R) = \mu_{Jn},$$

then create such node and connect it with an output node c_i, if $t_i = 1$, connect it with all input neurons and assign the corresponding weights $\mu_{J_i}^{i}$ to connections.
4. If there are still not-processed patterns in L and $k < k_{\max}$, then go to the step 1 and continue learning using next pattern, and otherwise stop.
5. Determine a rule base by one of three procedures:

   a. "Simple" rules learning: we leave the first k rules only (stop creation of rules, if it was created $k = k_{\max}$ rules).
   b. The "best" learning rules: we process patterns in L and accumulate activating of every rule neuron for every class of patterns which were entered into system NEFClass. If rule neuron R shows the greater accumulation of activating for a class C_j than for a class C_R, which was specified initially for this rule, then change implication of rule R from C_R to C_j, that means connect R with the output neuron c_j. We continue processing of patterns in L farther and calculate for every rule neuron the function:

$$V_R = \sum_{p \in L} a_R^{(p)} \cdot e_p \tag{5.5}$$

   where

$$e_p = \begin{cases} 1 & \textit{if pattern p is classified correctly} \\ -1 & \textit{otherwise} \end{cases}$$

   We leave k rule neurons with the greatest values of V_R and delete other rule neurons from the system NEFClass.
   c. The "best for every class" algorithm of learning: we operate as in the previous case, but leave for each class C_j only those best $\left[\frac{k}{m}\right]$ rules, the consequences of which relate to the class C_j (where $[x]$ is integer part from x).

Learning of Fuzzy Sets MF

Stage 2

On the second stage learning of parameters of membership functions (MF) of fuzzy sets. is performed A learning algorithm with teacher of the system NEFClass must adapt MF of fuzzy sets. The algorithm cyclic runs through all learning patterns of the sample L, executing the following steps, until one of stop criteria will be fulfilled [1–3].

Steps:

1. Choose a next pattern (p, t) from sample L, enter it into FNN NEFclass and determine an output vector c.
2. For every output neuron c_i calculate the value δ_{C_i}

$$\delta_{C_i} = t_i - a_{C_i},$$

where t_i is a desired output,a_{c_i} is an real output of neuron c_i.

3. For every rule neuron R, for which output is $a_R > 0$ execute:

   a. determine a value δ_R, equal

$$\delta_R = a_R \cdot (1 - a_R) \cdot \sum_{C \in U_3} W(R, C)\delta_C \tag{5.6}$$

   b. Find such x', that

$$W(x', R)(a_{x'}) = \min_{x \in U_1} \{W(x, R)(a_x)\}. \tag{5.7}$$

   c. For fuzzy sets $W(x', R)$ determine displacement (shift) of parameters of MF $\Delta_a, \Delta_b, \Delta_c$, using learning speed $\sigma > 0$:

$$\Delta_b = \sigma \cdot \delta_R \cdot (c - a) \cdot \operatorname{sgn}(a_{x'} - b), \tag{5.8}$$

$$\Delta_a = -\sigma \cdot \delta_R \cdot (c - a) + \Delta_b, \tag{5.9}$$

$$\Delta_c = \sigma \cdot \delta_R \cdot (c - a) + \Delta_b. \tag{5.10}$$

   and execute the changes of $W(x', R)$.

   d. Calculate an rule error:

$$E = a_R \cdot (1 - a_R) \cdot \sum_{c \in U_3} (2 \cdot W(R, c) - 1) \cdot |\delta_c|. \tag{5.11}$$

End of iteration. Repeat the described iterations until condition of stop will be fulfilled. It is possible to use as criteria of stop, for example, such:

1. An error has not decreased during n iterations.
2. Stop learning after achievement of the defined (desirably close to the zero) error value.

5.3 Analysis NEFClass Properties. The Modified System NEFClassM

FNN NEFClass has several obvious advantages, distinguishing it among the other classification systems. The most important are: easiness of implementation, high-speed algorithms of learning, as well as that is the most important, high accuracyof data classification—at the level of the best systems in this area. However, the basic system NEFClass has some shortcomings:

1. formulas used for parameters learning are empirical in nature, in addition,
2. it is not clear how to choose in the learning algorithm the speed parameter σ. Therefore, these shortcomings were deleted in the modification of basic system —so-called system NEFClass-M (modified) developed in [5].

Randomization and careful selection rate constants learning σ are inherent properties of the system NEFCLASS-M. These properties have been designed to mitigate the impact some of the shortcomings of the original model and have made it possible to achieve a significant improvement in the quality of classification.

Randomization. Because of the nature of the training algorithm "simple" rules base and learning algorithm of fuzzy sets, the outcome of the training network for these algorithms are highly dependent on order, in which samples are represented in a learning sample. If, for example, the samples will be sorted by classes, the system will better classify the patterns of one class and substantially worse—the patterns of the other class. Ideally, the patterns in the training sample must be randomly mixed, in order to avoid the negative effect.

Implementation of the system NEFClassM [5] avoids this complexity by "randomization" of patterns order in a learning sample after its boot. Moreover, such an accidental "randomization" occurs before each iteration of learning algorithm. As show further experiments, this allows to achieve a more stable and, often, the better classification results, which do not depend on the order in which patterns in a learning sample has been submitted by a user.

Choice of speed training. In the learning algorithm of fuzzy sets in the model NEFCLASS is used parameter training speed σ. As experiments had shown, carried out in the course of developing the NEFClass M, this parameter plays a vital role in the success of the training.

The experiments had shown that, under other parameters being equal, for each specific task training there exists a certain value σ, which ensures a minimum percentage of erroneous classification after the training. Unfortunately, to obtain analytical dependence for optimal parameter value is very difficult because learning algorithm NEFCLASS as a whole is empirical; however, using search and try method it was found that for many tasks optimal value σ lies in the narrow range [0.06–0.1], in particular it may be equal to 0.07. This value has been set for the program which implements a modified model NEFClass M [5].

The Modified Model NEFCLASS

Consider the basic shortcomings in the NEFCLass learning algorithm.

The analysis of the drawbacks of NEFCLASS has shown that their principal cause lies mostly in an imperfect learning algorithm of fuzzy sets. Therefore, a natural approach, aimed to correct the situation, was the replacement of empirical learning algorithm by the strict optimization algorithm with all the ensuing consequences for network architecture and algorithms.

Both the original and modified model NEFCLASS are based on the architecture of a fuzzy perceptron [1, 5, 6]. Architectural differences of the original and the modified model is in the form of membership functions of fuzzy sets, function t-norm for calculation rules activations of neurons, as well as aggregating function (t-conorms), determining the activation of output neurons.

The application of numerical optimization methods requires differentiability of the membership functions of fuzzy sets—condition to which the triangular membership functions don't satisfy. Therefore the modified model of fuzzy sets uses the Gaussian membership functions, described as

$$\mu(x) = \exp\left\{ -\frac{(x-a)^2}{2b^2} \right\}.$$

This membership function is defined by two parameters—a and b. The requirement of differentiability also dictates the choice of t-norms (intersections) for calculating neuron activation rules. In the system NEFCLASS for this operation is used minimum; in the modified system NEFCLASS-M-product of the corresponding values.

Finally, the kind of aggregate function (t-conorm) for modified model is limited only by the weighted sum. The reason consists in the fact that the maximum function which is used in the original system also does not satisfy the condition of differentiability. The main change is obviously relates to a learning algorithm of fuzzy sets. The objective function in the modified system NEFClass is minimization of the mean squared error on the training sample by analogy with the classical (clear) neural networks:

$$\min E = \frac{1}{N}\sum_{p=1}^{N} \left\| a_c^{(p)} - a_c^{(p)*} \right\|^2$$

where the N—number of patterns in the training sample, $a_c^{(p)}$ is an activation vector of neurons in the output layer for the next training sample p, $a_c^{(p)*}$ is a target value of this vector for the pattern p. The components of the target vector for the pattern p are equal:

$$a_{ij}^{(p)*} = \begin{cases} 0, & i \neq j \\ 1, & i = j \end{cases}$$

where j is a number of the real class to which this pattern p belongs, i is classification of pattern p by NEFClass. The argument of numerical optimization aimed at reducing MSE for the training set is the aggregate vector of parameters a and b of FNN. As a specific training method can be used any method of unconstrained optimization such as the gradient method or the conjugate gradient method, these both methods were implemented in this investigation.

5.4 Experimental Studies. Comparative Analysis of FNN NEFClass and NEFClassM in Classification Problems

Experiments were conducted on the classification of the two sets of data IRIS and WBC [5, 6]. Selection of IRIS and WBC test kits was dictated by two considerations: firstly, these sets can be considered standard for classification problems, and secondly, in the original works of authors NEFCLASS model was tested on these data sets [1, 2]. This allows to compare the results of the base system NEFCLASS with a modified NEFCLASS_M and estimate the effect of introduced improvements.

IRIS Data Set
IRIS set contains 150 samples belonging to three different classes (Iris Setosa, Iris Versicolour, and Iris Virginica), 50 samples of each class. Each sample is characterized by four properties [1–3].

IRIS is the only one set by classification simplicity for which even a simple strategy of rules selection gives good results.

In the first experiment, in a modified model NEFClass-M "simple" rules learning algorithm was used, and their number was limited to 10 with 3 fuzzy sets per variable (all other parameters were set to the default values). As a result, the system has created 10 rules and achieved only 4 classification errors of the 150 (i.e. 97.3 % correct) patterns.

The best result, which was managed to achieve with the "simple" rules learning algorithm is three rules with two essential variables, x_3 and x_4, and the same order of misclassification (4 errors) [5]:

R1: IF (any, any, large, large) THEN Class 3
R2: IF (any, any, medium, medium) THEN Class 2
R3: IF (any, any, small, small) THEN Class 1

The same result was achieved for the "better" and "best in class" rules learning algorithms. However, for the last two algorithms it's possible further reduction in the number of fuzzy sets for variable x_3 and x_4 under the following rules (6 erroneous classification):

R1: IF (any, any, small, small) THEN Class 1
R2: IF (any, any, large, small) THEN Class 2
R3: IF (any, any, large, large) THEN Class 3

The authors of model NEFCLASS obtained the similar results, except that in their experiments, they used three fuzzy sets (linguistic values) for x_3 and x_4 [1, 2]. Thus, for a set of data IRIS it was managed to achieve better results than in the original works—exclusively simple set rules of two variables with only two decomposing sets for each variable.

Dataset WBC

The next test sample for classification was standard data sample Wisconsin Breast Cancer (WBC). When processing sample Wisconsin Breast Cancer using system NEFClass-M interesting results were obtained which didn't always coincide with the results of the basic model NEFCLASS.

Following the course of the experiments by the authors of NEFCLASS [1, 2] for system training rule base learning algorithm with the "best in the class" (three sets in the variable). was used with maximum 4 rules. The resulting error of misclassification obtained for the system NEFClass-M was 28 patterns of 663 (95.7 % correct) [5]. Very interesting is the fact that for model NEFClass for similar parameters correct classification value was only 80.4 % (135 misclassification).

This is a significant advantage of the modified system NEFClass-M which can be explained by suggested modifications that distinguish this model from basic NEFCLASS model, namely, the use of randomization algorithm, the choice of learning rate and application of numerical algorithm of optimization (gradient method for MF learning.

The best result that was managed to obtain for the data set WBC is the rule base of 8 rules with five essential variables x_1, x_2, x_4, x_6 and x_9 (misclassification—19 errors) [5]:

R1: IF (small, small, any, small, any, small, any, any, small) THEN Class 1
R2: IF (small, small, any, large, any, small, any, any, small) THEN Class 1
R3: IF (small, small, any, small, any, small, any, any, large) THEN Class 1
R4: IF (large, large, any, small, any, large, any, any, small) THEN Class 2
R5: IF (large, large, any, large, any, small, any, any, small) THEN Class 2
R6: IF (small, large, any, small, any, large, any, any, small) THEN Class 2
R7: IF (large, small, any, small, any, small, any, any, small) THEN Class 2
R8: IF (large, small, any, small, any, small, any, any, large) THEN Class 2

Comparable results (24 misclassification) were obtained with the use of a maximum of 2 rules ("the best in the class") with all the important variables, except x_5 and x_7:

R1: IF (small, small, small, small, any, small, any, small, small) THEN Class 1
R2: IF (large, large, large, small, any, large, any, large, small) THEN Class 2

Thus, the results obtained by NEFCLASS-M are superior over basic model NEFCLASS both in number of rules/significant variables and classification accuracy. This confirms the efficiency of the modifications made to the model NEFClass: randomization, the correct choice of speed training and application of numerical optimization algorithms.

5.5 Object Recognition on Electro-Optical Images Using Fuzzy Neural Networks

5.5.1 General Characteristics of the System

Today, remote sensing includes a variety of imaging techniques in all ranges of the electro-magnetic spectrum—from ultraviolet to the infrared. When viewing images obtained from meteorological geostationary satellites, which cover almost the entire hemisphere, it can be concluded that remote sensing—is the main method of finding qualitative relevant information on the state of the Earth's surface, as well as of objects located on it.

Remote probing method, usually includes a set of measurements. Using them not measured parameters of objects, and some value associated with them are detected. For example, we need to assess the condition of agricultural crops. However, the satellite apparatus only responds to the light intensity from these objects to certain sections of the optical range. In order to "decipher" these data it's necessary to carry out preliminary investigations, which include a variety of experiments to study crop conditions by contact methods; to study on reflectance in electro-magnetic spectrum (water, soil, plants) at different locations of the satellite at a different brightness and parameter measuring device. Next, you must determine what kind of other objects are displayed in the image, and only then determine the state of the surface (crop, water).

The width of the surface scanning inspection is a characteristic feature of satellite methods for studying the Earth, they enable to obtain high-quality images in a relatively short period of time.

Information about objects studies comes from the satellite, usually in digital form, and remote sensing system composed of multiple sensors that are sensitive to the incoming light or sound waves from the object; and a computer that constantly reads the data from the sensor and saves them to the hard drive. Such a sensor is usually mounted in a moving object, so that at rotation to sense the necessary territory. As an example, the system can be powered with a digital camera to monitor the sea waves. With this system it is possible to obtain the following types of data: the depth and morphology of the ocean, and the reasonability of using land, type of flora. These data are invaluable in monitoring the processes such as erosion of the mountains, natural changes in the soil, construction of cities and changing the type of green cover of the Earth.

Remote sensing makes it possible to carry out rapid shooting and accumulating an extensive archive of aerial-imagery. Today, virtually the entire surface of the Earth's land (and a significant portion of the water surface) is fixed with space shooting at different viewing conditions (time of year, time of day, and so on.).

5.5.2 *The Concept of Multi-spectral Electro-Optical Systems*

Basic Definitions
The concept of multi-spectral system is shown in Fig. 5.2. The system captures an image of the Earth's surface using a sensor with a two-dimensional detector grid. The resulting data are substantially continuous spectrum for each picture element (pixel). Schematically, the system components can be represented as a cube: 3D three-dimensional dataset—the images which are characterized by three features, and coordinates and wavelength (spectrum) [6, 9].

The spectral curve shows the relationship between the wavelength and the reflectivity of the object. The shape of the spectral curve can distinguish objects. For example, the vegetation is highly reflective in the near-infrared region and low in average, compared with the soil. An example is shown in Fig. 5.3.

Using the on-board computers generated image is transmitted to terrestrial stations for subsequent processing. Image processing algorithms include calibration, mapping fixes geometric curvature and classification of surfaces. In some cases, the algorithm is used for data processing in real time.

5.5.3 *Types of Sensors. Multispectral and Hyperspectral Systems*

Multispectral system—is a system installed on a board of the aircraft or satellite (sensor) that takes pictures in a certain spectral range and has from 2 to 15 spectral frequencies. Frequency bands—defined as a part of the spectrum (color) with a certain width, for example 10 or 50 nm.

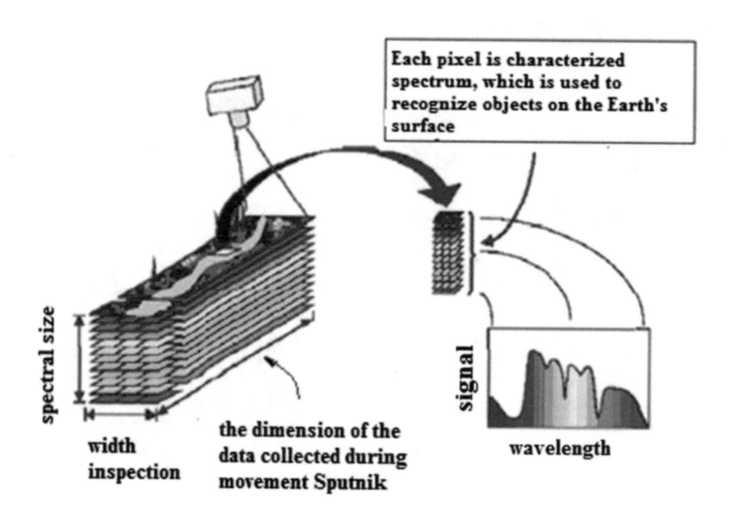

Fig. 5.2 The components of remote sensing as a 3D cube

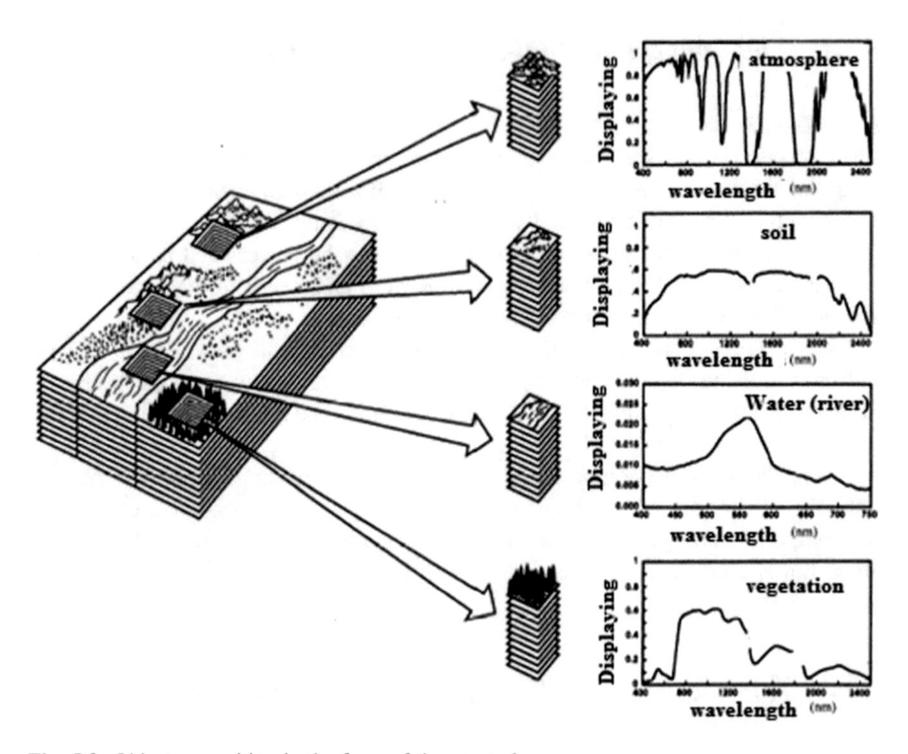

Fig. 5.3 Object recognition in the form of the spectral curve

Figure 5.4 shows an example of the spectral range of multispectral sensor.

Multispectral data set is characterized by 5–10 strips of relatively high spectral width of 70–400 nm. Typically, in a multispectral systems, frequency band images are not close to each other, so they are not contiguous. They can be broad or narrow, but a significant number of very narrow (about 10 nm). Modern multispectral

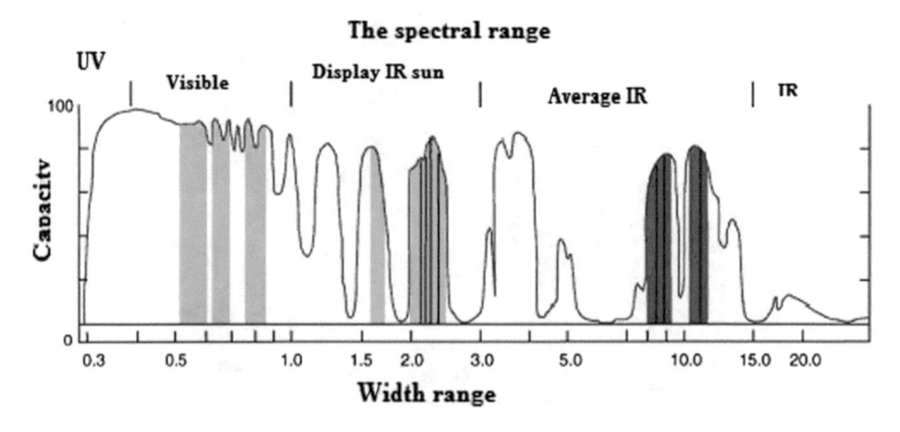

Fig. 5.4 Frequency band of multispectral sensor

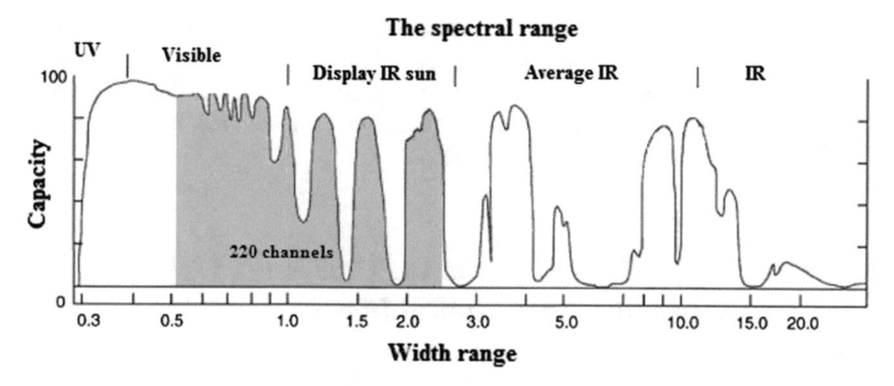

Fig. 5.5 The frequency band of hyperspectral sensor

images have a high spectral extension. For example, at the sensor Aster spectral extension for low range is from 0.07 to 0.17 μm. The narrower is the spectral range to which the sensor is sensitive, the narrower is the spectral extension.

Hyperspectral system operation principle is the same as multispectral, but differ in the number of spectral frequencies. They have tens of hundreds of narrow adjacent frequency bands. Generally, such systems have 100–200 frequencies on a relatively small spectral width (5–10 nm). Figure 5.5 is an example of the spectral range of hyperspectral sensor.

The spectral curves of the image, usually starting with 400 nm from the blue end of the visible spectrum. These systems measure the range of up to 1100 or even 2500 nm.

5.5.4 *Principles of Imaging Systems*

In the multispectral and hyperspectral system, electromagnetic radiation is split by a prism into a plurality of narrow compatible strips. The energy of each band is

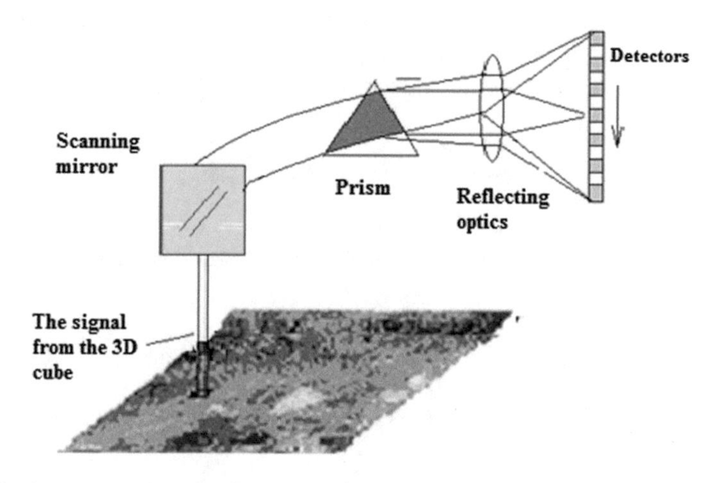

Fig. 5.6 The concept of constructing the image sensor—spectrometer

measured by a separate detector.. The schematic diagram of the construction of the image sensor—spectrometer is shown in the Fig. 5.6.

Using a scanning mirror, a prism and a set of optical lenses an image enters in the detector array from which the image is fed into a computer, recorded on the hard disk and transferred for further processing.

5.6 Application of NEFClass in the Problem of Objects Recognition at Electro-Optical Images

Using multi-spectral electro system operating in three ranges—red, green and blue images were obtained of the ocean and the coastal surface. It was required to recognize objects in the form of geometric shapes on water surface and on the sand. For these purposes, accounting the complexity of the problem as well as a large level of noise it was suggested to use fuzzy neural networks, in particular NEFClass. In order to organize the training of FNN NEFClass a number of learning algorithms were developed—gradient, conjugate gradient and genetic ones and their efficiency was investigated and compared to the basic training algorithm of the system NEFClass [7].

Gradient Learning Algorithm for NEFClass

For the first stage of the algorithm—learning rule base the first phase of the basic algorithm NEFClass is used. The second stage uses a gradient algorithm for training the feedforward neural network, which is described below [7].

Let the criterion of training fuzzy neural network, which has 3 layers (one hidden layer), be as follows:

$$e(W) = \sum_{i=1}^{M} (t_i - NET_i(W))^2 \rightarrow \min, \tag{5.12}$$

where t_i—the desired value of the ith output of neural network;

$NET_i(W)$—the actual value of the ith neural network output for the weight matrix $W = [W^I, W^O]$, $W^I = W(x, R) = \mu_j(x)$, $W^O = W(R, C)$.

Let activation function for the hidden layer neurons (neurons rules) be such:

$$O_R = \prod_{i=1}^{N} \mu_{j_i}^{(i)}(x_i), \quad j = 1, \ldots q_i, \tag{5.13}$$

where $\mu_{ji}(x)$—membership function, which has the form (Gaussian):

$$\mu_{j_i}^{(i)}(x) = e^{-\frac{(x-a_{ji})^2}{b_{ji}^2}}, \tag{5.14}$$

and the activation function of neurons in the output layer (weighted sum):

$$O_C = \frac{\sum_{R \in U_2} W(R,C) \cdot O_R}{\sum_{R \in U_2} W(R,C)}, \tag{5.15}$$

or maximum function:

$$O_C = \max\ W(R,C) \cdot O_R. \tag{5.16}$$

Consider the gradient learning algorithm of fuzzy perceptron.

1. Let $W(n)$—be the current value of the weights matrix. The algorithm has the following form:

$$W(n+1) = W(n) - \gamma_{n+1} \nabla_w e(W(n)), \tag{5.17}$$

   where γ_n—the step size at nth iteration;
   $\nabla_w e(W(n))$—gradient (direction), which reduces the criterion (5.12).
2. At each iteration, we first train (adjust) the input weight W, which depend on the parameters a and b (see the expression 5.14)

$$a_{ji}(n+1) = a_{ji}(n) - \gamma_{n+1} \frac{\partial e(W)}{\partial a_{ji}}, \tag{5.18}$$

$$b_{ji}(n+1) = b_{ji}(n) - \gamma'_{n+1} \frac{\partial e(W)}{\partial b_{ji}}, \tag{5.19}$$

   where γ'_{n+1}—step size for parameter b.

$$\frac{\partial e(W)}{\partial a_{ji}} = -2 \sum_{k=1}^{M} ((t_k - NET_k(w)) \cdot W(R,C)) \cdot O_R \cdot \frac{(x - a_{ji})}{b_{ji}^2}, \tag{5.20}$$

$$\frac{\partial e(W)}{\partial b_{ji}} = -2 \sum_{k=1}^{M} ((t_k - NET_k(w)) \cdot W(R,C)) \cdot O_R \cdot \frac{(x - a_{ji})^2}{b_{ji}^3}. \tag{5.21}$$

3. We find (train) output weight:

$$\frac{\partial e(W^O)}{\partial W(R,C_k)} = -\left(t_k - NET_k\left(W^O\right)\right) O_R, \tag{5.22}$$

$$W_k^O(n+1) = W_k^O(n) - \gamma''_{n+1} \frac{\partial e(W^O)}{\partial W(R,C_k)}. \tag{5.23}$$

4. $n := n+1$ and go to the next iteration.

The gradient method is the first proposed learning algorithm, it is easy to implement, but has the disadvantages:

1. converges slowly;
2. only finds a local extremum.

Conjugate Gradient Method for the System NEFClass

Conjugate gradient algorithm, as well as more general algorithm of conjugate directions, was used in the field of optimization thanks to a wide class of problems for which it ensures the convergence to the optimal solution for a finite number of steps. Its description is considered in the Chap. 1 and isn't described here.

Genetic Method for Training System NEFClass

Consider the implementation of a genetic algorithm to train NEFCLASS. This algorithm is a global optimization algorithm. It uses the following mechanisms [8]:

1. crossing-over pairs of parents and generation of descendants;
2. mutation (random effects of the action);
3. the natural selection of the best (selection).

The purpose of training—to minimize the mean square error:

$$E(W) = \frac{1}{M}\sum_{k=1}^{M} (t_k - NET_k(W))^2, \tag{5.30}$$

where M is the number of classes; t_k is the desired classification;
$NET_k(W)$—classification result of NEFCLASS;
$W = [W_I, W_O]$, $W_I = \left\| w_{ij}^I \right\|$ are inputs weights, $W_O = \left\| w_{ij}^O \right\|$—output weights.
Any individual (specimen) is described by the appropriate vector of weights W.
Set the initial population of N individuals $[W_I(0), \ldots, W_i(0), \ldots, W_N(0)]$.
Calculate the index of fitness (FI), and evaluate the quality of recognition:

$$FI(W_i) = C - E(W_i) \to \max, \tag{5.31}$$

where C—a constant.

Next step is the crossing of parental pairs. When selecting parents a probabilistic mechanism is used. Let P_i be the probability of selecting the ith parent

$$P_i = \frac{FI(W_i(0))}{\sum_{i=1}^{N} FI(W_i(0))}, \tag{5.32}$$

Then the crossing of selected pairs is performed.

It's possible to apply different mechanisms of crossing. For example: for the first offspring even components of the vector of the first parent and the odd components of the vector of the other parent are taken, and for the second on the contrary:

$$W_i(0) \oplus W_k(0) = W_i(1) + W_k(1) \tag{5.33}$$

$$w_{ij}(1) = \begin{cases} w_{ij}(0), & if \quad j = 2m \\ w_{kj}(0), & if \quad j = 2m - 1 \end{cases}$$

$$w_{kj}(1) = \begin{cases} w_{kj}(0), & if \quad j = 2m \\ w_{ij}(0), & if \quad j = 2m - 1 \end{cases} \tag{5.34}$$

where $W_i = \left[w_{ij}\right]_{j=1,R}$, $m = R/2$.

Choose $\frac{N}{2}$ pairs of parents and generate N descendants.

After generating offsprings, the mutation acts on the new population:

$$w'_{ij}(n) = w_{ij}(n) + \xi(n) \tag{5.35}$$

where $a = const \in [-1; +1]$;

$\xi(n) = ae^{-\alpha n}$, α—mutation rate of extinction;
α—is selected randomly from the interval [0, 1].

Then, after the effect of mutation selection procedure is performed in a population, which allows to choose the "fittest" individuals. Different mechanisms of selection may be used.

1. Complete replacement of the old to the new population.
2. N Selecting the best of all existing species by the criterion of maximum FI $N_{par} + N_{ch}$.

After the crossing, mutation and selection the current iteration ends. The iterations are repeated until one of the stop criteria will be fulfilled.

5.6.1 Experiments to Recognize Objects in the Real Data

For images processing the electro-optical imaging system ENVI was used and its ability to map, that is, to combine the images of the check points, obtained from the different spectral cameras [7]. This enables to get a multispectral image. In the Fig. 5.7 initial data for mapping are shown.

After selecting the 15 control points in the images in different spectrum (this function is not automated) images are merged and we get the so-called multispectral cube. The result is shown in Fig. 5.8.

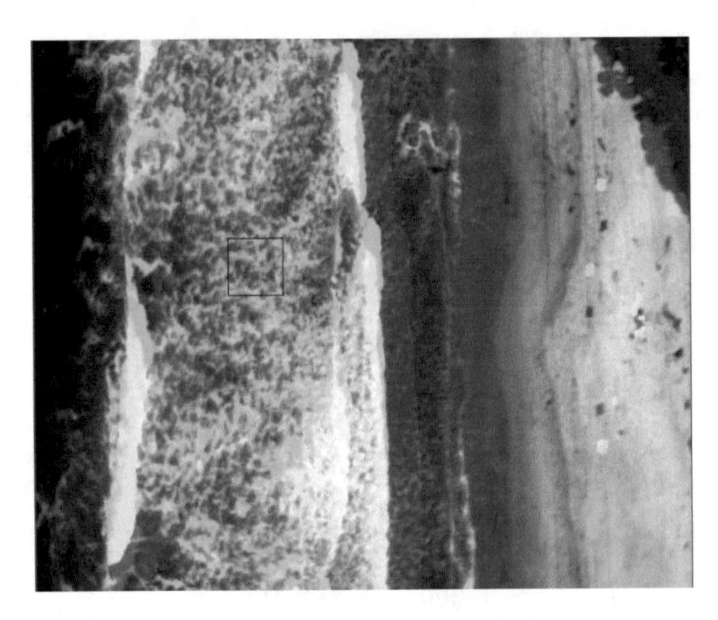

Fig. 5.7 Initial data

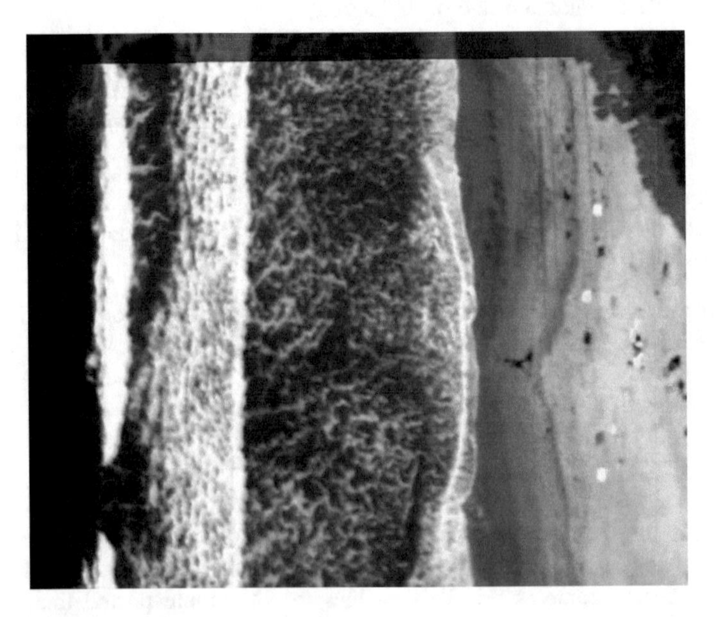

Fig. 5.8 Multispectral image

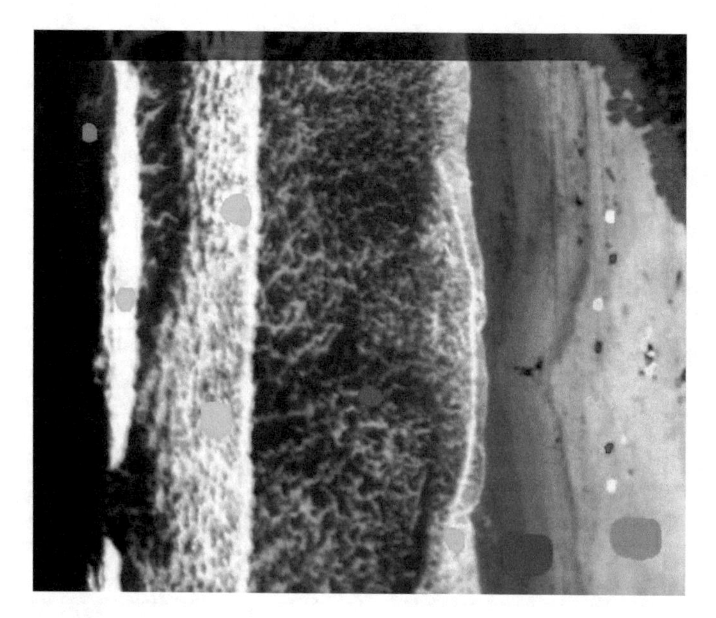

Fig. 5.9 Image of ROI

On the images there were nine different types of surfaces that need to be classified. For analysis and processing, so-called ROI (Region of Interest) on images were used. On the image homogeneous region was determined, for example, sand, water, foam, target red target white color and so on. The result of this detection can be seen in Fig. 5.9.

Next, using a processing system the mean value and the variance of the selected region were received. The data obtained were later tabulated.

These data characterize the nine classes of surface areas:

- white target; red target; green target; blue target; yellow target; foam; water; dry sand; wet sand.

For classification of objects it was suggested to use FNN NEFClass_M [5, 7]. These types of surfaces correspond to nine output nodes in the system NEFClass_M.

The total number of features used to classify the kinds of surfaces is four, namely:

- the brightness in the red spectrum (RS);
- the brightness in the blue spectrum (BS);
- the brightness in the green spectrum (GS);
- brightness in the infrared spectrum (IS).

The total number of data patterns is 99, 11 for each class.

Present the main statistical characteristics of the data set obtained by multispectral system "Mantis" (Tables 5.1 and 5.2) [9].

Table 5.1 Statistical characteristics of multispectral system "Mantis"

Evidence	Minimum	Maximum	Average	Pattern deviation	The correlation between the symptoms and the class
Brightness in the RS	28.81	255.00	165.40	76.14	−0.46
Brightness in the BS	72.93	255.00	165.43	68.62	−0.32
Brightness in the GS	44.34	254.89	121.57	57.64	−0.52
Brightness in the IS	17.03	255.00	140.84	81.58	−0.49

Table 5.2 The correlation between the features

	Brightness in the RS	Brightness in the BS	Brightness in the GS	Brightness in the IS
Brightness in the RS	1	0.7	0.58	0.95
Brightness in the BS		1	0.77	0.7
Brightness in the GS			1	0.59
Brightness in the IS				1

To explore the effectiveness of various learning algorithms in the problem of electro-optical image recognition using NEFClass software kit was developed named NEFClass—BGCGG (Basic, Gradient, Conjugate Gradient Genetic) [6].

Further experiments were carried out with the software kit NEFClass-BGCGG. According to the basic principle of model investigation experiments were carried out by changing only one parameter each time. Of the available 99 patterns 54 patterns served as a training sample. The other 45 patterns were used for testing. The values of the basic parameters of the simulation algorithm were set to the starting positions (see Table 5.3).

During the process of training 15 rules was generated presented in Table 5.4.

The dependence of the quality of training on the number of rules that are generated in the first stage was investigated. For an objective assessment of the results testing on the test sample was performed. For this purpose we varied the number of rules, starting from 9 to 14. The results are shown in the Table 5.5.

The obtained result is natural, the more rules, the better the results of the test classification.

We have investigated the effect of the terms number in features on the quality of classification. Comparative table is given below (see Table 5.6).

Very interesting result was obtained in this series of experiments [9].

Table 5.3 The values of the parameters for the program

Parameter	Value
Algorithm generation rules	The best for the class
The learning algorithm	Classic
Number of generating rules	Maximum
The aggregation function	Weighted sum
The number of terms (values) for each feature	5 for all
Limiting the intersection of fuzzy sets	1
Speed training for weight coefficients between the input nodes and the rule nodes	$\sigma_a = 0,1$ $\sigma_b = 0,1$ $\sigma_c = 0,1$
Speed training for weight coefficients between the rules layer and the output layer	$\sigma = 0,1$
The maximum number of epochs	50

Table 5.4 The rule base of a fuzzy classifier

№ of rule	№ Feature 1 value	№ Feature 2 value	№ Feature 3 value	№ Feature 4 value	Class №
1	4	4	4	4	0
2	4	0	1	4	1
3	4	0	0	4	1
4	4	1	0	4	1
5	2	3	1	1	2
6	1	0	1	0	3
7	4	4	1	4	4
8	3	4	3	3	5
9	3	3	2	3	5
10	4	4	3	3	5
11	0	0	0	0	6
12	3	2	1	2	7
13	1	0	0	1	8
14	1	1	0	1	8
15	1	0	0	0	8

Table 5.5 The dependence of the quality of classification on the number of rules

Number of rules	MSE	True classification %
9	13.071009	24
10	9.545608	15
11	9.910701	15
12	9.705482	15
13	4.769655	4
14	4.739224	4
15	4.751657	4

Table 5.6 The dependence of the quality of classification on the number of terms

Number of terms	MSE	True classification %
4	5.928639	4
5	4.626252	4
6	4.957257	4
7	5.228448	4
8	5.633563	4
9	6.797175	4
10	7.897521	7

From the Table 5.6 it follows that there exists an optimal number of terms that can be used to describe a collection of data during training. When the number of terms exceeds this value the number of misclassified samples increases, that is, by increasing the complexity of the model error increases.

System training using classical algorithm with the optimal number of terms in the features was performed. Forms of membership functions for each feature are shown in Fig. 5.10.

The total sum of squared errors was 2.852081, the number of erroneous classifications—zero in the training set, while for the test sample MSE was equal to 4.6252, which is not bad result.

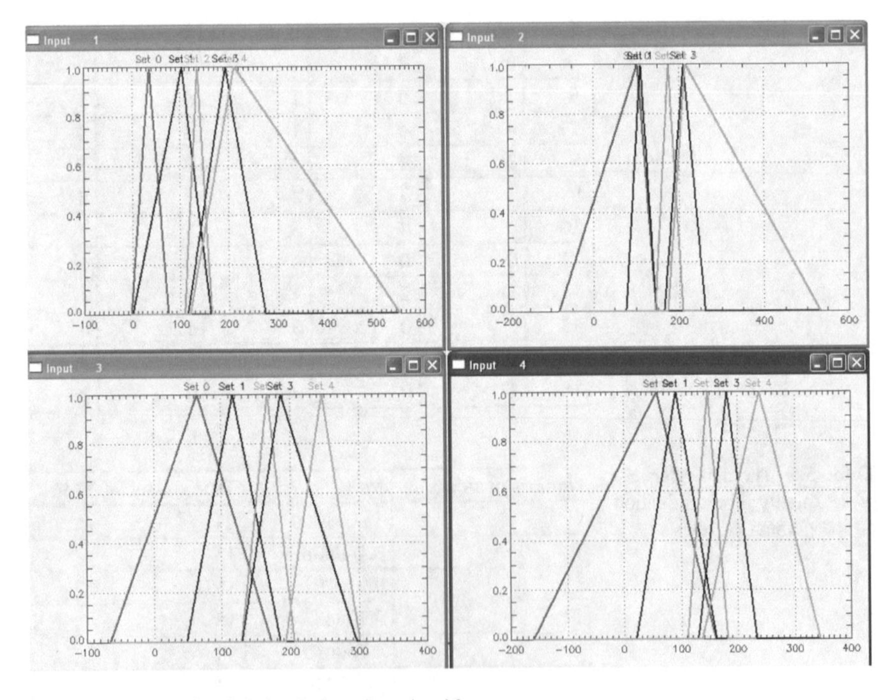

Fig. 5.10 The result of a classic learning algorithm

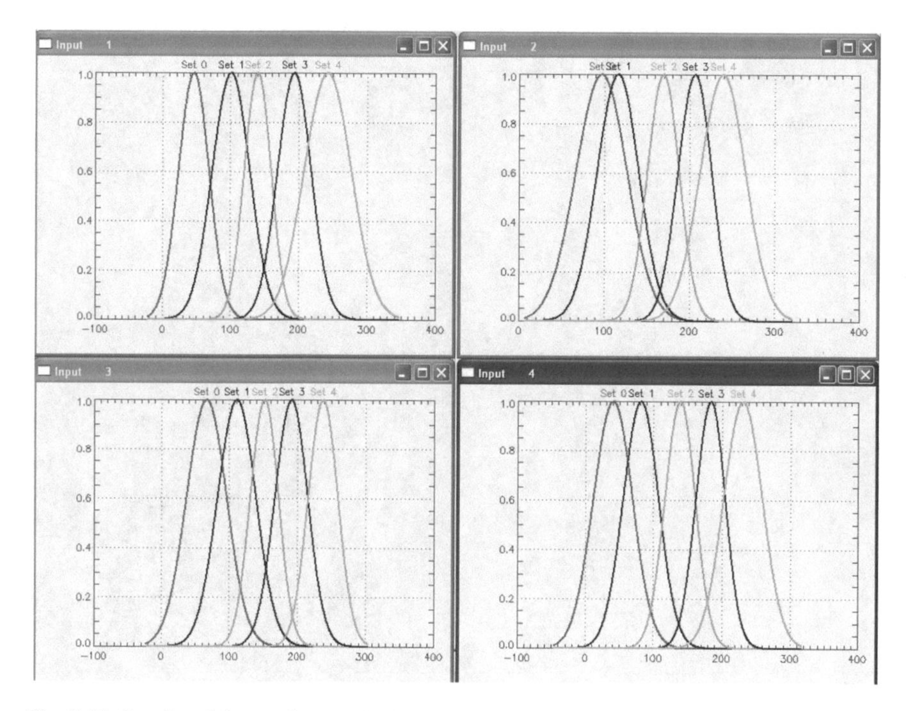

Fig. 5.11 Results of the gradient method

Experiments with the gradient algorithm. The results are shown in Fig. 5.11 (MF of fuzzy sets for each of the four features).

An error at the end of the training was 2.042015, that a little bit better than for classical method. When testing MSE was 3.786005, and the portion of misclassification was 4 %.

Further, the option automatic speed adjustment of MF parameters was included, that is, we used the algorithm "golden section" for step value optimization. The results are shown below (Fig. 5.12).

The same experiments were carried out with a conjugate gradient algorithm. The results are shown in Fig. 5.13.

Further the method of golden section was added to training algorithm. The results can be seen in Fig. 5.14.

Finally, experiments with a genetic algorithm with different MF—triangular and Gaussian were carried out [9].

The results of learning using different algorithms are presented in the comparative charts (Fig. 5.15) and Table 5.7. Note that for the training sample excellent results by the criterion of the percentage of misclassification were obtained for all algorithms.

For all algorithms, this criterion is zero. However, on the test sample, the results were worse: at least two samples were misclassified. Also the sum of squared error

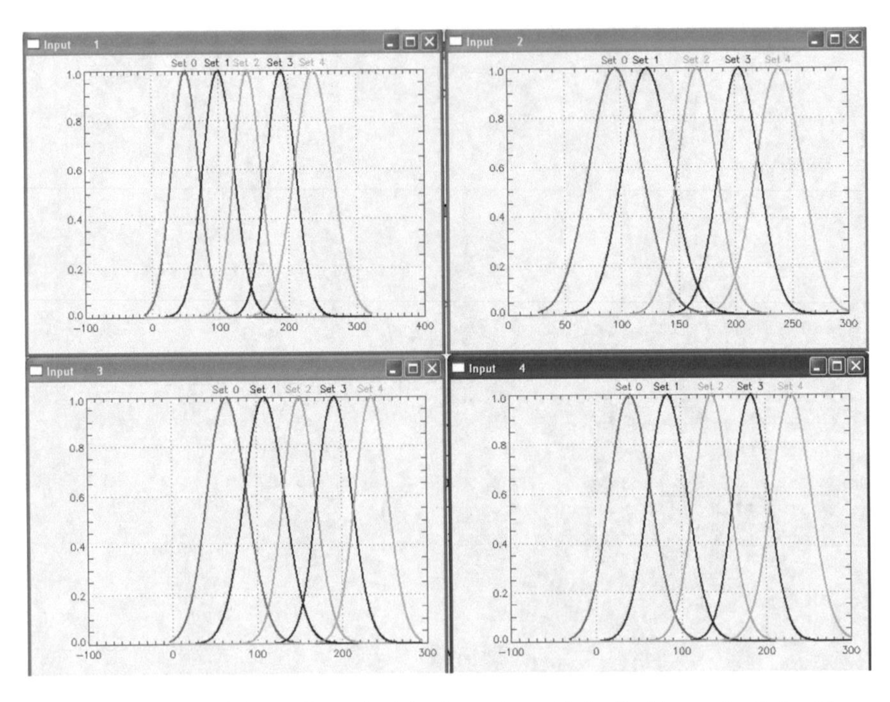

Fig. 5.12 The result of the gradient algorithm in tandem with the "golden section" algorithm

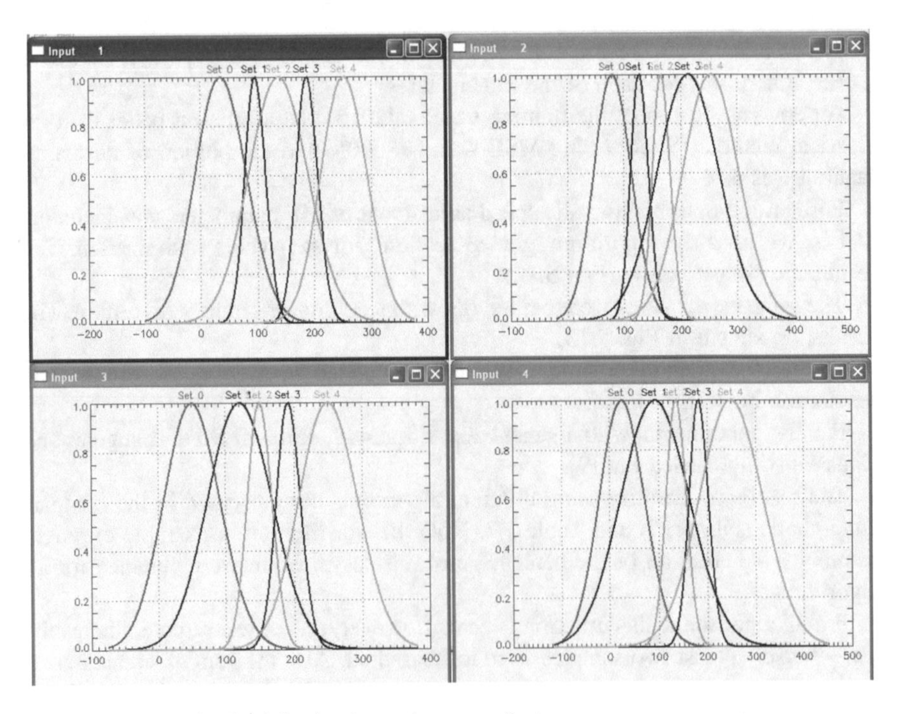

Fig. 5.13 The result of training by the conjugate gradient

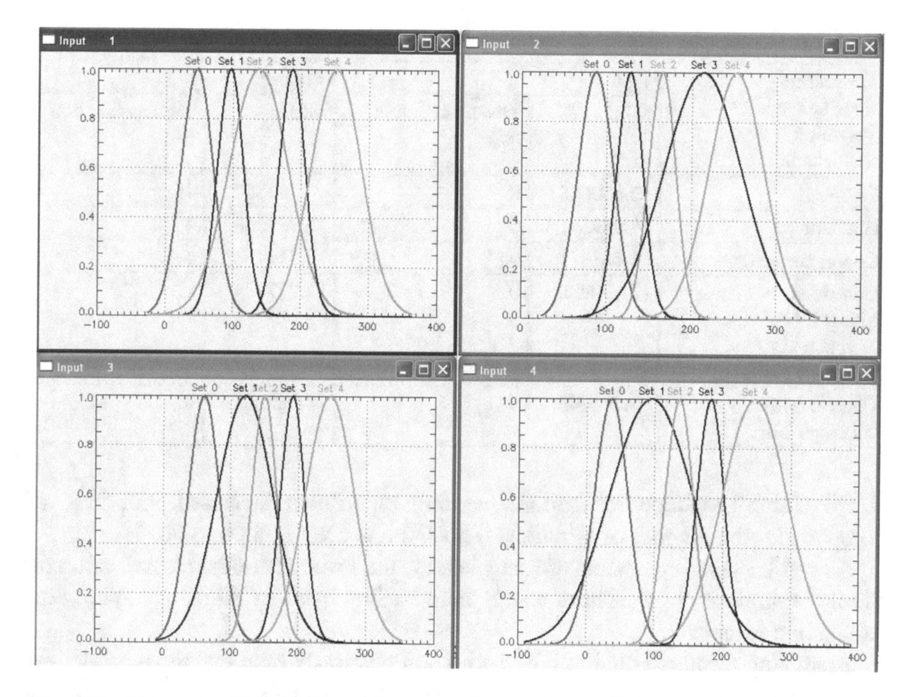

Fig. 5.14 The result of training by conjugate gradient method with the selection step by "golden section" algorithm

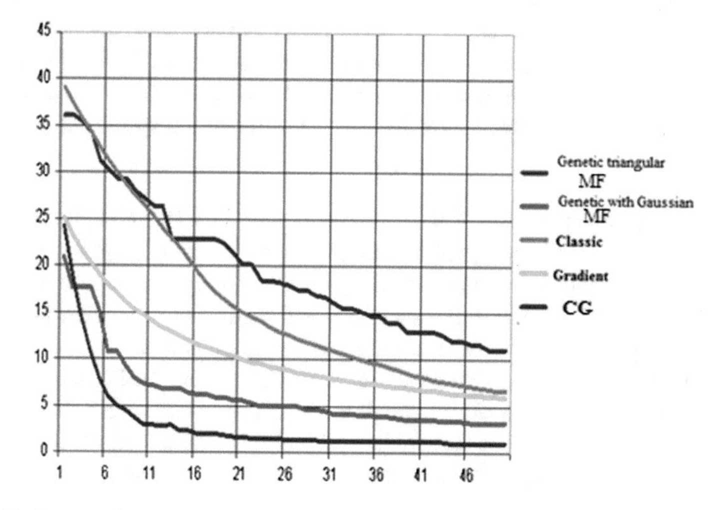

Fig. 5.15 Comparative curves of the convergence rate to the optimal classification of different learning algorithms

Table 5.7 Comparison table for different learning algorithms

The learning algorithm of weighting coefficients	Training		Testing	
	MSE	Misclassification (%)	MSE	Misclassification (%)
Classic	6.650668	0	7.285827	4
Gradient	5.9893	0	6.829068	4
Conjugate gradient	1.132871	0	3.314763	4
Genetic with triangular membership functions (MF)	11.110936	0	13.677424	4
Genetic with Gaussian MF	3.204446	0	4.568338	4

(MSE) for all, without exception, learning algorithms increased. For ease of comparison, the number of iterations (epochs) has been limited to 50.

As can be seen, the results are satisfactory, the level of correct classification on the test sample is 96 %. These results may be improved by forming a more representative sample.

Analyzing the curves in the Fig. 5.15 it can be clearly seen that the best method for the rate of convergence is the conjugate gradient method. Then the next is a genetic algorithm with Gaussian function. Less effective is the gradient method. Next by rate of convergence is classical algorithm used in the system NEFClass. And at the end of row the least effective is genetic method with a triangular membership function.

However, the MSE criterion by which the curves were plotted, displays ambiguously classification quality. An important criterion for evaluation of methods efficiency is the minimum number of misclassified samples. From Table 5.7 one can see that all algorithms show the same results with respect to this criterion.

Conclusions

1. The chapter describes the FNN NEFClass for the classification. Its advantages and disadvantages are analyzed. A modified system NEFClass_M is considered that differs by randomization of learning sample, configuration parameters—the speed of learning and the use of Gaussian MF instead of triangular, allowing to implement gradient learning algorithm and its modifications.
2. Comparative experimental study of FNN NEFClass and NEFClass_M demonstrated the benefits of the latest in quality classification (% correct classification).
3. Investigations FNN NEFClass in the problem of classification of objects of electro-optical imaging, which led to the following conclusions:

   - In the test data the effect of various parameters on the system efficiency was explored. The results showed that the quality of the training system NEFClass depends on the rate of learning. An increase in the parameter of

speed training for weights between input and rule nodes, may improve the speed of convergence and attain much closer to the minimum point of error function. However, during experiments, it was observed setting too large speed is not appropriate, as the phenomenon of "oscillation" occurs.
- Best algorithm for rules generation is "The best for the class."

5.7 Recognition of Hand-Written Mathematical Expressions with Application of FNN

Mathematical expressions are widely-used in scientific and technical literature but their entering in PC and I-pads with help of such devices as mouse and key board is rather difficult. The possibility of entering mathematical expression with help of stylus and sensor screen enables to make this process more natural, convenient and simple for user. Therefore last years the problem has arisen of hand-written mathematical expression recognition with application of FNN.

Goal of presented the work was to develop a method and software for hand-written mathematical expressions recognition which are entered in computer in on-line mode.

The problems to be solved were the following [7]:

1. The investigation of properties of hand-written mathematical symbols and expressions.
2. Development and investigation of approach to selection of informative features describing the properties of hand-written symbols and utilized for constructing data base of fuzzy neural classifier.
3. Investigation and analysis of training algorithms of fuzzy classifier NEFCLASS.
4. Development of algorithm of hand-written mathematical expressions recognition using FNN NEFCLASS.
5. Development of software kit implementing the suggested algorithms.
6. Structural analysis of hand-written mathematical expressions
7. The design of information technology of pattern recognition of hand-written mathematical expressions in on-line mode

Pattern Recognition of Hand-Written Mathematical Expression in On-line Mode

Problem Statement

While pattern recognition on-line it's possible to track the trajectory of writing symbol and define the point of touching pen (an initial symbol point) and a point of pen non-touching (final point) when writing. The input data for pattern recognition task is sequence of pen coordinates called segment consisting of touching point, non-touching point and all points lying between them. The pattern recognition problem consists of two stages:

Table 5.8 Table of symbols used for writing hand-written mathematical expressions

Cifer	1	2	3	4	5	6	7	8	9	0																
Latin letters	a	b	c	d	e	f	g	h	i	j	k	l	m	n	o	p	q	r	s	t	u	v	w	x	y	z
brackets	[	]	(	)	{	}																				
Greek letters	α	β	γ	δ	ε	θ	λ	μ	ξ	π	ω	ϕ	∂	Ω												
Other symbols	∅	∞	Δ	∇	∑	∏	√	∫	∧	∨	.	,	;	:												
	=	≠	≈	~	<	>	≥	≤																		
	+	-	/	\	÷	%	!	±	||																	
	∀	→	∉	∈																						

At the stage of pattern recognition symbols consisting of one part are considered

At the stage of structural analysis are reconstructed symbols consisting of several parts,

Cifer	1	2	3	4	5	6	7	8	9	0		
Latin letters	a	b	c	d	e	g	h	k	m	n		
	o	p	q	r	s	u	v	w	y	z		
Brackets	[	]	(	)	{	}						
Greek letters	α	β	γ	δ	ε	λ	μ	ξ	ω	ϕ	∂	Ω
Other symbols	∞	Δ	∇	∑	∏	√	∫	∧	∨	.	,	
	+	-	/	\	|	~	<	>				
Auxillary	ȷ	l	ʃ									

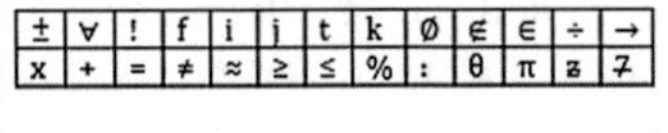

±	∀	!	f	i	j	t	k	∅	∉	∈	÷	→
x	+	=	≠	≈	≥	≤	%	:	θ	π	ƶ	7

1. Hand-written symbols recognition;
2. Structural analysis which allows to determine the space relations among symbols.

The division of problem into two stages has some advantages, in particularly, sub-problems may be solved independently one from another that enables to analyze the obtained results and make some improvements at each stage preserving by this the integrity of the whole system (the list of symbols to be recognized is shown in Table 5.8).

An Approach to Constructing Informative Features Using Method of Approximating Points

The first problem to be solved at hand-written mathematical expressions pattern recognition is selection or construction of informative features of hand-written symbols. An approach to constructing informative features was suggested based on extracting so-called approximating points of linearized line approximating a symbol curve [10].

For finding approximating points algorithm of Ramer–Douglas–Pecker was suggested which enables to cut by several degrees the total number of points approximating the symbol curve preserving its form with maximal accuracy.

The algorithm finds the point which is maximal distant from the line passing through the first and the last points on the curve. If this point locates at the distance lesser than ε, then the obtained line approximates the curve with accuracy not

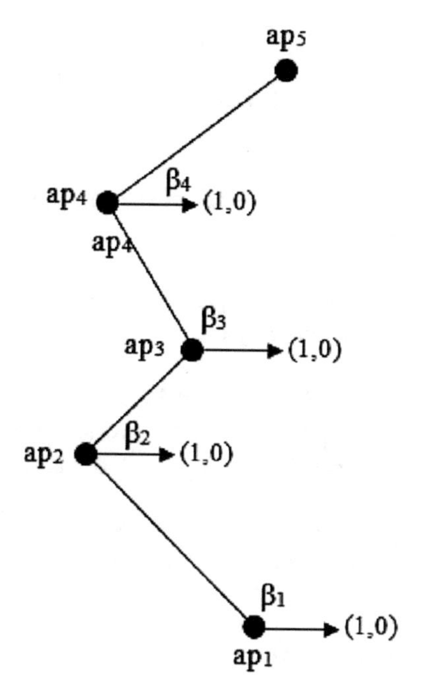

Fig. 5.16 A set of informative features for symbols pattern recognition

greater than ε. If this distance is greater than ε, then algorithm recurrently repeats at the pair initial point-obtained point and the pair obtained point—final point. Experimentally was found optimal coefficient of smoothing accuracy by which the optimal ratio between the number of left points and maximal preserving the form of initial symbol curve.

Informative Features

For pattern recognition of hand-written symbols was suggested the set of informative features which describe the properties of hand-written symbols [10] (see Fig. 5.16)

1. Maximal angle
   Find angles β_j, which lie between segments of symbol divided by approximating points and axis X: $\beta_j = \arccos\left(\vec{x} \cdot \frac{\vec{v}_j}{\|\vec{v}_j\|}\right)$
   where $\vec{x} = (1,0)$, vector v_j is defined as: $v_j = ap_i - ap_{i-1},\ 2 \le i \le k(ap) - 1$, $k(ap)$—total number of approximating points.
   Feature maximal angle is defined by formula $max\ \beta_j$
2. Average value of angles
   Calculate the difference between sequentially located angles β_j:
   where $2 \le i \le k(ap)$,
   Then the feature average angle value can be calculated by formula:

$$a_{aver} = \frac{1}{k(ap) - 2} \sum_{n=1}^{k(ap)-2} \varphi_n$$

3. Linearity
   The value of feature linearity can be calculated as a ratio of number of all angles such $\varphi_n \leq |\varepsilon|$ to a total number of angles φ_n.
4. Topology
   Total number of cases where the sequence of angles φ_n, changes the sign from plus to minus and from minus to plus.
5. Histogram of angles
   Divide coordinate plane on 8 sections by 45 grades each. Every angle β_j, lies in one of the sections. For determination of histogram of angles in *i*th section divide the number of angles located in it by the total number of angles:

$$HA_i = \frac{number\ of\ angles\ in\ the\ section\ i}{total\ number\ of\ angles}, \quad i = 1, \ldots, 8$$

6. Square of segments
   For calculating this feature find vectors:

$$\vec{u}_i = ap_{i+1} - ap_1$$

$$\vec{v}_i = ap_{i+2} - ap_1$$

$$\vec{a}_i = ap_{i+2} - ap_{1+1}$$

where $1 \leq i \leq k(ap) - 2$

The square of segments may be calculated by formula (see Fig. 5.17):

$$S = \sum_{i=1}^{k(ap)-2} \frac{(\vec{u}_i \times \vec{v}_i) \cdot \mathrm{sgn}(\vec{u}_i \times \vec{v}_i)}{\|\vec{a}_i\|}$$

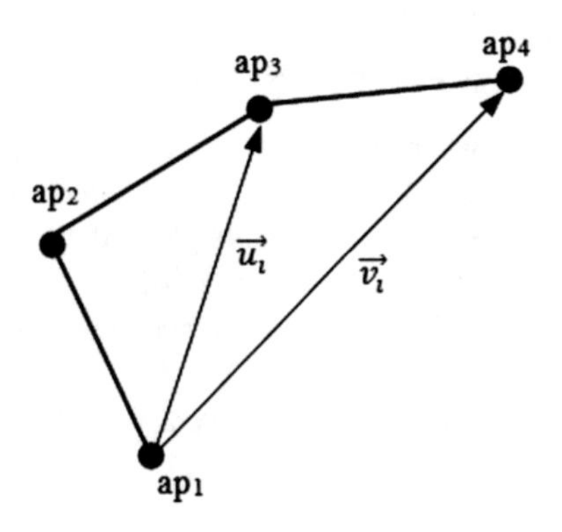

Fig. 5.17 Example of segments

The example of finding informative features on base of two methods: approximating points and dominant points is shown below (Fig. 5.18).

Comparative analysis of methods approximating points and dominant points was performed consisting of the following steps:

1. Test sample was constructed so: it was created ten random variants of writing each of 70 symbols.
2. For each symbol values of thirteen features based approximating and dominant points were found.
3. For analysis of obtained results the average deviation from the mean value was used

$$d = \frac{1}{n}\sum_{i=1}^{n} x_i - \bar{x}$$

where $\bar{x}$ is a mean arithmetic for results of ten experiments for each informative feature.

The experimental results of methods of dominant points and approximating points for 70 symbols are shown in the Fig. 5.19 and presented in the Table 5.9.

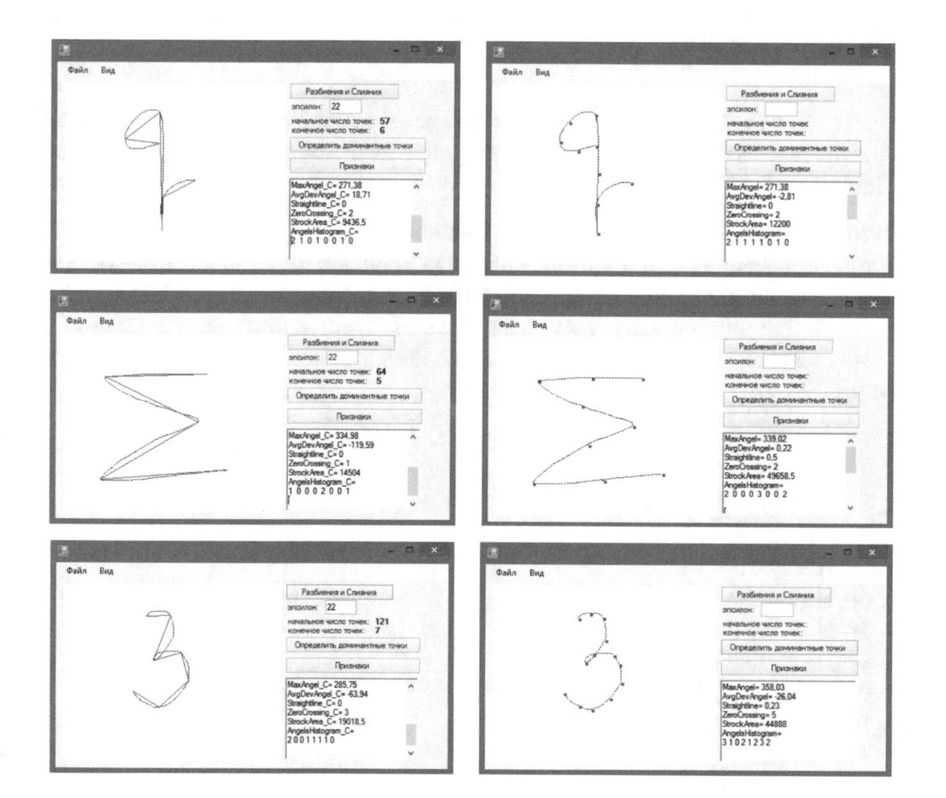

Fig. 5.18 Finding of informative features for methods of dominant and approximating points

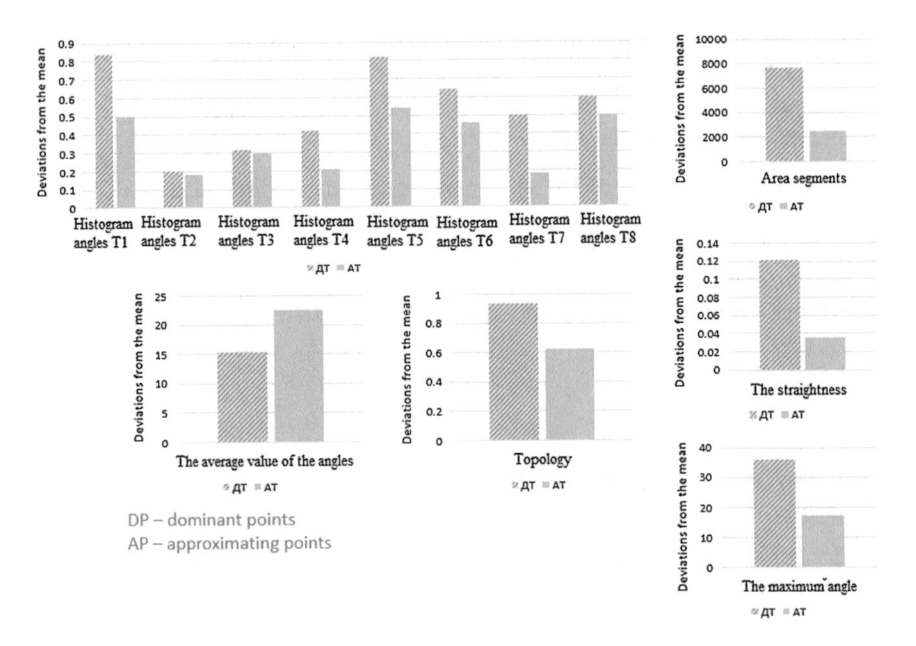

Fig. 5.19 The experimental results of dominant points and approximating points methods application for recognition

For all 70 symbols the average percentage of improvement average deviation from the mean value using approximating points method was 32.85 % as compared with DP-method.

Recognition of Hand-Written Math. Symbols

For the efficiency estimation of the suggested approach for mathematical symbols recognition the experiments were carried out with application of FNN NEFCLASS with different training algorithms and various MF.функций принадлежности.

As inputs of FNN NEFCLASS were used the suggested 13 informative features [7, 9]:

x_1—Histogram of angles_T1
x_2—Histogram of angles_T2
x_3—Histogram of angles_T3
x_4—Histogram of angles_T4
x_5—Histogram of angles_T5
x_6—Histogram of angles_T6
x_7—Histogram of angles_T7
x_8—Histogram of angles_T8
x_9—Average angle value
x_10—Maximal angle
x_11—Linearity
x_12—Square of segments

Table 5.9 Deviations from the mean for methods of approximating points and dominant points

Dominant point	Deviations from the mean	Approximating point	Deviations from the mean	% variation from the mean deviation
Histogram angles T1	0.84	Histogram angles T1	0.5	40.47619048
Histogram angles T2	0.2	Histogram angles T2	0.18	10
Histogram angles T3	0.32	Histogram angles T3	0.3	6.25
Histogram angles T4	0.42	Histogram angles T4	0.21	50
Histogram angles T5	0.82	Histogram angles T5	0.54	34.14634146
Histogram angles T6	0.64	Histogram angles T6	0.46	28.125
Histogram angles T7	0.5	Histogram angles T7	0.18	64
Histogram angles T8	0.6	Histogram angles T8	0.5	16.66666667
The average value of the angles	15.3673	The average value of the angles	22.6521	−47.40455383
The maximum angle	36.0874	The maximum angle	17.2534	52.18996104
The straightness	0.122	The straightness	0.03584	70.62295082
Area segments	7670.32	Area segments	2458.68	67.94553552
Topology	0.94	Topology	0.62	34.04255319
			Av.% improvement	32.85081887

x_13—Topology
Outputs were 70 symbols to be recognized.

Learning sample: 280 hand-written symbols,
Testing sample: 140 hand-written symbols

Comparative analysis of different training algorithms of NEFCLASS (Fig. 5.20).

For training Gaussian MF was used: $\mu(x) = \exp\left[-\frac{(x-c)^2}{2\sigma^2}\right]$

For membership functions parameters training the following algorithms were investigated [11]:

- Gradient algorithm;
- Conjugated gradient algorithm;
- Genetic algorithm.

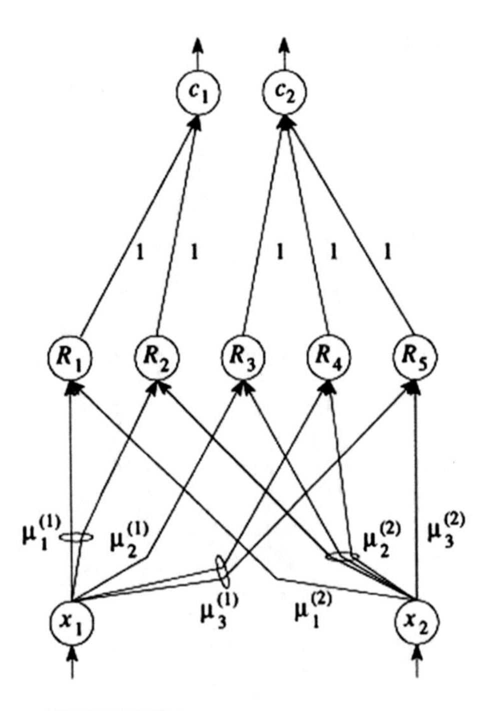

Fig. 5.20 Architecture of FNN NEFClass

The experimental investigations of training algorithms were carried out. On the Fig. 5.21 the dependence "percentage error on number of iterations" for all training algorithms is presented.

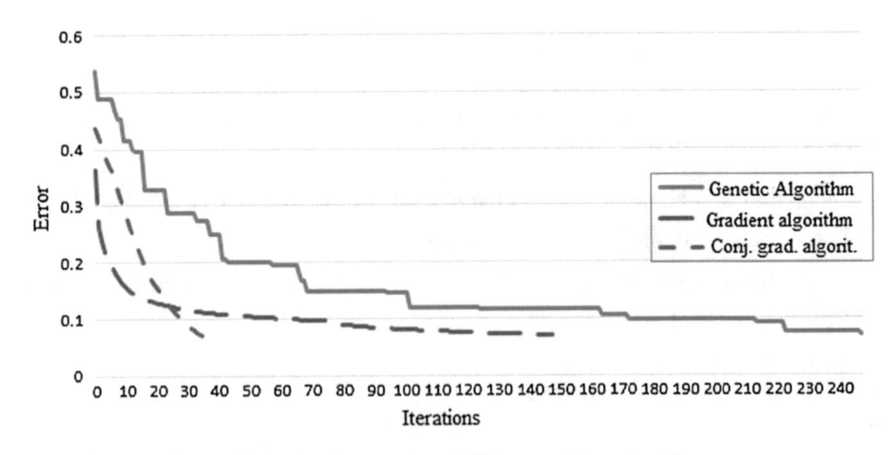

Fig. 5.21 Experimental investigations results of different training algorithms

Comparative analysis of different training algorithms of NEFCLASS using misclassification error (%) for different mathematical symbols are presented below (Figs. 5.22, 5.23, 5.24 and 5.25).

Structural Analysis of Mathematical Expressions
After the stage of symbols recognition the main problem is the determination of the space relations among the components of mathematical expression. The approach for structural analysis of mathematical expressions was suggested which consists of 3 stages [7].

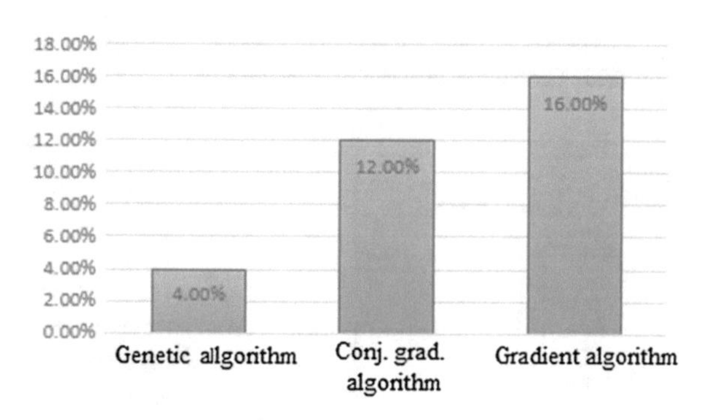

Fig. 5.22 Classification results for symbol "∑"

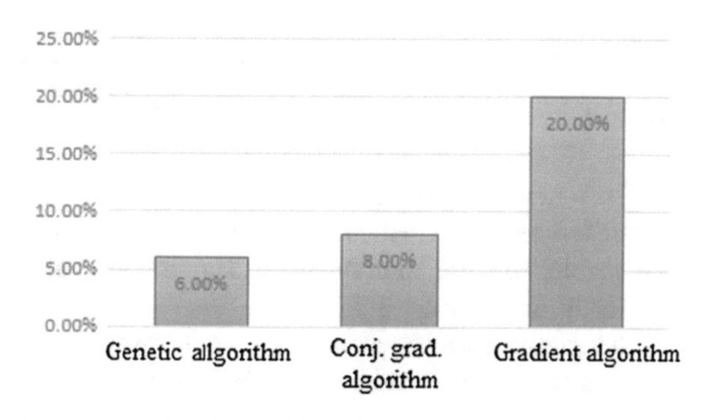

Fig. 5.23 Classification results for symbol "7"

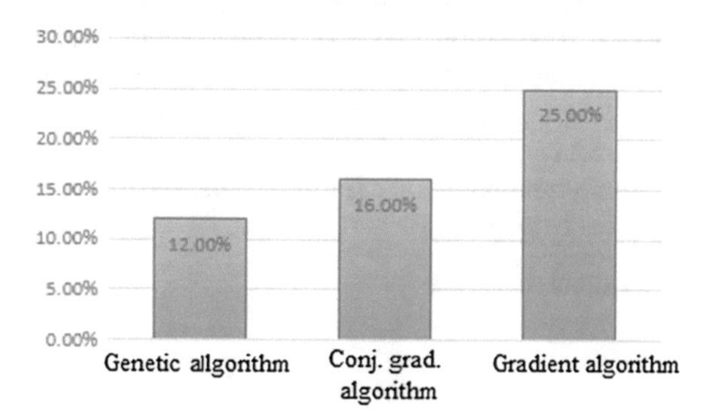

Fig. 5.24 Classification results for symbol "∫"

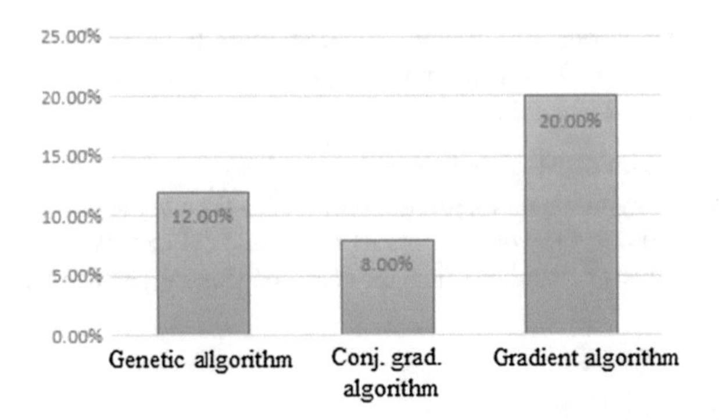

Fig. 5.25 Classification results for symbol "S"

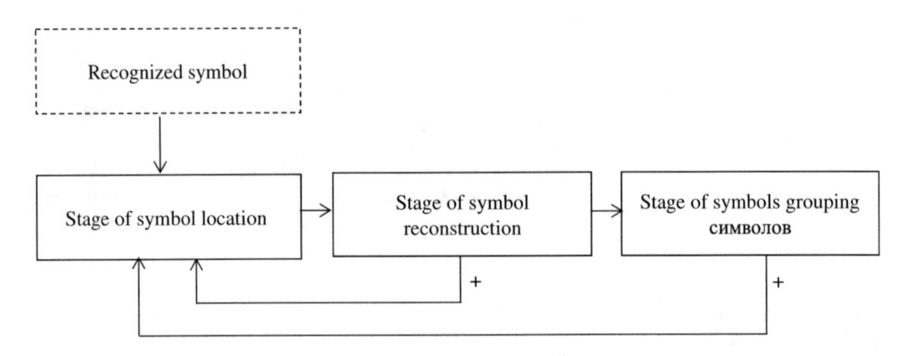

Fig. 5.26 Block-schema of symbols structural analysis

The schema of structural analysis is presented in the figure below (Fig. 5.26).

The written mathematical expression is looked through after each entered change that enables to write components of expression in any order and introduce changes in already written expression.

The symbols are divided into several groups depending on feasible and obligatory positions of location the other symbols relating to them.

The *feasible position* is acceptable but not obligatory for filling symbol position.

Obligatory position is position mandatory for filling symbol position. In the Table 5.10 are shown feasible and obligatory positions (Fig. 5.27).

For solution of the problem a dynamic base of heuristic rules is used based on knowledge on writing order and space relations between symbols shown in the Table 5.11.

Some symbols allow to group several separate symbols into one group. To these symbols belong

- Various brackets,
- Fractional feature,
- Sum $\sum$,
- Product $\prod$,
- Integral,

Table 5.10 Feasible and obligatory positions for different math. symbols

Group of symbols	Feasible positions	Obligatory positions
0–9, π, ∞	1, 2, 6	
$\int$, Π, Σ	4, 6, 8	2
$\sqrt{}$	1, 6, 7	2
Thrigonom. functions	1, 6	2
Logarithmic functions	1, 3, 6	2
Letters	1, 2, 3, 6, 8	
Opening brackets	6	2
Closing brackets	2	6

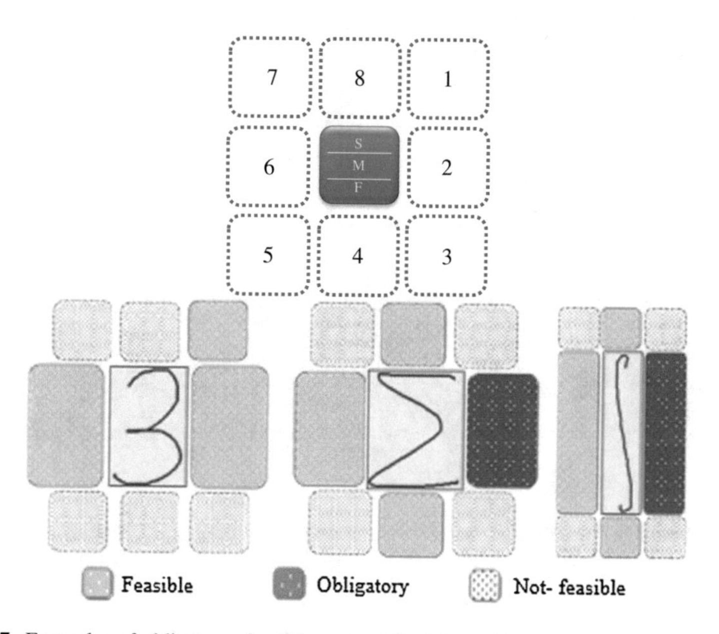

Fig. 5.27 Examples of obligatory, feasible and not-feasble positions

- Arithmetic root,
- Point, comma.

Symbol group is considered as united symbol, for each symbol exist feasible and obligatory positions of locating the neighbor symbols (see example in Fig. 5.28).

For testing were selected 50 different mathematical expressions including trigonometric and logarithmic functions, operations of root extracting, division and so on. The experimental results are presented in the Table 5.12.

In a result of testing the total value of correctly structured mathematical expressions was obtained. It equals to 78 % of test sample.

Table 5.11 Writing order and space positions betweeen symbols

Current symbol *Si*	Symbol S_{i-1}	Position of s_i relating to s_{i-1}	Symbol S_{i-2}	Position of s_{i-1} relating to s_{i-2}	Result
–	∨	M			∀
–	+	4			±
.	l	8			i
–	l	S			t
–	ſ	S			f
.	J	8			j
/	\	M			x
\	/	M			x
>	–	2 or F			→
|	–	F or M	|	S	π
l	–	F or M	J	S	π
l	~	F or M	J	S	π
n	1	6	s	6	sin
s	0	6	c	6	cos
m	i	6	1	6	lim

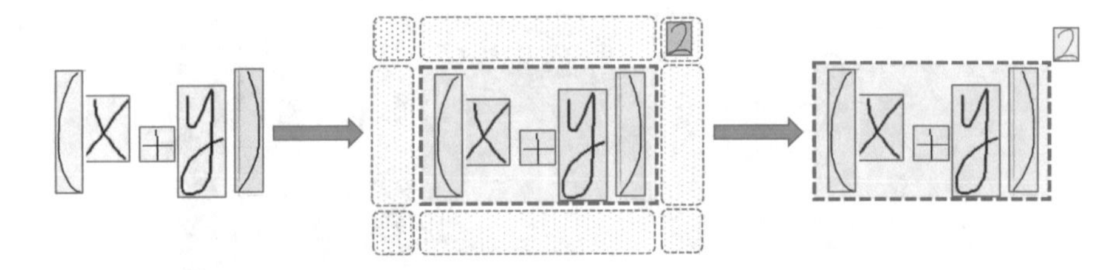

Fig. 5.28 Examples of obligatory and feasible positions for locating neighbor symbol

Table 5.12 Experimental results of structuring mathematical expressions

General number of math expressions	Number of erroneously structured math. expressions	% of erroneously structured math. expressions
50	11	22

The information system for hand-written math. expressions recognition was developed based on the suggested methods of recognition and structural analysis [12].

1. Information system is implemented as client-server architecture and contents a set of software and hardware enabling gathering, storage, processing and transmission of obtained information.

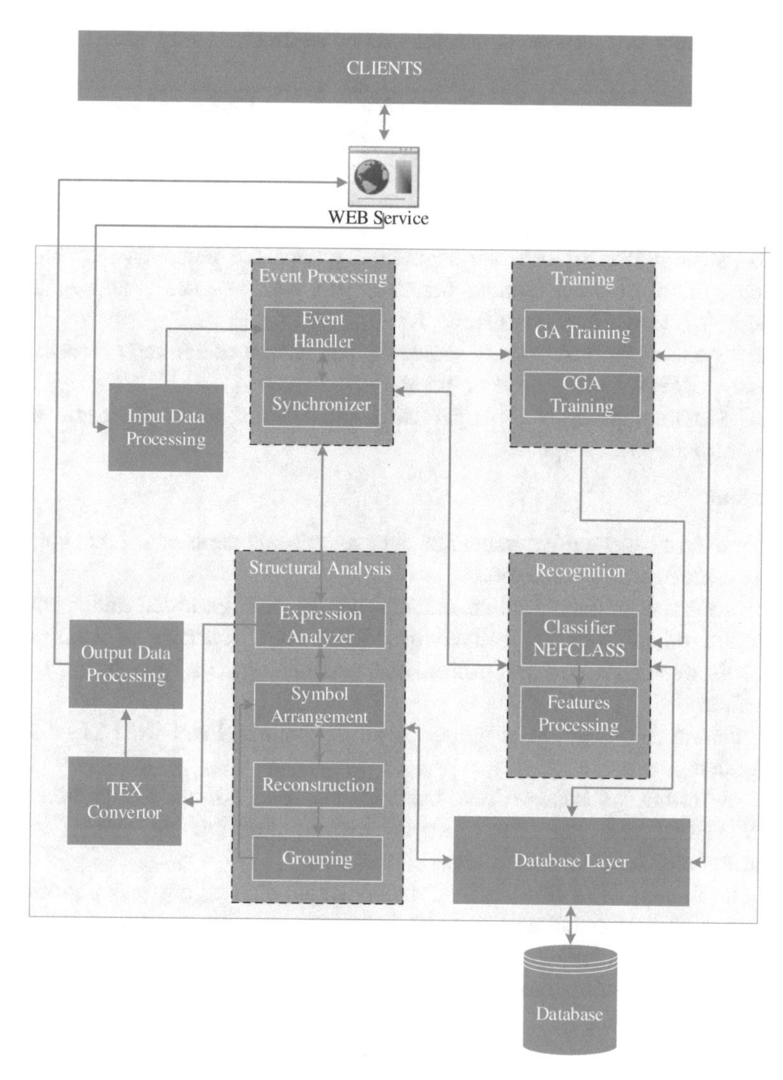

Fig. 5.29 Architecture of information system for hand-written math. expressions recognition

2. Information system is used for hand-written mathematical expression recognition in on-line mode based on suggested methods and algorithms.
3. It is represented in Fig.5.29.

General Results of Experimental Investigations

The total test sample consisted of 140 mathematical expressions.
Total time of recognition of each symbol after training is 150 ms.
The results of final experiments are presented in the following Table 5.13.

Table 5.13 Results of experimental investigations of hand-written mathematical expressions recognition

Recognition of separate segments	Symbols reconstruction (for correctly recognized segments)	Structural analysis of mathematical expressions
91.54 %	95.32 %	71.29 %

Field "Recognition of separate segments" shows the percentage of correctly classified symbols written without break of pen with one script and enables to estimate recognition efficiency (Table 5.13).

Field "Symbols reconstruction" shows the percentage of correctly reconstructed symbols of correctly recognized segments.

Field "Structural analysis of mathematical expressions" shows the percentage of correctly structured expressions.

Conclusion

1. The problem hand-written mathematical symbols and expressions recognition in on-line mode was considered.
2. The properties of hand-written math. symbols were explored and method of selection of informative features describing the properties of hand-written symbols was developed and utilized for constructing data base of fuzzy neural classifier.
3. For classification of hand-written math. symbols FNN NEFCLASS was suggested.
4. The algorithm of hand-written mathematical expressions recognition using FNN NEFCLASS was developed, the investigation and analysis of training algorithms of fuzzy classifier were carried out.
5. Structural analysis of hand-written mathematical expressions was performed.
6. Software kit implementing the suggested algorithms was developed and design of information technology of pattern recognition of hand-written mathematical expressions in on-line mode was performed.

References

1. Nauck, D., Kruse, R.: Generating classification rules with the neuro-fuzzy system NEFCLASS. In: Proceedings of Biennial Conference of the North American Fuzzy Information Processing Society (NAFIPS'96). Berkeley (1996)
2. Nauck, D., Kruse, R.: New learning strategies for NEFCLASS. In: Proceedings of Seventh International Fuzzy Systems Association World Congress IFSA'97, vol. IV, pp. 50–55. Academia Prague (1997)
3. Nauck, D., Kruse, R.: What are neuro-fuzzy classifiers? In: Proceedings of Seventh International Fuzzy Systems Association World Congress IFSA'97, vol. IV, pp. 228–233. Academia Prague (1997)

4. Nauck, D.: Building neural fuzzy controllers with NEFCON-I. In: Kruse, R., Gebhardt, J., Palm, R. (eds.) Fuzzy Systems in Computer Science. Artificial Intelligence, pp. 141–151. Vieweg, Wiesbaden (1994)
5. Zaychenko, Yu.P., Sevaee, F., Matsak, A.V.: Fuzzy neural networks for economic data classification. In: Informatic, Control and Computer Engineering, vol. 42, pp. 121–133. Vestnik of National Technical University of Ukraine "KPI" (2004) (rus)
6. Zaychenko, Yu.P.: Fuzzy Models and Methods in Intellectual Systems, 354 pp. Kiev Publishing House, "Slovo" (2008). Zaychenko, Yu.P.: Fuzzy group method of data handling under fuzzy input data. Syst. Res. Inf. Technol. (3), 100–112 (2007) (rus)
7. Zaychenko, Yu.P., Naderan, E.: Structural analysis of hand-written mathematical expressions. In: Informatics, Control and Computer Engineering, № 54, pp. 12–17. NTUU "KPI" Herald (2011) (rus)
8. Zaychenko, Yu.P.: Fundamentals of Intellectual Systems Design, 352 pp. Kiev Publishing House, "Slovo" (2004) (rus)
9. Zaychenko, Yu.P., Petrosyuk, I.M., Jaroshenko, M.S.: The investigations of fuzzy neural networks in the problems of electro-optical images recognition. Syst. Res. Inf. Technol. (4), 61–76 (2009) (rus)
10. Naderan, E., Zaychenko, Yu.: The extraction of informative features of hand-written mathematical symbols. In: Informatics, Control and Computer Engineering, № 57, pp. 124–128. Herald of NTUU "KPI" (2012) (rus)
11. Zaychenko, Yu.P., Naderan, E.: The application of fuzzy classifier NEFClass to the problem of on-line recognition of hand-written mathematical expressions. Herald of Cherkassy State Technical University, № 1, pp. 3–7 (2012) (rus)
12. Naderan, E., Zaychenko, Yu.: Approach to development of the hand-written mathematical expressions recognition system which are entered in on-line mode. In: Informatics, Control and Computer Engineering, № 58, pp. 56–60. Herald of NTUU "KPI" (2013) (rus)

Chapter 6
Inductive Modeling Method (GMDH) in Problems of Intellectual Data Analysis and Forecasting

6.1 Introduction

This chapter is devoted to the investigation and application of fuzzy inductive modeling method known as Group Method of Data Handling (GMDH) in problems of intellectual data analysis (Data Mining), in particularly its application in the forecasting problem in macroeconomy and financial sphere.

The problem consists in forecasting *models* construction and finding unknown functional dependence between given set of macroeconomic indices and forecasted variable using experimental data.

The advantage of inductive modeling method GMDH is a possibility of constructing adequate model directly in the process of algorithm run. The specificity of fuzzy GMDH is getting the interval estimates for forecasting variables.

In this chapter the review of main results concerning GMDH and fuzzy GMDH is presented, analysis of application of various membership functions (MF) and perspectives of fuzzy GMDH application for forecasting in macroeconomy and financial sphere are estimated.

The Sect. 6.2 contains the problem formulation.

In the Sect. 6.3 main principles and ideas of GMDH are considered. In the Sect. 6.4 the generalization of GMDH in case of uncertainty—new method fuzzy GMDH suggested by authors is described which enables to construct fuzzy models almost automatically. The Sect. 6.5 contains the algorithm of fuzzy GMDH. In the Sect. 6.6 the fuzzy GMDH with Gaussian and bell-wise membership functions MF are considered and their similarity with triangular MF is shown. In the Sect. 6.7. fuzzy GMDH with different partial descriptions in particular orthogonal polynomials of Chebyshev and Fourier are considered.

In the Sect. 6.8 the problem of adaptation of fuzzy models obtained by GMDH is considered and the corresponding adaptation algorithms are described. The Sect. 6.9 contains the results of numerous experiments of GMDH and fuzzy

M.Z. Zgurovsky and Y.P. Zaychenko, *The Fundamentals of Computational Intelligence: System Approach*, Studies in Computational Intelligence 652,
DOI 10.1007/978-3-319-35162-9_6

GMDH application for forecasting share prices and Dow Jones index at New York stock exchange (NYSE). The extension and generalization of fuzzy GMDH in case of fuzzy inputs is considered and its properties are analyzed in the Sect. 6.11.

6.2 Problem Formulation

A set of initial data is given inclusive: input variables $\{X(1), X(2), \ldots, X(N)\}$ and output variables $\{Y(1), Y(2), \ldots, Y(N)\}$, where $X = [x_1, x_2, \ldots, x_n]$ *is* n-tuple vector, N is a number of observations.

The task is to synthesize an adequate forecasting model $Y = F(x_1, x_2, \ldots, x_n)$, and besides, the obtained model should have the minimal complexity. In particularly, while solving forecasting problem as an output variable Y a forecasting model is used $X(N+K) = f(X(1), \ldots, X(N))$, where K is a value of a forecasting interval.

The constructed model should be adequate according to the initial set of data, and should have the least complexity (Fig. 6.1).

The distinguishing features of the problem are the following:

1. Form of functional dependence is unknown and only model class is determined, for example, polynomial of any degree or Fourier time series.
2. short data samples;
3. time series $x_i(t)$ in general case is non-stationary.

In this case the application of conventional methods of statistical analysis (e.g. regressional analysis) is impossible and it's necessary to utilize methods based on computational intelligence (CI). To this class belongs Group Method of Data Handling (GMDH) developed by acad. Ivakhnenko [1] and extended by his colleagues. GMDH is a method of inductive modeling. The method inherits ideas of biological evolution and its mechanisms:

1. crossing-over of parents and offsprings generation;
2. selection of the best offsprings.

GMDH method belongs to self-organizing methods and allows to discover internal hidden laws in the appropriate object area.

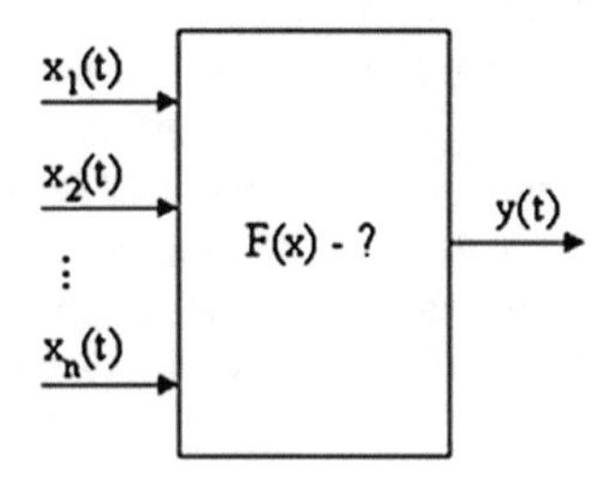

Fig. 6.1 Graphical representation of the problem

The advantages of GMDH algorithms is the possibility of constructing optimal models with a small number of observations and unknown dynamics among variables. This method doesn't demand to know the model structure a priori, the model is constructed by algorithm itself in the process of its run.

6.3 The Basic Principles of GMDH

Let's remind the fundamental principles of GMDH [1, 2]. The full interconnection between input $X(i)$ and output $Y(i)$ in the class of polynomial models may be presented by so-called generalized polynomial of Kolmogorov-Gabor:

$$Y = a_0 + \sum\nolimits_{i=1}^{n} a_i x_i + \sum_{j=1}^{n} \sum_{i \le j} a_{ij} x_i x_j \sum_{i=1}^{n} \sum_{j \le i} \sum_{k \le j} a_{ijk} x_i x_j x_k + \cdots \tag{6.1}$$

where all the coefficients a_0, a_i, a_{ij}, are unknown.

While constructing model (search coefficients values) as a criterion of adequacy the so-called regularity criterion (mean squared error—MSE) is used

$$\overline{\varepsilon^2} = \frac{1}{N} \cdot \sum_{i=1}^{N} (y_i - f(X_i))^2, \tag{6.2}$$

where N is a sample size (number of observations).

It' s demanded to find minimum $\bar{\varepsilon}^2$.

GMDH method is based on the following principles [1].

The principle of multiplicity of models. There is a great number of models providing zero error on a given sample. It's enough simply to raise the degree of the polynomial model. If N nodes of interpolation are available, then it's possible to construct the family of models each of which gives zero error on experimental points $\bar{\varepsilon}^2 = 0$.

The principle of self-organization. Denote S as model complexity. The value of an error depends on the complexity of a model. As the level of complexity S grows the error first drops, attains minimum value and then begins to rise (see Fig. 6.2).

We need to find such level of complexity for which the error would be minimal. In addition if to take into account the action of noise we may make the following conclusions concerning ε:

1. With the increase of noise the optimal complexity $s_0 = \arg\min \bar{\varepsilon}^2$ shifts to the left.
2. With the increase of noise level the value of optimal criterion $\min \bar{\varepsilon}^2(s)$ grows.

Theorem of incompleteness by Geodel: In any formal logical system there are some statements which cannot be proved or refuted using the given system of

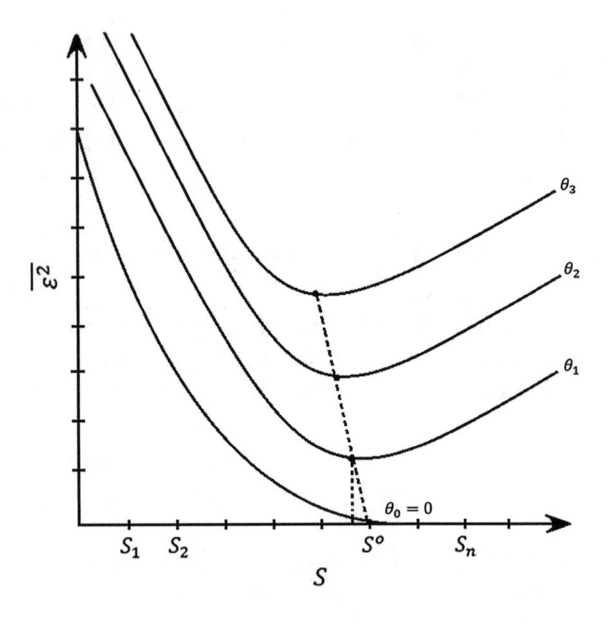

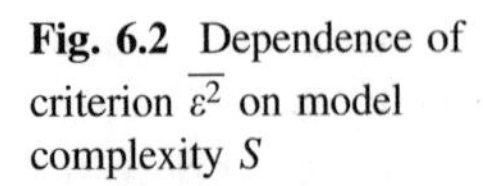
Fig. 6.2 Dependence of criterion $\overline{\varepsilon^2}$ on model complexity S

axioms and staying in the margins of this system. That to prove or refute such statement one need go out this system and use some external information (meta information) which is called "external complement". In our case as external information stands additional sample of data which wasn't used for the finding unknown coefficients of the model.

So one way to overcome incompleteness of sample is to use **principle of external complement** which means that the whole sample should be divided into two parts—training subsample and test subsample. The search of optimal model is performed in such a way:

- at the training sample N_{train} the estimates $\bar{a}_0, \bar{a}_i, \bar{a}_{ij}$, are determined.
- at the test sample N_{test} the best models are selected.

The Principle of Freedom of Choice

For each pair of inputs x_i and x_j partial descriptions are being built (all in all C_n^2) of the form:

- $\bar{Y}_s = \varphi(x_i, x_j) = a_0 + a_i x_i + a_j x_j, \quad s = 1\ldots C_n^2(\text{linear});$ (6.3)

- or $\bar{Y}_s = \varphi(x_i, x_j) = a_0 + x_i + a_j x_j + a_{ii} x_i^2 + a_{ij} x_i x_j + a_{jj} x_j^2,$
$s = 1\ldots C_n^2(\text{quadratic}).$

1. Determine the coefficients of these model using LSM (least square method) at the training sample (i.e. find estimates $\bar{a}_0, \bar{a}_1, \ldots, \bar{a}_j, \ldots, \bar{a}_N, \bar{a}_{11}, \ldots, \bar{a}_{ij}, \ldots, \bar{a}_{NN}$.
2. Further at the test sample for each of these models calculate the value of regularity criterion:

$$\bar{\delta}_s^2 = \frac{1}{N_{test}} \cdot \sum_{k=1}^{N_{testx}} [Y(k) - \bar{Y}_s(k)]^2 \tag{6.4}$$

(where $Y(k)$ is real output value of the kth point of test; $\bar{Y}_s(k)$ is a value of this criterion on kth point obtained by model, N_{test} is a number of points at the test sample); as alternate criterion "unbiasedness" criterion may be used:

$$N_{ub} = \frac{1}{N_1 + N_2} \sum_{k=1}^{N} \left(y_k^* - y_k^{**}\right)^2, \tag{6.5}$$

where the sample is also divided in two parts N_1 and N_2, y_k^* are outputs of the model built on the subsample N_1, y_k^{**} are outputs of model built on subsample N_2, $N = N_1 + N_2$.
3. Determine F (this number is called a freedom of choice) best models using one of these criteria. The selected models y_i are then transferred to the second row of model construction. We search coefficients of new partial descriptions:

$$z_I = \varphi^{(2)}(y_i, y_j) = a_0^{(2)} + a_1^{(2)} y_i + a_2^{(2)} y_j + a_3^{(2)} y_i^2 + a_4^{(2)} y_i y_j + a_5^{(2)} y_j^2.$$

The process at the second row runs in the same way. The selection of the best models is carried out similarly, but $F_2 < F_1$. The process of rows construction repeats more and more till MSE (regularity criterion) falls. If at the mth layer the increase of the error $\bar{\varepsilon}^2$ occurs, the algorithm stops. In this case we find the best model at the preceding layer and then moving backward by its connections find models of preceding layer and successfully passing all the used connections at the end we' ll reach the first layer and find the analytical form of the optimal model (with minimal complexity).

6.4 Fuzzy GMDH. Principal Ideas. Interval Model of Regression

The drawbacks of GMDH are following:

1. GMDH utilizes least squared method (LSM) for finding the model coefficients but matrix of linear equations may be close to degenerate and the corresponding

solution may appear non-stable and very volatile. Therefore, the special methods for regularisation should be used;
2. after application of GMDH point-wise estimations are obtained but in many cases it's needed find interval value for coefficient estimates;
3. GMDH doesn't work in case of incomplete or fuzzy input data.

Therefore, in the last 10 years the new variant of GMDH—fuzzy GMDH was developed and refined which may work with fuzzy input data and is free of classical GMDH drawbacks [3].

In works [3, 4] the linear interval model regression was considered:

$$Y = A_0Z_0 + A_1Z_1 + \cdots + A_nZ_n, \tag{6.6}$$

where A_i is a fuzzy number of triangular form described by pair of parameters $A_i = (\alpha_i, c_i)$, where α_i is interval center, c_i is its width, $c_i \geq 0$.

Then Y is a fuzzy number, parameters of which are determined as follows:

the interval center

$$\alpha_y = \sum \alpha_i z_i = \alpha^T \cdot z, \tag{6.7}$$

the interval width

$$c_y = \sum c_i \cdot |z_i| = c^T |z|. \tag{6.8}$$

In order that the interval model be correct it's necessary that real value of output should belong to the interval of uncertainty described by the following constraints:

$$\begin{cases} \alpha^T z - c^T \cdot |z| \leq y \\ \alpha^T z + c^T \cdot |z| \geq y \end{cases}. \tag{6.9}$$

For example, for the partial description of the kind

$$f(x_i, x_j) = A_0 + A_1x_i + A_2x_j + A_3x_ix_j + A_4x_i^2 + A_5x_j^2 \tag{6.10}$$

it's necessary to assign in the general model (6.6)

$$z_0 = 1\ z_1 = x_i\ z_2 = x_j\ z_3 = x_ix_j\ z_4 = x_i^2\ z_5 = x_j^2$$

Let the training sample be $\{z_1, z_2, \ldots, z_M\}$, $\{y_1, y_2, \ldots, y_M\}$. Then for the model (6.10) to be adequate its necessary to find such parameters (α_i, c_i) $i = \overline{1, n}$, which satisfy the following inequalities:

$$\begin{cases} \alpha^T z_k - c^T \cdot |z_k| \leq y_k \\ \alpha^T z_k + c^T \cdot |z_k| \leq y_k \end{cases}, \quad k = \overline{1, M}. \tag{6.11}$$

Let's formulate the basic requirements for the linear interval model of partial description of a kind (6.10).

It's necessary to find such values of the parameters (α_i, c_i) of fuzzy coefficients for which:

1. real values of the observed outputs y_k would drop in the estimated interval for Y_k;
2. the total width of the estimated interval for all sample points would be minimal.

These requirements lead to the following linear programming problem:

$$\min\left(C_0 \cdot M + C_1 \sum_{k=1}^{M} |x_{ki}| + C_2 \sum_{k=1}^{M} |x_{kj}| + C_3 \sum_{k=1}^{M} |x_{ki}x_{kj}| + \right. \\ \left. + C_4 \sum |x_{ki}^2| + C_5 \sum_{k=1}^{M} \left|x_{kj}^2\right|\right) \tag{6.12}$$

under constraints:

$$a_0 + a_1 x_{ki} + a_2 x_{kj} + a_3 x_{ki} x_{kj} + a_4 x_{ki}^2 + a_5 x_{kj}^2 - \left(C_0 + C_1 |x_{ki}| + C_2 |x_{kj}| \right. \\ \left. + C_3 |x_{ki} x_{kj}| + C_4 |x_{ki}^2| + C_5 \left|x_{kj}^2\right|\right) \leq \mathrm{y_k} \tag{6.13}$$

$$a_0 + a_1 x_{ki} + a_2 x_{kj} + a_3 x_{ki} x_{kj} + a_4 x_{ki}^2 + a_5 x_{kj}^2 + \left(C_0 + C_1 |x_{ki}| + C_2 |x_{kj}| + \right. \\ \left. + C_3 |x_{ki} x_{kj}| + C_4 |x_{ki}^2| + C_5 \left|x_{kj}^2\right|\right) \geq \mathrm{y_k} \tag{6.14}$$

$$k = \overline{1, M}, \quad C_p \geq 0, \quad p = 0, 1, \ldots, 5$$

where k is an index of a point.

As we can easily see the task (6.12)–(6.14) is linear programing (LP) problem. However, the inconvenience of the model (6.12)–(6.14) for the application of standard LP methods is that there are no constraints of non-negativeness for variables α_i. Therefore for its solution it's reasonable to pass to the dual LP problem by introducing dual variables $\{\delta_k\}$ and $\{\delta_{k+M}\}$, $k = \overline{1, M}$. Using simplex-method for the dual problem and after finding the optimal values for the dual variables $\{\delta_k\}$ the optimal solutions $(\alpha_i,\ c_i)$ of the initial direct problem will be also found.

6.5 The Description of Fuzzy Algorithm GMDH

Let's present the brief description of the algorithm FGMDH [2, 3].

1. Choose the general model type by which the sought dependence will be described.
2. Choose the external criterion of optimality (criterion of regularity or non-biasedness).
3. Choose the type of partial descriptions (for example, linear or quadratic one).
4. Divide the sample into training N_{train} and test N_{test} subsamples.
5. Put zero values to the counter of model number k and to the counter of rows r (iterations number).
6. Generate a new partial model f_k (6.10) using the training sample. Solve the LP problem (6.12)–(6.14) and find the values of parameters α_i, c_i.
7. Calculate using test sample the value of external criterion ($N_{ub}^{(r)}$ or $\delta_k^{(2)}(r)$).
8. $k = k + 1$. If $k > C_N^2$ for $r = 1$ or $k > C_F^2$ for $r > 1$, then $k = 1$, $r = r + 1$ and go to step 9, otherwise go to step 6.
9. Calculate the best value of the criterion for models of rth iteration. If $r = 1$, then go to step 6 otherwise, go to step 10.
10. If $|N_{ub}(r) - N_{ub}(r-1)| \leq \varepsilon$ or $\delta^{(2)}(r) \geq \delta^{(2)}(r-1)$, then go 11, otherwise select F best models and assigning $r = r + 1$, $k = 1$, go to step 6 and execute $(r + 1)$th iteration.
11. Select the best model out of models of the previous row (iteration) using external criterion.

6.6 Analysis of Different Membership Functions

In the first papers devoted to fuzzy GMDH [2, 3] the triangular membership functions (MF) were considered. But as fuzzy numbers may also have the other kinds of MF it's important to consider the other classes of MF in the problems of modeling using FGMDH. In the paper [4] fuzzy models with Gaussian and bell-shaped MF were investigated.

Consider a fuzzy set with MF of the form:

$$\mu_B(x) = e^{-\frac{1}{2}\cdot\frac{(x-a)^2}{c^2}}, \tag{6.15}$$

B is a fuzzy number with Gaussian MF. Such fuzzy number is given by pair $\beta = (a, c)$, where a-is a center and c is a value determining the width of the interval (Fig. 6.3).

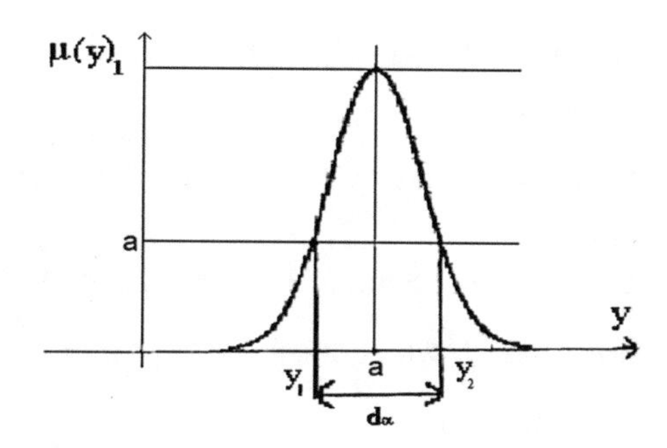

Fig. 6.3 Subset of α level

Let the linear interval model for partial description of FGMDH take the form (6.15). Then the problem is formulated as follows: find such fuzzy numbers B_i, with parameters (a_i, c_i), that:

1. the observation y_k would belong to a given estimate interval for the set $Y(k)$ with degree not less than α, $0<\alpha<1$;
2. the width of estimated interval of the degree α would be minimal;
   The width of the interval of degree α is equal (see Fig. 6.3):

$$d_\alpha = y_2 - y_1 = 2(y_2 - a),$$

where $(y_2 - a)$ is determined from the condition:

$$\alpha = \exp\left\{-\frac{1}{2}\cdot\frac{(y_2 - a)^2}{c^2}\right\} \tag{6.16}$$

Hence $d_\alpha = 2c \cdot \sqrt{-2\ln\alpha}$.

So the goal function may be written as follows:

$$\min\sum_{k=1}^{M} d_\alpha^k = \min\sum_{k=1}^{M} 2C_k\sqrt{-2\ln\alpha} = 2\sqrt{-2\ln\alpha}\min\sum_{k=1}^{M}\sum_{i=1}^{n} C_i\cdot|z_{ik}|. \tag{6.17}$$

As $2\cdot\sqrt{-2\ln\alpha}$ is a positive constant not influencing on the set of the values c_i, which minimize (6.17), it's possible to divide the goal function by this constant and transform the goal function to the initial form (6.12).

Now consider the first demand: $\mu(y_k) \geq \alpha$.

It's equivalent to $\exp\left\{-\frac{1}{2}\cdot\frac{(y_k - a_k)^2}{c_k^2}\right\} \geq \alpha$.

This inequality may be transformed to the following system of inequalities:

$$\begin{aligned} a_k + c_k\sqrt{-2\ln\alpha} \geq y_k \\ a_k - c_k\sqrt{-2\ln\alpha} \leq y_k \end{aligned}. \tag{6.18}$$

Taking into account $\alpha_k = \sum_{i=1}^{n} a_i z_{ki}$; $c_k = \sum_{i=1}^{n} c_i |z_{ki}|$, the problem of finding optimal fuzzy model will be finally transform to the following LP problem:

$$\begin{aligned} \min\Bigg(& C_0 \cdot M + C_1 \sum_{k=1}^{M} |x_{ki}| + C_2 \sum_{k=1}^{M} |x_{kj}| + C_3 \sum_{k=1}^{M} |x_{ki}x_{kj}| + \\ & + C_4 \sum_{k=1}^{M} |x_{ki}^2| + C_5 \sum_{k=1}^{M} \left|x_{kj}^2\right| \Bigg), \end{aligned} \tag{6.19}$$

under constraints:

$$\left.\begin{aligned} a_0 + a_1 x_{ki} + \cdots + a_5 x_{kj}^2 + (C_0 + C_1|x_{ki}| + \cdots + C_5\left|x_{kj}^2\right|) \cdot \sqrt{-2\ln\alpha} \geq y_k \\ a_0 + a_1 x_{ki} + \cdots + a_5 x_{kj}^2 - (C_0 + C_1|x_{ki}| + \cdots + C_5\left|x_{kj}^2\right|) \cdot \sqrt{-2\ln\alpha} \leq y_k \end{aligned}\right\}$$

$$k = \overline{1, M} \tag{6.20}$$

That to solve this problem like in the case with triangular MF pass to the dual LP problem of the form:

$$\max\left(\sum_{k=1}^{M} y_k \cdot \delta_{k+\mathrm{M}} - \sum_{k=1}^{M} y_k \cdot \delta_k\right), \tag{6.21}$$

with constraints of equalities and inequalities:

$$\begin{gathered} \sum_{k=1}^{M} \delta_{k+\mathrm{M}} - \sum_{k=1}^{M} \delta_k = 0, \\ \sum_{k=1}^{M} X_{ki} \cdot \delta_{k+M} - \sum_{k=1}^{M} X_{ki_i} \cdot \delta_k = 0 \\ \cdots \\ \sum_{k=1}^{M} X_{kj}^2 \cdot \delta_{k+M} - \sum_{k=1}^{M} X_{kj}^2 \cdot \delta_k = 0 \end{gathered} \tag{6.22}$$

$$\left.\begin{array}{c}\sum_{k=1}^{M}\delta_k+\sum_{k=1}^{M}\delta_{k+M}\le\frac{M}{\sqrt{-2\ln\alpha}}\\ \sum_{k=1}^{M}|X_{ki}|\cdot\delta_{k+M}+\sum_{k=1}^{M}|X_{ki}|\cdot\delta_k\le\frac{\sum_{k=1}^{M}|X_{ki}|}{\sqrt{-2\ln\alpha}}\\ \cdots\\ \sum_{k=1}^{M}\left|X_{kj}^2\right|\cdot\delta_{k+M}+\sum_{k=1}^{M}\left|X_{kj}^2\right|\cdot\delta_k\le\frac{\sum_{k=1}^{M}\left|X_{kj}^2\right|}{\sqrt{-2\ln\alpha}}\end{array}\right\}\tag{6.23}$$

$$\delta_k\ge 0, k=\overline{1,2M}.$$

Thus, fuzzy GMDH allows to construct fuzzy models and has the following advantages:

1. The problem of optimal model determination is transferred to the problem of linear programming, which is always solvable.
2. There is interval regression model built as the result of method work;
3. There is a possibility of the obtained model adaptation.

6.7 Fuzzy GMDH with Different Partial Descriptions

6.7.1 Investigation of Orthogonal Polynomials as Partial Descriptions

As it well-known from the general GMDH theory, models pretendents are generated on the base of so-called partial description-elementary models of two variables (e.g. linear polynomials) which utilize as arguments the partial descriptions of preceding stages for optimal model synthesis. But the choice of partial description form is not automated and is to be solved the expert.

Usually as partial descriptions linear or quadratic polynomials are used. The alternative to this class of models is application of orthogonal polynomials. The choice of orthogonal polynomials as partial descriptions is determined by the following advantages:

- Owing to orthogonal property the calculation of polynomial coefficients which approximates the simulated process goes faster than for non-orthogonal polynomials.
- The coefficients of polynomial approximating equation don't depend on the degree of initial polynomial model so if apriori the real polynomial degree is not known we may check the polynomials of various degrees and by this property the coefficients obtained for polynomials of lower degrees remain the same after transfer to higher polynomial degrees. This property is the most important during investigation of real degree of approximating polynomial. One of the properties of orthogonal polynomials widely used in this work is the *property of*

almost equal errors. This property means that approximation errors oscillates in the middle of estimation interval between two almost equal bounds (margins). Owing to this the very large errors do not happen, on the contrary, in most cases the error values are small. Therefore the damping of approximation errors occurs.

Main Concepts

Consider the approximating equation for one-dimensional system

$$\bar{y}(x) = b_0F_0(x) + b_1F_1(x) + \cdots + b_mF_m(x), \tag{6.24}$$

where $\bar{y}$ is initial (real) value which is estimated,

$F_v(x)$ is orthogonal polynomial of degree v, which has the orthogonal property, i.e.

$$\sum_{i=0}^{r} F_\mu(x_i)F_v(x_i) = 0, \quad \forall \mu \neq v, \tag{6.25}$$

r is a length of training sample.

In the general case (for continuous variables)

$$\int_a^b \omega(x)F_\mu(x)F_v(x)dx = 0, \tag{6.26}$$

where μ, ν are nonnegative integer numbers (polynomial degree);

$\omega(x)$ is a weighting coefficient.

Investigation of Chebyshev's Orthogonal Polynomials as Partial Descriptions

Chebyshev's orthogonal polynomials in general case have the following form [5]:

$$F_v(\xi) = T_v(\xi) = \cos(v \cdot \arccos \xi), \; -1 \leq \xi \leq 1, \tag{6.27}$$

These polynomials have the following orthogonality property

$$\int_{-1}^{1} \frac{T_\mu(\xi)T_v(\xi)d\xi}{\sqrt{1-\xi^2}} = \begin{cases} 0 & \text{if} \quad \mu \neq \xi; \\ \frac{\pi}{2} & \text{if} \quad \mu = \xi \neq 0;, \\ \pi & \text{if} \quad \mu = \xi = 0. \end{cases} \tag{6.28}$$

where $\sqrt{1-\xi^2}$ is a weighting coefficient $\omega(\xi)$ in the Eq. (6.26).

Let's present several Chebyshev's orthogonal polynomials of lower degrees:

Table 6.1 Testing sample for BP plc (ADR)

Data	BP plc (ADR)
Nov 15, 2011	43.04
Nov 16, 2011	41.78
Nov 17, 2011	41.60

$$T_0(\xi) = 1$$
$$T_1(\xi) = \xi$$
$$T_2(\xi) = 2\xi^2 - 1$$

$$T_{v+1}(\xi) = 2\xi \cdot T_v(\xi) - T_{v-1}(\xi)$$

The approximating Chebyshev's orthogonal polynomial for $\bar{y}$ is obtained on the base of function S minimization.

$$S = \int_{-1}^{1} \omega(\xi)\left(y(\xi) - \sum_{i=0}^{m} b_i T_i(\xi)\right)^2 d\xi. \tag{6.29}$$

where from it follows:

$$b_k = \begin{cases} \frac{1}{\pi}\int\limits_{-1}^{1} \frac{y(\xi)}{\sqrt{1-\xi^2}} d\xi, & k = 0 \\ \frac{2}{\pi}\int\limits_{-1}^{1} \frac{y(\xi)T_k(\xi)}{\sqrt{1-\xi^2}} d\xi, & k \neq 0 \end{cases}. \tag{6.30}$$

Hence the approximating equation is obtained

$$\bar{y}(\xi) = \sum_{k=0}^{m} b_k T_k(\xi). \tag{6.31}$$

As it may be readily seen from the presented expressions coefficient b_k in the Eq. (6.30) doesn't depend on choice of degree m. Thus, the variable m doesn't demand recalculation of b_j, $\forall j \leq m$, while such recalculation is necessary for non-orthogonal approximation.

Determining the Best Degree of Approximating Polynomial

The best degree m^* of approximating may be obtained on the base of hypothesis that investigation results $y(i)$, $i = 1, 2, \ldots, r$, have independent Gaussian distribution in the bounds of some polynomial function $\bar{y}$ of definite degree for example, $m^* + \mu$, where

$$\bar{y}_{m^*+\mu}(x_i) = \sum_{j=0}^{m^*+\mu} b_j x_i^j, \tag{6.32}$$

and a dispersion σ^2 of distribution $y - \bar{y}$ doesn't depend on μ.

It's clear that for very small m (m = 0, 1, 2, …) σ_m^2 decreases as m grows.

As in accordance with previously formulated hypothesis dispersion doesn't depend on m, the best degree m^* is a minimal m, for which $\sigma_m \cong \sigma_{m+1}$.

For determining m^* it's necessary to calculate the approximating polynomials of various degrees. As coefficients b_j in the Eq. (6.32) don't depend on m, the determination of the best degree of polynomial is accelerated.

Constructing Linear Interval Model

Let we have the forecasted variable Y and input variables $x_1, x_2, \ldots x_n$. Let's search the relation between them in the following form

$$Y = A_1 f_1(x_1) + A_2 f_2(x_2) + \cdots + A_n f_n(x_n), \tag{6.33}$$

where A_i is a fuzzy number of triangular type given as $A_i = (\alpha_i, c_i)$, functions f_i are determined so [2]:

$$f_i(x_i) = \sum_{j=0}^{m_i} b_{ij} T_j(x_i). \tag{6.34}$$

The degree m_i of function f_i is determined using hypothesis defined in the preceding section. So if denote $z_i = f_i(x_i)$, we'll get linear interval model in its classical form.

6.7.2 *Investigation of Trigonometric Polynomials as Partial Descriptions*

General Concepts

Let we have periodic function with period P, $P > 0$:

$$g(x + P) = g(x) \tag{6.35}$$

In, general we may consider any periodic function which has period 2π:

$$f(x) = g\left(\frac{Px}{2\pi}\right); \quad f(x + 2\pi) = g\left(\frac{Px}{2\pi} + P\right) = f(x) \tag{6.36}$$

Let function $f(x)$ be periodic with period 2π defined at the interval $[-\pi, \pi]$, and its derivative $f'(x)$ is also defined at the interval $[-\pi, \pi]$. Then the following equality holds

$$S(x) = f(x);\ \forall x \in [-\pi; \pi] \tag{6.37}$$

where

$$S(x) = \frac{a_0}{2} + \sum_{j=1}^{\infty} (a_j \cos(jx) + b_j \sin(jx)). \tag{6.38}$$

Coefficients a_j, b_j are calculated by Euler formulas

$$\begin{aligned} a_j &= \frac{1}{\pi} \int_{-\pi}^{\pi} f(x) \cos(jx) dx;, \\ b_j &= \frac{1}{\pi} \int_{-\pi}^{\pi} f(x) \sin(jx) dx; \end{aligned} \tag{6.39}$$

Trigonometric Polynomials

Definition. A trigonometric polynomial of the degree M is called the following polynomial

$$T_M(x) = \frac{a_0}{2} + \sum_{j=1}^{M} (a_j \cos(jx) + b_j \sin(jx)). \tag{6.40}$$

The following theorem is true stating that exists such $\exists M,\ 2M < N$, which minimizes the expression:

$$\sum_{j=1}^{N} (f(x_i) - T_M(x_i))^2. \tag{6.41}$$

Hence the coefficients of corresponding trigonometric polynomial are determined by formulas:

$$\begin{aligned} a_j &= \frac{2}{N} \sum_{i=1}^{N} f(x_i) \cos(jx_i); \\ b_j &= \frac{2}{N} \sum_{i=1}^{N} f(x_i) \sin(jx_i). \end{aligned} \tag{6.42}$$

Constructing a Linear Interval Model

Let it be the variable Y to be forecasted and input variables $x_1, x_2, \ldots x_n$. Let's search the dependence among them in the form:

$$Y = A_1 f_1(x_1) + A_2 f_2(x_2) + \cdots + A_n f_n(x_n), \tag{6.43}$$

where A_i is a fuzzy number of triangular type given as $A_i = (\alpha_i, c_i)$, functions f_i are determined in such a way:

$$f_i(x_i) = T_{M_i}(x_i). \tag{6.44}$$

The degree M_i of function f_i is determined by the theorem described in the preceeding section.

Therefore if to assign $z_i = f_i(x_i)$, the linear interval model will be obtained in its classical form.

Note before start of modeling initial data should be normalized at the interval $[-\pi, \pi]$

6.7.3 The Investigation of ARMA Models as Partial Descriptions

General Concepts

The choice autoregression models with moving average (ARMA) for investigation is based on their wide applications for modeling and forecasting in economy.

Consider the classical ARMA (n, m)-model. As a rule it is described in the form of difference equation:

$$\begin{aligned} y(k) + a_1 y(k-1) + \cdots + a_n y(k-n) = \\ = b_1 u(k-1) + \cdots + b_m u(k-m) + v(k). \end{aligned} \tag{6.45}$$

or in the form:

$$\begin{aligned} y(k) = b_1 u(k-1) + \cdots + b_m u(k-m) - \\ - a_1 y(k-1) - \cdots - a_n y(k-n). \end{aligned} \tag{6.46}$$

In vector form this model may be presented, as follows

$$y(k) = \theta^T \psi(k) + v(k), \tag{6.47}$$

where $\theta^T = [a_1 \ldots a_n \, b_1 \ldots b_m]$ is vector of model parameters;

$\psi^T(k) = [-y(k-1) - y(k-2) - \cdots - y(k-n) \, u(k-1) \ldots u(k-m)]$ is measurement vector;

$y(k)$ is dependent variable, $u(k)$ is independent variable, $v(k)$ is random noise. Model parameter vector is estimated with any identification method.

Constructing Linear Interval Model

Let θ be a vector of fuzzy parameters (triangular form). Then in the Eq. (6.47) the noise is missing and linear interval model in its classical form is obtained

$$Y = A_0 z_0 + A_1 z_1 + \cdots + A_p z_p,$$

where $p = n + m$;

$$A_i = \begin{cases} a_i, & i = 1, \ldots, n \\ b_{i-n}, & i = n+1, \ldots, n+m \end{cases}.$$

For constructing the model of optimal complexity (in our case it's the determination of n, m and model parameters) the multilevel polynomial GMDH algorithm may be used.

At the first level a set fuzzy model-pretendents is generated in the class ARMA (i,j), where $i = 1, \ldots, n_0$; $j = 1, \ldots, m_0$, n_0, m_0 are varied. From this set using selection procedure the best models are selected and transferred to the next level of synthesis etc. Evidently, if to use ARMA—models as arguments in the next level model also ARMA-models will be obtained.

As a result of GMDH algorithm application ARMA-model $(n*, m*)$ will be obtained, where n* and m* are optimal values in the sense of GMDH.

6.8 Adaptation of Fuzzy Models Coefficients

While forecasting by self-organizing methods (fuzzy GMDH, in particular) the problem arises connected with necessity of huge amount of repetitive calculations in case of the training sample size increase or while forecasting in real time when it's needed to correct the obtained model in accordance with new available data. In this work for solution of this problem the following adaptation methods are suggested: stochastic approximation and recurrent LSM method. The possibility of choice one of several adaptation methods enables to carry out more wide explorations and make grounded recommendations concerning the properties and application sphere of each method.

Taking into account new information obtained while forecasting may be done by two approaches. The first one is to correct parameters of a forecasting model with new data assuming that model structure didn't change. The second approach consists in adaptation of not only model parameters but its optimal structure as well. This way demands the repetitive use of full GMDH algorithm and is connected with huge volume of calculations. The second approach is used if adaptation of parameters doesn't provide good forecast and the new real output values don't drop in the calculated interval for its estimate.

In our consideration the first approach is used based on adaptation of FGMDH model parameters with new available data. Here the recursive identification methods are preferably used, especially the recursive least squared method (LSM). In this method the parameters estimates on the next step are determined on the base of estimates on the previous step, model error and some information matrix which is modified during all estimation process and therefore contains data which may be used at the next steps of adaptation process.

Hence, model coefficients adaptation will be simplified substantially. If to store information matrix obtained while identification of optimal model using fuzzy GMDH then for model parameters adaptation it will be enough to fulfill only one iteration by recursive LSM method.

6.8.1 Application of Stochastic Approximation Method for Adaptation

Consider a discrete stationary system

$$y_k = \Phi(U_k, p), \tag{6.48}$$

where U_k is a vector of input model variables,

p is a parameters vector to be estimated,

Φ is a given function.

The vector estimate $\widehat{p}_{n+1}$ on the step$(n + 1)$ is defined in such a way:

$$\widehat{p}_{n+1} = \widehat{p} - \rho_n \Psi_n$$

where Ψ_n is a function which depends on U_n and y,

ρ_n is a sequence of scalar correcting coefficients.

Let $J_n(\hat{p}_n)$ be scalar quality index of identification which is defined so [4]:

$$J_n(\hat{p}_n) = \frac{1}{2}(y_n - \Phi(U_n, \hat{p}_n))^2 \tag{6.49}$$

Then

$$\Psi_n = \frac{\partial J_n(\widehat{p}_n)}{\partial p_n} = \begin{pmatrix} \dfrac{\partial J_n(\widehat{p}_n)}{\partial p_{n_1}} \\ \cdots \\ \dfrac{\partial J_n(\widehat{p}_n)}{\partial p_{n_m}} \end{pmatrix}. \tag{6.50}$$

Here $\widehat{p}_{n_1}, \ldots, \widehat{p}_{n_m}$ are components of a vector $\widehat{p}_n$.

From the expression (6.50) we obtain:

$$\frac{\partial J_n(\widehat{p}_n)}{\partial p_n} = (\Phi(U_n, \widehat{p}_n) - y_n) \cdot g(U_n, \widehat{p}_n), \tag{6.51}$$

where

$$g(U_n, \widehat{p}_n) = \frac{\partial \Phi(U_n, \widehat{p}_n)}{\partial \widehat{p}_n}. \tag{6.52}$$

In the case of linear interval model we have [6]:

$$\Phi(U_n, \widehat{p}_n) = \widehat{p}_n^T U_n \tag{6.53}$$

where $\widehat{p}_n = (\alpha_1\ \alpha_2 \ldots\ \alpha_m)$ or $\widehat{p}_n = (C_1\ C_2 \ldots\ C_m)$;
$U_n = (z_1\ z_2 \ldots\ z_m)$ or $U_n = (|z_1|\ |z_2| \ldots\ |z_m|)$;
for adaptation of α_i or C_i correspondingly.
Then

$$g(U_n, \widehat{p}_n) = \frac{\partial \Phi(U_n, \widehat{p}_n)}{\partial \widehat{p}_n} = U_n, \tag{6.54}$$

and we obtain:

$$\Psi_n = \left(\Phi(U_n, \widehat{p}_n) - y_n\right) \cdot U_n. \tag{6.55}$$

The first of scalar correcting coefficients is determined so:

$$\rho_1 = \frac{1}{\left\| g(U_n, \widehat{p}_n) \right\|^2}. \tag{6.56}$$

In the case of linear interval model:

$$\rho_1 = \frac{1}{\|U_n\|^2} = \frac{1}{U_n^T U_n}. \tag{6.57}$$

Thus, stochastic approximation represents the convenient tools for recursive identification of model parameters which may be linearized and applied for model coefficients adaptation obtained by fuzzy GMDH. Adaptation procedure applies to parameters α_i, as well as to C_i.

In software implementation coefficients adaptation is fulfilled by expert in step-mode forecasting or by automatic forecasting procedure for one or several steps ahead using the following principle: the adaptation takes place only in the case when the real value of output variable drops out of interval bounds for estimate obtained by a model.

The main disadvantage of this algorithm is that while constructing the first of the sequence of correcting coefficients the information obtained in the process of parameters estimation using fuzzy GMDH algorithm is not used.

Therefore it's desirable to modify the algorithm of parameters estimation so that output of fuzzy GMDH algorithm for optimal model altogether with the data accumulated during the process of estimation were used. Such approach is considered in the next section.

6.8.2 The Application of Recurrent LSM for Model Coefficients Adaptation

Consider the following model:

$$y(k) = \theta^T \Psi(k) + v(k), \tag{6.58}$$

where $y(k)$ is a dependent (output) variable,

$\Psi(k)$ is a measurements vector,
$v(k)$ are random disturbances,
θ—is a parameters vector to be estimated.

The parameters estimate θ at the step N is performed due to such formula [2, 4, 7]:

$$\widehat{\theta}(N) = \widehat{\theta}(N-1) + \gamma(N)[y(N) - \widehat{\theta}^T(N-1)\Psi(N)], \tag{6.59}$$

where $\gamma(N)$—is a coefficients vector which is determined by formula:

$$\gamma(N) = \frac{P(N-1)\Psi(N)}{1 + \Psi^T(N)P(N-1)\Psi(N)}, \tag{6.60}$$

where $P(N-1)$ is so-called an "information matrix", determined by formula:

$$P(N-1) = P(N-2) - \frac{P(N-2)\psi(N-1)\,\psi^T(N-1)P(N-2)}{1 + \psi^T(N-1)P(N-2)\psi(N-1)}. \tag{6.61}$$

Linear Interval Model Parameters Adaptation

As one can easily see in (6.61), the information matrix may be obtained independent on parameters estimation process and parallel to it. The adaptation of two parameter vectors $\theta_1^T = [\alpha_1, \ldots, \alpha_m]$; $\theta_2^T = [C_1, \ldots, C_m]$; is performed in such a way using the formulas [4]:

$$\begin{aligned}\hat{\theta}_1(N) &= \hat{\theta}_1(N-1) + \gamma_1(N)\left[y(N) - \hat{\theta}_1^T(N-1)\Psi_1(N)\right] \\ \hat{\theta}_2(N) &= \hat{\theta}_2(N-1) + \gamma_2(N)\left[y_c(N) - \hat{\theta}_2^T(N-1)\Psi_2(N)\right]\end{aligned}, \quad (6.62)$$

where $\Psi_1^T = [z_1, \ldots, z_m]$; $\Psi_2^T = [|z_1|, \ldots, |z_m|]$,

$$y_c(N) = |y(N) - \theta_1^T(N-1)\Psi_1(N)|.$$

6.9 The Application of GMDH for Forecasting at the Stock Exchange

Consider the application of GMDH and fuzzy GMDH for forecasting at the stock exchange NYSE. As input variables the following stock prices at NYSE were chosen: close prices of companies Hess Corporation, Repsol YPF, S.A. (ADR), Eni S.p.A. (ADR), Exxon Mobil Corporation, Chevron Corporation, and TOTAL S.A. (ADR) [5].

As an output variable close prices of British Petroleum—BP plc (ADR) were chosen.

The initial data for training were taken in the period from 20 September 2011 year till 14 November 2011 year. For testing were used the data of company BP plc (ADR) since 15 November 2011 to 17 November 2011 year presented in the Table 6.1.

The input variables were chosen after correlation matrix analysis.

The second problem was forecasting industrial index Dow-Jones Average. As the input variables in this problem were taken close stock prices of the following companies which form it: American Express Company, Bank of America, Coca-cola, McDonald's, Microsoft Corp., Johnson & Johnson, Intel Corp.

The training sample data were also taken in the period since 20 September to 14 November 2011 year. For test sample were taken data of Dow Jones Industrial Average since 15 November 2011 year to 17 November 2011 year (Table 6.2).

For experimental investigations classical GMDH and fuzzy GMDH were used. The comparison with cascade neo-fuzzy neural network [5] was also carried out during the experiments. For investigations the percentage of training sample was chosen 50 %, 70 %, 90 %. Freedom of choice was taken such $F = 5, 6$.

Table 6.2 Testing data for Dow Jones Industrial Average

Data	Dow Jones Industrial Average
Nov 15 2011	12096.16
Nov 16 2011	11905.59
Nov 17 2011	11770.73

Table 6.3 Forecast criteria for shares of BP plc (ADR)

Step of forecast	Criterion	Percentage of training sample		
		50 %	70 %	90 %
1	MSE test	0.851172	0.195661	0.222304
	MSE	0.556938	0.285995	0.269439
	MAPE test	1.828592	0.952620	1.122068
	MAPE	1.405313	1.034865	0.994513
	DW	0.604306	2.171417	2.222481
	R-square	0.650841	0.969484	0.906763
	AIC	2.185818	1.590313	1.514696
	BIC	−3.023414	−2.349042	−2.244520
	SC	2.235525	1.640021	1.564404
2	MSE test	1.734720	0.433501	1.052307
	MSE	0.974776	0.425016	0.391586
	MAPE test	2.199273	1.274355	1.718034
	MAPE	1.412474	1.227474	1.090535
	DW	0.587306	1.913641	2.219508
	R-square	0.706036	1.046596	0.906839
	AIC	2.223931	1.874082	1.514855
	BIC	−3.059720	−2.698769	−2.244745
	SC	2.273639	1.923790	1.564562
3	MSE test	2.981124	1.093896	3.238221
	MSE	1.723194	0.675693	0.839414
	MAPE test	2.411189	1.788249	2.352890
	MAPE	1.571035	1.496220	1.241341
	DW	0.587306	1.952832	2.219508
	R-square	0.706036	1.023470	0.906839
	AIC	2.223931	1.923849	1.514855
	BIC	−3.059720	−2.754263	−2.244745
	SC	2.273639	1.973556	1.564562

For fuzzy GMDH triangular, Gaussian and bell-shaped membership functions were used. For Gaussian and bell-shaped membership functions the following level values were taken $a = 0.3; 0.5; 0.7; 0.9$.

For constructing models the following four partial descriptions were used:

- a linear model of the form:

$$f(x_i, x_j) = A_{01} + A_{11} * x_i + A_{12} * x_j,$$

Table 6.4 Forecast criteria for Dow Jones Industrial Average

Step of forecast	Criterion	Percentage of training sample		
		50 %	70 %	90 %
1	MSE test	40494.427565	33109.754270	40286.725033
	MSE	26900.763617	62553.804951	26432.081096
	MAPE test	1.462066	1.363148	1.405372
	MAPE	1.149183	1.809130	1.191039
	DW	1.917430	1.013536	1.862203
	R-square	0.990922	0.804774	0.849329
	AIC	12.753417	13.551054	12.729001
	BIC	−7.360808	−7.482613	−7.356959
	SC	12.796073	13.593710	12.771656
2	MSE test	41546.293300	31602.995442	43907.693743
	MSE	27793.341807	61328.007065	32746.387552
	MAPE test	1.481603	1.374860	1.521950
	MAPE	1.167355	1.819287	1.280730
	DW	1.917020	1.013536	1.669879
	R-square	0.989976	0.804774	0.811488
	AIC	12.753849	13.551054	12.935378
	BIC	−7.360875	−7.482613	−7.389255
	SC	12.796504	13.593710	12.978034
3	MSE test	58651.409938	34923.660449	56028.016943
	MSE	37306.640205	61736.892393	34792.320367
	MAPE test	1.584580	1.394577	1.534979
	MAPE	1.230185	1.830302	1.288520
	DW	1.917430	1.013536	1.669879
	R-square	0.990922	0.804774	0.811488
	AIC	12.753417	13.551054	12.935378
	BIC	−7.360808	−7.482613	−7.389255
	SC	12.796073	13.593710	12.978034

- a squared model:

$$f(x_i, x_j) = A_{01} + A_{11} * x_i + A_{12} * x_j + A_{21} * x_i^2 + A_{22} * x_j^2 + A_{23} x_i * x_j,$$

- Fourier polynomial of the first degree:

$$f(x_i, x_j) = (A_{01})/2 + A_{11} * \cos(x_i) + A_{12} * \sin(x_j),$$

- Chebyshev's polynomial of the second degree:

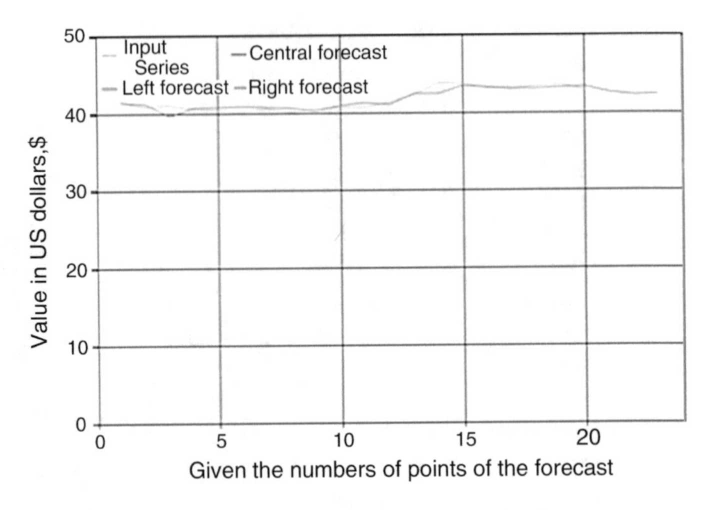

Fig. 6.4 Forecast for share prices of BP plc (ADR) at 3 steps ahead

Table 6.5 Forecast criteria for shares BP plc (ADR) closing prices for fuzzy GMDH for each forecast step

Step of forecast	Criterion	Percentage of training sample		
		50 %	70 %	90 %
1	MSE test	1.248355	0.793523	0.317066
	MSE	0.748864	0.612827	0.485599
	MAPE test	2.041366	2.065155	1.386096
	MAPE	1.452656	1.723657	1.505965
	DW	0.686478	1.763043	1.839065
	R-square	1.064526	0.938058	1.008042
	AIC	2.456985	2.228267	2.053752
	BIC	−3.268440	−3.063808	−2.892253
	SC	2.506693	2.277974	2.103459
2	MSE test	1.256828	1.871883	2.440575
	MSE	0.728499	0.989650	0.795699
	MAPE test	2.072085	2.367006	2.426847
	MAPE	1.468987	1.845798	1.656084
	DW	0.686478	1.795021	1.839065
	R-square	1.064526	0.874163	1.008042
	AIC	2.456985	2.147717	2.053752
	BIC	−3.268440	−2.986448	−2.892253

(continued)

Table 6.5 (continued)

Step of forecast	Criterion	Percentage of training sample		
		50 %	70 %	90 %
	SC	2.506693	2.197424	2.103459
3	MSE test	1.317349	5.408313	6.385595
	MSE	0.788241	2.388737	1.625459
	MAPE test	2.129226	3.095381	3.169584
	MAPE	1.496548	2.149238	1.834261
	DW	0.686478	1.762895	1.839065
	R-square	1.064526	0.945517	1.008042
	AIC	2.456985	2.229614	2.053752
	BIC	−3.268440	−3.065077	−2.892253
	SC	2.506693	2.279321	2.103459

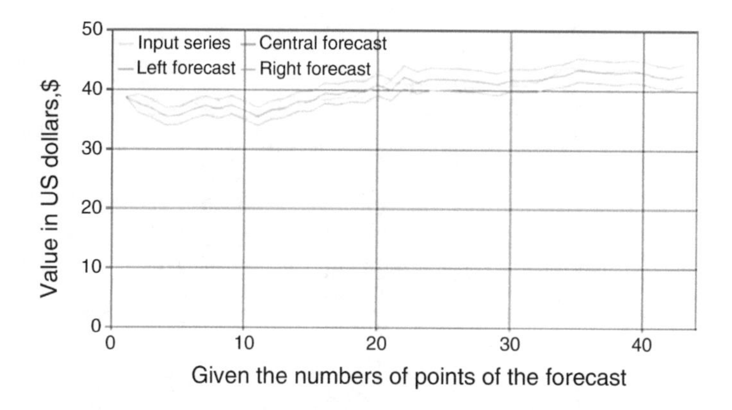

Fig. 6.5 Forecasts flow charts for fuzzy GMDH with linear partial descriptions, 3 steps ahead

$$f(x_i, x_j) = (A_{01}) + A_{11} * x_i + A_{12} * (x_j^2 - 1),$$

where A_i is a fuzzy number with triangular, Gaussian and bell-shaped membership function.

Experimental Investigations of Inductive Modeling Methods

1. In the first experiments the investigations of classical GMDH were carried out.

As partial descriptions were taken linear model, squared model, Chebyshev's polynomial and Fourier polynomial.

All the forecasts were carried out with change of training sample size, namely 50 %, 70 %, 90 %. The freedom of choice varied so $F = 5, 6, 7$. But as the results of experiments had shown the fluctuations of forecast values differed only at the 3rd

Table 6.6 Forecast criteria for each forecast step Dow Jones Industrial Average for fuzzy GMDH

Step of forecast	Criterion	Percentage of training sample		
		50 %	70 %	90 %
1	MSE test	37900.205497	37035.864923	37221.489676
	MSE	25676.116414	62162.341245	43498.250784
	MAPE test	1.379057	1.207524	1.425909
	MAPE	1.137631	1.705561	1.627755
	DW	1.910164	1.265273	1.338295
	R-square	0.920628	0.928216	0.769091
	AIC	12.708769	13.569508	13.219191
	BIC	−7.353765	−7.485345	−7.432833
	SC	12.751424	13.612164	13.261847
2	MSE test	37541.139628	39456.166212	38884.456009
	MSE	25793.209060	61020.777256	43549.202548
	MAPE test	1.379904	1.230402	1.457562
	MAPE	1.143661	1.711295	1.628665
	DW	1.910164	1.265273	1.338295
	R-square	0.920628	0.928216	0.769091
	AIC	12.708769	13.569508	13.219191
	BIC	−7.353765	−7.485345	−7.432833
	SC	12.751424	13.612164	13.261847
3	MSE test	44442.485544	44478.491577	79402.073988
	MSE	29782.905175	63478.044185	49364.835751
	MAPE test	1.430837	1.289476	1.678009
	MAPE	1.176753	1.719494	1.651875
	DW	1.910164	1.265273	1.338295
	R-square	0.920628	0.928216	0.769091
	AIC	12.708769	13.569508	13.219191
	BIC	−7.353765	−7.485345	−7.432833
	SC	12.751424	13.612164	13.261847

or 4th digit after comma with various F. Thefore, in the succeeding experiments the freedom of choice was taken equal to $F = 6$.

For forecasts accuracy analysis the errors were calculated and below in the Tables 6.3 and 6.4 the following criteria of forecast quality are presented: MSE for test sample, MSE for full sample, MAPE for full and test sample, Durbin-Wattson criterion (DW), R-squared, Akaike criterion (AIC), Bayes information criterion (BIC), and Shwarz criterion (SC). MAPE criterion was taken in percentages.

In the Fig. 6.4 the forecast of BP plc (ADR) shares prices at 3 steps for F = 6 and 70 % of training sample is presented (Table 6.4).

Table 6.7 Comparative forecasting results for BP plc (ADR) share prices

Step of forecast	Criteria	Neo-fuzzy neural network	Forecast results for GMDH Partial description (PD)				Forcast results for fuzzy GMDH Partial description			
			Linear	Quadratic	Fourier polynomial	Chebyshev's polynomial	Linear	Quadratic	Fourier polynomial	Chebyshev's polynomial
1	MSE	5.231	0.285	1.905	0.859	0.365	0.481	0.130	1.691	0.757
	MAPE	4.521	1.034	1.965	1.624	1.114	1.374	0.813	2.960	1.459
2	MSE	6.111	0.425	3.090	1.094	0.366	0.498	0.150	1.742	1.029
	MAPE	5.398	1.227	2.916	1.814	1.115	1.481	0.818	2.977	1.584
3	MSE	7.490	0.675	4.978	2.144	0.523	0.572	0.308	2.183	1.505
	MAPE	6.521	1.496	4.434	2.050	1.320	1.494	0.908	3.024	1.681

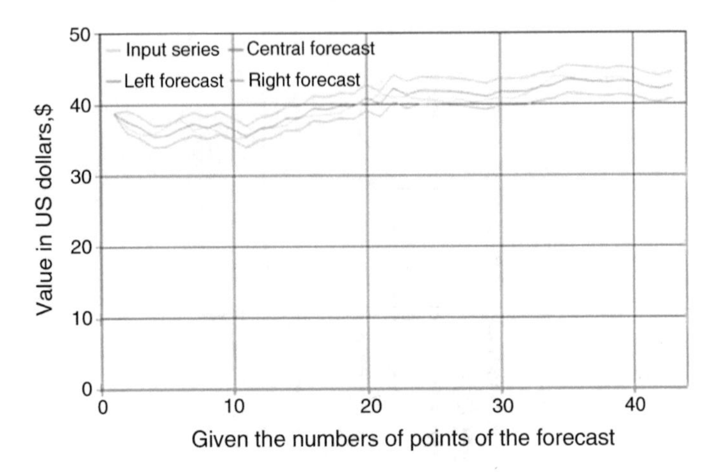

Fig. 6.6 Index Dow Jones I. A. forecast results at 3 steps ahead with FGMDH

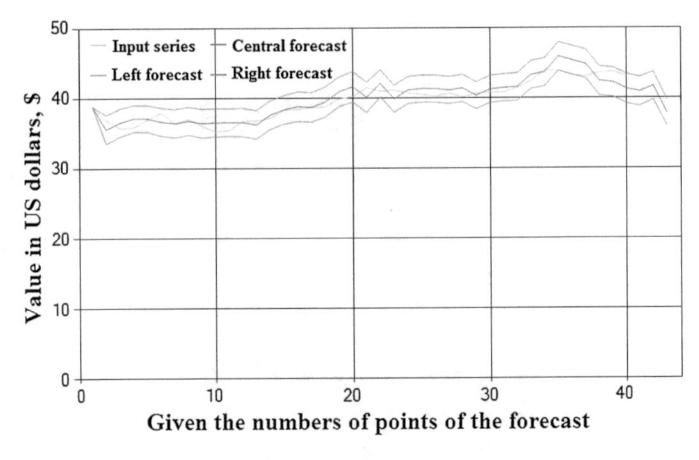

Fig. 6.7 Forecast results for BP plc (ADR shares by FGMDH with quadratic partial descriptions, 3 steps ahead

Fuzzy GMDH

In this section the forecasting results for fuzzy GMDH are presented. As partial descriptions linear model, squared model, Chebyshev's polynomial and Fourier polynomial were taken. As membership functions triangular, Gaussian and bell-shaped membership function were used. All the forecasts were carried out with change of training sample size, namely 50 %, 70 %, 90 %.

In the Fig. 6.5 the flow charts of forecasts made by fuzzy GMDH are presented for 3 steps ahead with F = 6, training sample size—70 %, and Gaussian MF.

Significance level $\alpha = 0.7$.

Table 6.8 Comparative forecasting results for index Dow Jones I. A

Step of forecast	Criteria	Neo-fuzzy neural network	Forecast results for GMDH Partial description (PD)				Forcast results for fuzzy GMDH Partial description			
			Linear	Quadratic	Fourier polynomial	Chebyshev's polynomial	Linear	Quadratic	Fourier polynomial	Chebyshev's polynomial
1	MSE	514561	26900	38225	40142	23818	25176	21332	42205	24464
	MAPE	5.231	1.149	1.298	1.445	1.111	1.137	1.046	1.487	1.125
2	MSE	584371	27793	39460	40930	23978	25793	223491	59059	24767
	MAPE	5.992	1.167	1.322	1.445	1.119	1.143	1.098	1.614	1.144
3	MSE	624501	37306	50471	41720	27337	29782	38291	63900	24910
	MAPE	6.179	1.230	1.386	1.460	1.157	1.176	1.099	1.623	1.160

Table 6.9 MSE comparison for different methods of experiment 1

	GMDH	FGMDH	FGMDH with fuzzy inputs, Triangular MF	FGMDH with fuzzy inputs, Gaussian MF
MSE	0.1129737	0.0536556	0.055557	0.028013

For forecasts accuracy analysis the errors were calculated and in the Tables 6.5 and 6.6 the following criteria of forecast quality are presented: MSE for test sample, MSE for full sample, MAPE for full and test sample, Durbin-Wattson criterion, R-squared, Akaike criterion, Bayes information criterion (BIC), and Shwartz criterion. Criteria were calculated for each forecast step using a test sample. The results for BP plc (ADR) shares are presented in the Table 6.5 and for Dow Jones Industrial Average in the Table 6.6.

On the Figs. 6.6 and 6.7 the flow charts of forecasts for shares BP plc (ADR) and Dow Jones I. A obtained by fuzzy GMDH are presented for 3 steps ahead with $F = 6$, a training sample size 70 % and Gaussian MF.

Significance level $a = 0.7$.

Further experiments for forecasting share prices of BP plc (ADR) and Dow Jones I. A. were carried out with application of GMDH and fuzzy GMDH with different partial descriptions: linear model, squared model, Chebyshev's polynomials and Fourier polynomials and with application of cascade neo-fuzzy neural networks as well. [6].

The final experimental results of forecasts at 1, 2 and 3 steps ahead with aforesaid methods for share prices of British Petroleum—BP plc (ADR) are presented in the Table 6.7, and for index Dow Jones Industrial Average—in the Table 6.8.

Judging from presented criteria estimates the best results were obtained with fuzzy GMDH with quadratic partial descriptions (PD), 70 % training sample and Gaussian membership function. The worst results were obtained using Fourier polynomials as PD. The both GMDH methods, classical and fuzzy, have shown the high forecast accuracy. If to compare the accuracy of both methods with linear partial descriptions then linear model by GMDH has shown more accurate results. But with all used PD more accurate forecasts were obtained using fuzzy GMDH with quadratic partial descriptions. At the same time cascade neo-fuzzy neural network gave the worse results than both GMDH methods.

The best results were obtained with fuzzy GMDH using quadratic partial descriptions with bell-shaped membership functions and 50 % training sample size. The worst results were obtained with Fourier polynomial as partial descriptions. The use of Chebyshev's polynomial as PD in classical GMDH has shown the best results while in fuzzy GMDH the most accurate estimates were obtained with linear and quadratic MF.

6.10 FGMDH Model with Fuzzy Input Data

Let's consider the generalization of fuzzy GMDH for case when input data are also fuzzy. Then a linear interval regression model takes the following form [8–10]:

$$Y = A_0 Z_0 + A_1 Z_1 + \cdots + A_n Z_n, \tag{6.63}$$

where A_i is a fuzzy number, which is described by three parameters $A_i = (\underline{A_i}, \breve{A}_i, \overline{A_i})$, $\breve{A}_i$ is an interval center, $\bar{A}_i$ is an upper border of the interval, A_i is a lower border of the interval, and Z_i is also a fuzzy number, which is determined by parameters $\left(\underline{Z}_i, \breve{Z}_i, \bar{Z}_i\right)$, $\underline{Z}_i$ is a lower border, $\breve{Z}_i$ is a center, $\bar{Z}_i$ is an upper border of a fuzzy number.

Then Y is a fuzzy number, which parameters are defined as follows (in accordance with L-R numbers multiplying formulas):

Center of interval $\breve{y} = \sum \breve{A}_i * \breve{Z}_i$,

Deviation in the left part of the membership function:

$$\breve{y} - \underline{y} = \sum \left(\left|\breve{A}_i\right| * (\breve{Z}_i - \underline{Z}_i) + (\breve{A}_i - \underline{A}_i) * \left|\breve{Z}_i\right|\right), \tag{6.64}$$

Lower border of the interval

$$\underline{y} = \sum \left(\breve{A}_i * \breve{Z}_i - \left|\breve{A}_i\right| * (\breve{Z}_i - \underline{Z}_i) - (\breve{A}_i - \underline{A}_i) * \left|\breve{Z}_i\right|\right)$$

Deviation in the right part of the membership function:

$$\bar{y} - \breve{y} = \sum \left(\left|\breve{A}_i\right| * (\bar{Z}_i - \breve{Z}_i) + \left|\breve{Z}_i\right| * (\bar{A}_i - \breve{A}_i)\right). \tag{6.65}$$

Upper border of the interval

$$\bar{y} = \sum \left(\left|\breve{A}_i\right| * (\bar{Z}_i - \breve{Z}_i) + \left|\breve{Z}_i\right| * (\bar{A}_i - \breve{A}_i) + \breve{A}_i * \breve{Z}_i\right). \tag{6.66}$$

For the interval model to be correct, the real value of input variable Y should lay in the interval got by the method application.

So, the general requirements to estimation linear interval model are to find such values of parameters $A_i = (\underline{A_i}, A_i, \overline{A_i})$ of fuzzy coefficients, which allow [8–10]:

1. observed values y_k should lay in the estimation interval for Y_k;
2. total width of the estimation interval should be minimal.

Input data for this task is $Z_k = [Z_{ki}]$-input training sample, and output values y_k

are known, $k = \overline{1,M}$, M is the number of observation points.
Triangular membership functions are considered:
Quadratic partial descriptions were chosen:

$$f(x_i, x_j) = A_0 + A_1x_i + A_2x_j + A_3x_ix_j + A_4x_i^2 + A_5x_j^2.$$

6.10.1 FGMDH with Fuzzy Input Data and Triangular Membership Functions

There are more than ten typical forms of curves to specify the membership functions. But the most wide spread are triangular, trapezoidal and Gaussian membership functions. Triangular membership function is described by three numbers (a, b, c), and its value in point x is computed as the following expression shows [6]:

$$\mu(x) = \begin{cases} 1 - \dfrac{b-x}{b-a}, & a \le x \le b \\ \dfrac{x-c}{b-c}, & b \le x \le c \\ 0, & \textit{in other cases} \end{cases}$$

If $b - a = c - b$, then it is the case of symmetric triangular membership function.

Triangular membership function is a special case of trapezoidal membership function, which is described by four parameters (a, b, c, d) while $b = c$. Its value in point x is computed as follows [6]:

$$\mu(x) = \begin{cases} 1 - \dfrac{b-x}{b-a}, & a \le x \le b \\ 1, & b \le x \le c \\ 1 - \dfrac{x-c}{d-c}, & c \le x \le d \\ 0, & \textit{in other cases} \end{cases}$$

Now let's consider the linear interval regression model:

$$Y = A_0Z_0 + A_1Z_1 + \cdots + A_nZ_n, \tag{6.67}$$

where A_i is a fuzzy number of triangular shape, which is described by three parameters $A_i = (\underline{A}_i, \breve{A}_i, \overline{A}_i)$, where a_i is a center of the interval, $\overline{A}_i$—its upper border, $\underline{A}_i$—its lower border.

Consider the case of symmetrical membership function for parameters A_i, so they can be described by the pair of parameters (a_i, c_i).

$\underline{A}_i = a_i - c_i$, $\overline{A}_i = a_i + c_i$, c_i—interval width, $c_i \ge 0$,

Z_i is also a fuzzy number of triangular shape, which is defined by parameters $(\underline{Z_i}, \breve{Z}_i, \overline{Z_i})$, $\underline{Z_i}$ is a lower border, $\breve{Z}_i$ is a center, $\overline{Z_i}$—an upper border of fuzzy number.

Then Y is a fuzzy number, which parameters are defined as follows:

the center of the interval $\breve{y} = \sum a_i * \breve{Z}_i$,
the deviation in the left part of the membership function:

$$\breve{y} - \underline{y} = \sum \left(a_i * (\breve{Z}_i - \underline{Z}_i) + c_i \left|\breve{Z}_i\right|\right),$$

thus, the lower border of the interval

$$\underline{y} = \sum \left(a_i * \underline{Z}_i - c_i \left|\breve{Z}_i\right|\right). \tag{6.68}$$

Deviation in the right part of the membership function:

$$\bar{y} - \breve{y} = \sum \left(a_i * (\overline{Z}_i - \breve{Z}_i) + c_i \left|\breve{Z}_i\right|\right) = \sum a_i \overline{Z}_i - a_i \breve{Z}_i + c_i \left|\breve{Z}_i\right|, \text{ so}$$

The upper border of the interval

$$\bar{y} = \sum \left(a_i * \bar{Z}_i + c_i \left|\breve{Z}_i\right|\right). \tag{6.69}$$

For the interval model to be correct, the real value of input variable Y should lay in the interval obtained by the method FGMDH.

It can be described in such a way:

$$\begin{cases} \sum \left(a_i * \underline{Z}_{ik} - c_i \left|\breve{Z}_{ik}\right|\right) \le y_k \\ \sum \left(a_i * \bar{Z}_{ik} + c_i \left|\breve{Z}_{ik}\right|\right) \ge y_k \end{cases}, \quad k = \overline{1, M}, \tag{6.70}$$

where $Z_k = [Z_{ki}]$ is an input training sample, y_k is known output value, $k = \overline{1, M}$, M—number of observation points.

So, the general requirements to a linear interval model are the following: to find such values of parameters (a_i, c_i) of fuzzy coefficients, which ensure [6]:

1. observed values y_k should locate in the estimation interval for Y_k;
2. total width of the estimation interval be minimal.

These requirements may be redefined as a task of linear programming:

$$\min_{a_i, c_i} \sum_{k=1}^{M} \left(\sum \left(a_i * \bar{Z}_i + c_i \left|\breve{Z}_i\right|\right) - \sum \left(a_i * \underline{Z}_i - c_i \left|\breve{Z}_i\right|\right) \right) \tag{6.71}$$

under conditions

$$\begin{cases} \sum (a_i * \underline{Z}_{ik} - c_i \left| \breve{Z}_{ik} \right|) \leq y_k \\ \sum (a_i * \bar{Z}_{ki} + c_i \left| \breve{Z}_{ik} \right|) \geq y_k, k = \overline{1, M} \end{cases} \tag{6.72}$$

Let's consider partial description of the form

$$f(x_i, x_j) = A_0 + A_1 x_i + A_2 x_j + A_3 x_i x_j + A_4 x_i^2 + A_5 x_j^2 \tag{6.73}$$

Rewriting it in accordance with the model (6.63) needs such substitutions

$$z_0 = 1 \; z_1 = x_i \; z_2 = x_j \; z_3 = x_i x_j \; z_4 = x_i^2 \; z_5 = x_j^2$$

Then math model takes the following form

$$\begin{aligned} &\min_{a_i, c_i}(2Mc_0 + a_1 \sum_{k=1}^{M} (\bar{x}_{ik} - \underline{x}_{ik}) + 2c_1 \sum_{k=1}^{M} \left| \breve{x}_{ik} \right| + a_2 \sum_{k=1}^{M} (\bar{x}_{jk} - \underline{x}_{jk}) + 2c_2 \sum_{k=1}^{M} \left| \breve{x}_{jk} \right| + \\ &+ a_3 \sum_{k=1}^{M} (\left| \breve{x}_{ik} \right| (\bar{x}_{jk} - \underline{x}_{jk}) + \left| \breve{x}_{jk} \right| (\bar{x}_{ik} - \underline{x}_{ik})) + 2c_3 \sum_{k=1}^{M} \left| \breve{x}_{ik} \breve{x}_{jk} \right| + 2a_4 \sum_{k=1}^{M} \left| \breve{x}_{ik} \right| (\bar{x}_{ik} - \underline{x}_{ik}) + \\ &+ 2c_4 \sum_{k=1}^{M} \breve{x}_{ik}^2 + 2a_5 \sum_{k=1}^{M} \left| \breve{x}_{jk} \right| (\bar{x}_{jk} - \underline{x}_{jk}) + 2c_5 \sum_{k=1}^{M} \breve{x}_{jk}^2) \end{aligned} \tag{6.74}$$

with the following constraints

$$\begin{aligned} &a_0 + a_1 \underline{x}_{ik} + a_2 \underline{x}_{jk} + a_3 (-\left| \breve{x}_{ik} \right| (\breve{x}_{jk} - \underline{x}_{jk}) - \left| \breve{x}_{jk} \right| (\breve{x}_{ik} - \underline{x}_{ik}) + \breve{x}_{ik} \breve{x}_{jk}) + \\ &+ a_4 (-2 \left| \breve{x}_{ik} \right| (\breve{x}_{ik} - \underline{x}_{ik}) + \breve{x}_{ik}^2) + a_5 (2 \left| \breve{x}_{jk} \right| (\breve{x}_{jk} - \underline{x}_{jk}) + \breve{x}_{jk}^2) - \\ &- c_0 - c_1 \left| \breve{x}_{ik} \right| - c_2 \left| \breve{x}_{jk} \right| - c_3 \left| \breve{x}_{ik} \breve{x}_{jk} \right| - c_4 \breve{x}_{ik}^2 - c_5 \breve{x}_{jk}^2 \leq y_k \end{aligned}$$

$$\begin{aligned} &a_0 + a_1 \bar{x}_{ik} + a_2 \bar{x}_{jk} + a_3 (\left| \breve{x}_{ik} \right| (\bar{x}_{jk} - \breve{x}_{jk}) + \left| \breve{x}_{jk} \right| (\bar{x}_{ik} - \breve{x}_{ik}) - \breve{x}_{ik} \breve{x}_{jk}) + \\ &+ a_4 (2 \left| \breve{x}_{ik} \right| (\bar{x}_{ik} - \breve{x}_{ik}) - \breve{x}_{ik}^2) + a_5 (2 \left| \breve{x}_{jk} \right| (\bar{x}_{jk} - \breve{x}_{jk}) - \breve{x}_{jk}^2) + c_0 + \\ &+ c_1 \left| \breve{x}_{ik} \right| + c_2 \left| \breve{x}_{jk} \right| + c_3 \left| \breve{x}_{ik} \breve{x}_{jk} \right| + c_4 \breve{x}_{ik}^2 + c_5 \breve{x}_{jk}^2 \geq y_k \end{aligned}$$

$$c_l \geq 0, l = \overline{0, 5}.$$

As one can see, this is the linear programing problem, but there are still no constraints for non-negativity of variables a_i, so it's reasonable to pass to a dual problem, introducing dual variables $\{\delta_k\}$ and $\{\delta_{k+M}\}$.

6.11 Experimental Investigations of GMDH and FGMDH

Experiment 1. Forecasting of RTS index (opening price)
Let's consider the results of experiments of forecasting RTS index using data from Russian stock-exchange.

RTS index is an official indicator of RTS stock exchange. It is calculated during trade session with every change of price of instruments, which are included in the list of its calculation. First value of index is the opening price, and the last is the closing price. RTS index is calculated as a ratio of total capitalization of securities included in list of its calculation to total market capitalization of these securities on the initial date and then multiplying this value by the value of the index on the initial date and correction coefficient:

$$I_n = Z_n * I_1 * \frac{MC_n}{MC_1},$$

where MC_n—sum of market capitalizations of shares on current time, USD.

$MC_n = \sum_{i=1}^{N} W_i * P_i * Q_i * C_i$, where W_i—correction coefficient, which takes into account the number of securities of ith type in free circulation;
C_i—coefficient, which limits share of capitalization of securities of ith type;
Q_i—quantity of securities of particular type, which were created by issuer till the current date;
P_i—price of ith security in USD on the calculation moment t;
N—number of types of securities in list, which is used for index calculation.

The list of securities, which is used for calculation of RTS index, consists of the most liquid shares of Russian companies chosen by Informational Committee and based on expert judgment. The number of securities may not exceed 50.

Information about index is covered on RTS web-site http://www.rts.ru/rtsindex. There are also the methods of index calculation, the list of securities for index calculation, index value, information about every security allotment into total capitalization of all securities in list.

The experiment contains 5 fuzzy input variables, which are the stock prices of leading Russian energetic companies included into the list of RTS:

- index LKOH—shares of "ЛУКОЙЛ" joint-stock company,
- EESR—shares of "РАО ЕЭС России" joint-stock company,
- YUKO—shares of "ЮКОС" joint-stock company,
- SNGSP—privileged shares of "Сургутнефтегаз" joint-stock company,
- SNGS—common shares of "Сургутнефтегаз" joint-stock company.
- Output variable is the value of RTS index (opening price) of the same period (03.04.2006–18.05.2006).

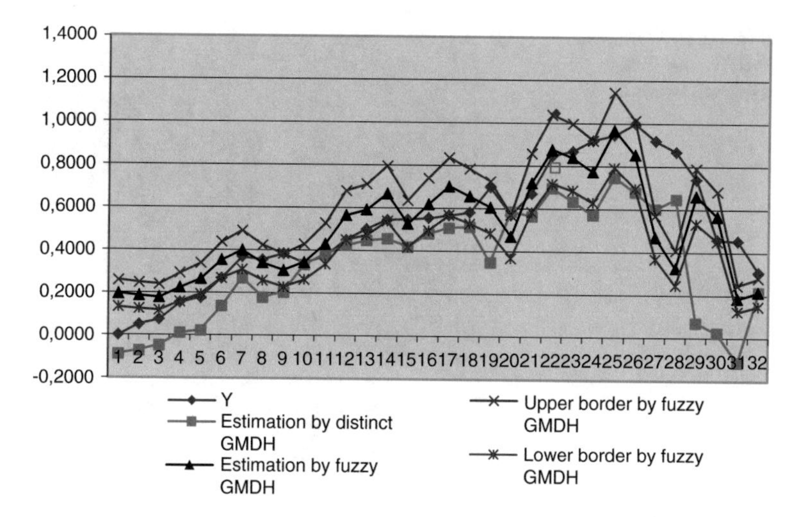

Fig. 6.8 Experiment 1 results using GMDH and FGMDH

Table 6.10 Comparison of different methods in experiment 2

	GMDH	FGMDH	FGMDH with fuzzy inputs, triangular MF	FGMDH with fuzzy inputs, Gaussian MF
MSE	0.051121	0.063035	0.061787	0.033097

Sample size contains 32 values. Training sample size has 18 values (optimal size of the training sample for current experiment). The following results were obtained [6].

Experiment 1 For normalized input when using Gaussian MF in group method of data handling with fuzzy input data the results of experiment are presented in Fig. 6.8 and Table 6.9: MSE for GMDH = 0.1129737, MSE for FGMDH = 0.0536556.

As the results of experiment 1 show, fuzzy group method of data handling with fuzzy input data gives more accurate forecast than GMDH and FGMDH. In case of triangular MF FGMDH with fuzzy inputs gives a little worse forecast than FGMDH with Gaussian MF (Fig. 6.9).

Experiment 2 RTS-2 index forecasting (opening price)

Sample size—32 values. Training sample size—18 values

For normalized input data when using triangular MF in FGMDH with fuzzy input data the experimental results are presented in Fig. 6.10.

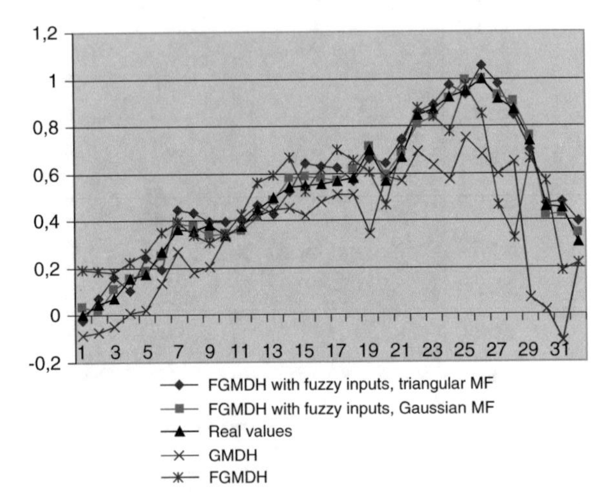

Fig. 6.9 GMDH, FGMDH, and FGMDH with fuzzy inputs comparison result

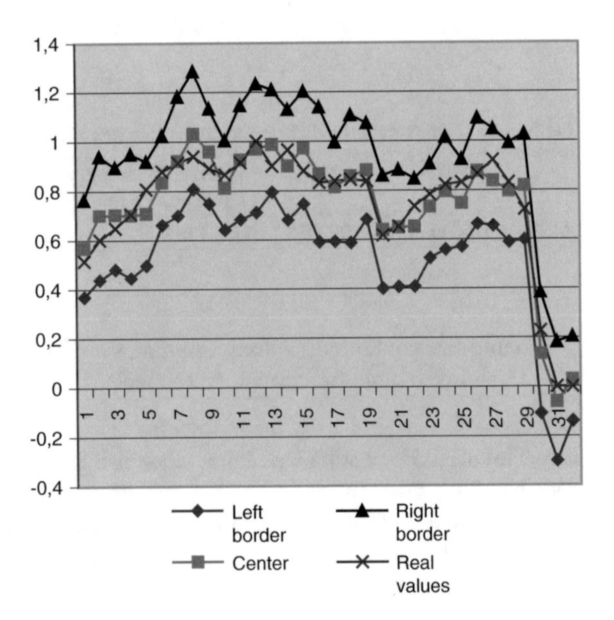

Fig. 6.10 Experiment 2 results for triangular MF

For normalized input data using Gaussian MF in FGMDH with fuzzy input data the following experimental results were obtained (see Fig. 6.11).

MSE for GMDH = 0.051121, MSE for FGMDH = 0.063035.

As the results of the experiment 2 show (Table 6.10), fuzzy group method of data handling with fuzzy input data gives better result than GMDH and FGMDH in case of both Gaussian and triangular membership function (Fig. 6.12).

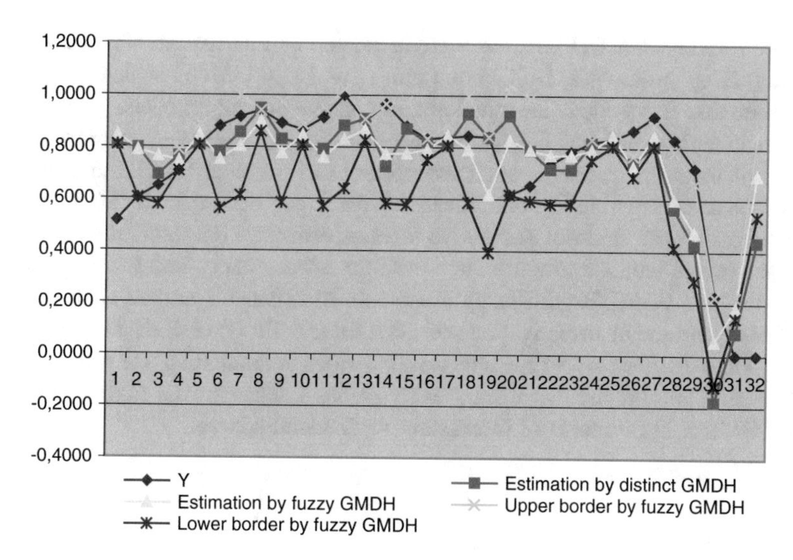

Fig. 6.11 Experiment 2 results using GMDH and FGMDH

Table 6.11 Comparison of different methods in experiment 2

	GMDH	FGMDH	FGMDH with fuzzy inputs, triangular MF	FGMDH with fuzzy inputs, Gaussian MF
MSE	0.051121	0.063035	0.061787	0.033097

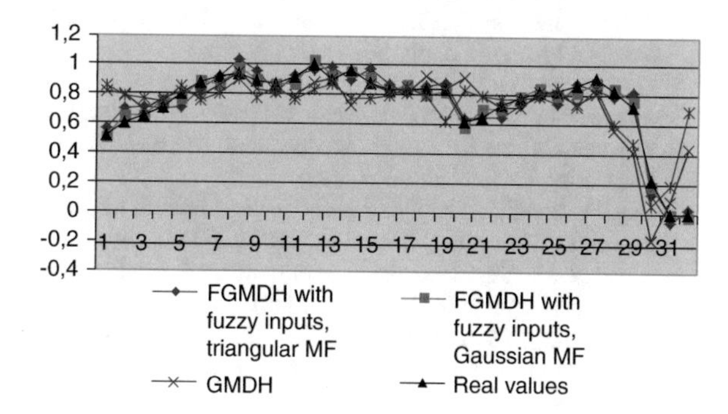

Fig. 6.12 GMDH, FGMDH and FGMDH with fuzzy inputs (center of estimation) results

6.12 Conclusions

In this chapter inductive modeling method GMDH is considered.

The main principles of GMDH are presented. This method enables to construct complex process models using experimental data. Opposite to classical identification methods GMDH has advantage to synthesize model structure automatically in the process of algorithm run.

Two different GMDH versions were presented and discussed: classical GMDH developed by prof. A.G. Ivakhnenko and new fuzzy GMDH which works with indefinite and fuzzy input information and constructs fuzzy models.

The advantage of fuzzy GMDH lies herein it doesn't use least square method for search of unknown model coefficients opposite to classical GMDH and therefore the problem of possible ill-conditioned matrix doesn't exist for it. Besides, fuzzy GMDH enables to find not point-wise forecast estimates but interval estimates for forecast values which allow to determine forecast accuracy. The generalization of fuzzy GMDH FGMDH with fuzzy inputs was also considered and analyzed.

The experimental investigations of GMDH and fuzzy GMDH in problems of share prices forecast at NYSE and Russian stock market RTS are presented and discussed. The comparative results analysis has confirmed the high accuracy of fuzzy GMDH in problems of forecasting in financial sphere.

References

1. Ivakhnenko, A.G., Zaychenko, Yu.P., Dimitrov, V.D.: Decision-Making on the Basis of Self-Organization, 363 pp. Moscow Publishing House "Soviet Radio"" (1976) (rus)
2. Zaychenko, Yu.P., Zayets, I. O. Ю.: П. Synthesis and adaptation of fuzzy forecasting models on the basis of self-organization method. Scientific papers of NTUU "KPI", № 3, pp. 34–41 (2001) (rus)
3. Zaychenko, Yu.P., Kebkal, A.G., Krachkovsky, V.F.: Fuzzy group method of data handling and its application for macro-economic indicators forecasting. Scientific papers of NTUU "KPI", № 2, pp. 18–26 (2000) (rus)
4. Zaychenko, Yu.P., Zayets, I.O., Kamotsky, O.V., Pavlyuk, O.V.: The investigations of different membership functions. In: Fuzzy Group Method of Data Handling. Control Systems and Machines, № 2, pp. 56–67 (2003) (rus)
5. Zaychenko, Yu.: Fuzzy group method of data handling in forecasting problems at financial markets. Int. J. Inf Models Anal. **1**(4), 303–317 (2012) (rus)
6. Zaychenko, Yu.P.: Fuzzy Models and Methods in Intellectual Systems, 354 pp. Kiev Publishing House "Slovo" (2008).
7. Zaychenko, Yu.: The fuzzy group method of data handling and its application for economical processes forecasting. Scientific Inquiry **7**(1), 83–98 (2006)
8. Zaychenko, Yu.: The fuzzy group method of data handling with fuzzy input variables. Scientific Inquiry, **9**(1), 61–76 (2008)
9. Zaychenko, Yu.P., Petrosyuk, I.M., Jaroshenko, M.S.: The investigations of fuzzy neural networks in the problems of electro-optical images recognition. Syst. Res. Inf. Technol. (4), 61–76 (2009) (rus)
10. Zaychenko, Yu.P.: Fuzzy group method of data handling under fuzzy input data. Syst. Res. Inf. Technol. (3), 100–112 (2007) (rus)

Chapter 7
The Cluster Analysis in Intellectual Systems

7.1 Introduction

Term cluster analysis (introduced by Tryon, 1939 for the first time) actually includes a set of various algorithms of classification without teacher [1]. The general question asked by researchers in many areas is how to organize observed data in evident structures, i.e. to develop taxonomy. For example, biologists set the purpose to divide animals into different types that to describe distinctions between them. According to the modern system accepted in biology, the person belongs to primacies, mammals, vertebrate and an animal. Notice that in this classification, the higher is aggregation level, the less is the similarity between members in the corresponding class. The person has more similarity to other primacies (i.e. with monkeys), than with the "remote" members of family of mammals (for example, dogs), etc.

The clustering is applied in the most various areas. For example, in the field of medicine the clustering of diseases, treatments of diseases or symptoms of diseases leads to widely used taksonomy. In the field of psychiatry the correct diagnostics of clusters of symptoms, such as paranoia, schizophrenia, etc., is decisive for successful therapy. In archeology by means of the cluster analysis researchers try to make taxonomy of stone tools, funeral objects, etc. Broad applications of the cluster analysis in market researches are well known. Generally, every time when it is necessary to classify "mountains" of information to groups, suitable for further processing, the cluster analysis is very useful and effective. In recent years the cluster analysis is widely used in the intellectual analysis of data (Data Mining), as one of the principal methods [1].

The purpose of this chapter is the consideration of modern methods of the cluster analysis, crisp methods(a method of C-means, Ward's method, the next neighbor, the most distant neighbor), and fuzzy methods, robust probabilistic and possibilistic clustering methods. In the Sect. 7.2 problem of cluster analysis is formulated, main criteria and metrics are considered and discussed. In the Sect. 7.3 classification of

M.Z. Zgurovsky and Y.P. Zaychenko, *The Fundamentals of Computational Intelligence: System Approach*, Studies in Computational Intelligence 652,
DOI 10.1007/978-3-319-35162-9_7

cluster analysis methods is presented, several crisp methods are considered, in particular hard C-means method and Ward's method. In the Sect. 7.4 fuzzy C-means method is described. In the Sect. 7.5 the methods of initial location of cluster centers are considered: peak and differential grouping and their properties analyzed. In the Sect. 7.6 Gustavsson-Kessel's method of cluster analysis is considered which is a generalization of fuzzy C-means method when metrics of distance differs from Euclidian.

In the Sect. 7.7 adaptive robust clustering algorithms are presented and analyzed which are used when initial data are distorted by high level of noise, or by outliers. In the Sect. 7.8 robust probabilistic algorithms of fuzzy clustering are considered.

Numerous results of pilot studies of fuzzy methods of a cluster analysis are presented in the Sect. 7.9 among them is a problem of UN countries clustering by indicators of sustainable development.

7.2 Cluster Analysis, Problem Definition. Criteria of Quality and Metrics

Let the set of observations c_1 be given, where $X_i = \{x_{ij}\},\ j = \overline{1, N}$. It is required to divide a set X into not intersected K subsets—clusters $S_1, \ldots, S_K$ so that to provide extremum of some criterion (functional of quality), that is:

to find such $S = (S_1, \ldots, S_K)$ that $f(S) \rightarrow min(max)$.

Different types of criteria (functional) of splitting are possible. It's worth to note that this task is closely connected with definition of some metrics in a feature space.

Consider the most widely used functionals of splitting quality [2]:

1. Coefficient of splitting F which is defined as follows:

$$F = \sum\nolimits_{j=1}^{K} \sum\nolimits_{i=1}^{n} \frac{w_{ij}^2}{n}, \tag{7.1}$$

where $w_{ij} \in [0; 1]$—some degree of membership of the ith object to the jth cluster. Change range is $F \in \left[\frac{1}{k}; 1\right]$, where n—number of objects, K—number of clusters.

2. Non-fuzziness index:

$$NFI = \frac{KF - 1}{k - 1}, \quad NFI \in [0; 1], \tag{7.2}$$

where K—number of classes (clusters); F—splitting coefficient.

3. Entropy of splitting:

$$H = -\sum_{j=1}^{K}\sum_{i=1}^{n}\frac{w_{ij}ln(w_{ij})}{n},\ H \in (0; \ln K). \quad (7.3)$$

4. The normalized entropy of splitting:

$$H_1 = \frac{H}{1 - K/n},\ H_1 \in \left(0; \frac{n \ln K}{n - K}\right), \quad (7.4)$$

where n is a number of points.

5. The modified entropy:

$$H_2 = \frac{H}{\ln K},\ H_2 \in (0, 1) \quad (7.5)$$

6. Second functional of Rubens:

$$F_2 = \frac{1}{2}\left(\frac{1}{n}\sum_{i=1}^{n}\max_j W_{ij} + \min_i \max_j W_{ij}\right),$$

$$F_2 \in \left(\frac{1}{K}, 1\right). \quad (7.6)$$

7. Third functional of Rubens (second index of Non-fuzziness) :

$$NF2I = \frac{KF2 - 1}{K - 1},\ NF2I \in (0, 1). \quad (7.7)$$

As initial information is set in the form of a matrix X, there is a metrics choice problem. Metrics choice—the most important factor influencing results of a cluster analysis. Depending on type of features various measures of distance (metrics) are used.

Table 7.1 Clustering metrics

No	Name of a metrics	Type of features	Formula for an assessment of a measure of proximity (metrics)
1	Euclidean distance	The quantitative	$d_{ik} = \left(\sum_{j=1}^{N} (x_{ij} - x_{kj})^2\right)^{1/2}$
2	Measure of similarity of Hamming	Nominal (qualitative)	$\mu_{ij}^{H} = \frac{n_{ik}}{N}$, where n_{ik}—number of coinciding features in samples X_i and X_K
3	Measure of similarity of Rogers-Tanimoto	Nominal scales	$\mu_{ij}^{R-T} = n''_{ik}(n'_i + n''_k - n''_{ik})$ where n''_{ik}—number of coinciding unit features at samples X_i and X_K; n'_i, n''_k—total number of unit features at samples X_i and X_K respectively
4	Manhattan metrics	The quantitative	$d_{ik}^{(1)} = \sum_{j=1}^{N} \lvert x_{ij} - x_{kj}\rvert$
5	Makhalonobis's distance	The quantitative	$d_{ik}^{M} = (x_{ij} - x_{kj})^T W^{-1} (x_{ij} - x_{kj})$ W—covariance matrix of sample $X = \{X_1, X_2, \ldots X_n\}$
6	Zhuravlev's distance	The mixed	$d_{ik} = \sum_{j=1}^{N} I_{ik}^{j}$, where $I_{ik}^{j} = \begin{cases} 1, & if \lvert x_{ij} - x_{kj}\rvert < \varepsilon \\ 0, & otherwise \end{cases}$

Let be samples X_i and X_K in N-dimensional feature space.

The main metrics of clustering are given in the Table 7.1

There is a large number of clustering algorithms which use various metrics and criteria of splitting.

Classification of Algorithms of Cluster Analysis

When performing a clustering it is important to know, how many clusters contains an initial sample It is supposed that the clustering has to reveal natural local grouping of objects. Therefore the number of clusters is the parameter which is often significantly complicates an algorithm if it is supposed to be unknown and significantly influencing quality of result if it is known.

The problem of a choice of clusters number is very nontrivial. It is enough to tell that for obtaining the satisfactory theoretical decision often it is required to make in advance very strong assumptions of properties of some family of distributions. But about what assumptions one can make when, especially at the beginning of research, of data practically it isn't known? Therefore algorithms of a clustering usually are constructed as some way of search clusters number and determination of its optimum value in the course of search.

The number of methods of splitting a set of objects into clusters is quite great. All of them can be subdivided on hierarchical and not hierarchical.

In not hierarchical algorithms their work and conditions of stop need to be regulated in advance often with large number of parameters that is sometimes difficult, especially at the initial stage of investigation. But in such algorithms big flexibility in a variation of a clustering is reached and usually the number of clusters is defined.

On the other hand, when objects are characterized by a large number of features (parameters), a task of grouping features is important. Initial information contains in a square matrix of features interconnections, in particular, in a correlation matrix. Basis of the successful solution of a grouping task is the informal hypothesis of a small number of the hidden factors which define structure of an interconnection between features.

In hierarchical algorithms one actually refuses to define a number of clusters, building a full tree of the enclosed clusters (so-called dendrogram). The number of clusters is defined from the assumptions, in principle, which aren't relating to work of algorithms, for example on dynamics of change of a threshold of splitting (merge) of clusters. Difficulties of such algorithms are well studied: choice of measures of proximity of clusters, problem of inversions of indexation in the dendrogramms, inflexibility of hierarchical classifications which is sometimes undesirable. Nevertheless, representation of a clustering in the form of a dendrogramm allows to gain the most complete display of structure of clusters.

Hierarchical algorithms are connected with dendrogramms construction and divided on:

1. agglomerative, characterized by consecutive merge of initial elements and the corresponding reduction of number of clusters (creation of clusters from below to top);
2. divisional (divided) in which the number of clusters increases, starting with one cluster therefore the sequence of the splitting groups is constructed (creation of clusters from top to down).

7.2.1 Hierarchical Algorithms. Agglomerative Algorithms

On the first step all the set of objects is represented as a set of clusters:

$$c_1 = \{i_1\}, c_2 = \{i_2\}, \ldots, c_m = \{i_m\}$$

On the following step two closest one to another clusters are chosen (for example, C_p and C_q) and unite in one joint cluster. The new set consisting already of m − 1 of clusters will be such:

$$c_1 = \{i_1\}, c_2 = \{i_2\}, \ldots, c_p = \{i_p, i_q\}, \ldots, c_p = \{i_m\}$$

Repeating process, we obtain step by step the consecutive sets consisting of *(m − 2), (m − 3), (m − 4)* and etc. clusters.

At the end of procedure the cluster consisting of m of objects and coinciding with an initial set *I* will be obtained.

For determination of distance between clusters it is possible to choose different metrics. Depending on it algorithms with various properties exist.

There are some methods of recalculation of distances with use of old values of distances for the united clusters differing in coefficients in a formula [1]:

$$d_{rs} = \alpha_p d_{ps} + \alpha_q d_{qs} + \beta d_{pq} + \gamma \left| d_{ps} - d_{qs} \right|$$

If clusters p and q unite in one cluster of r and it is required to calculate distance from a new cluster to cluster say, s, application of this or that method depends on a way of determination of distance between clusters, these methods differ with values of coefficients $\alpha_p, \alpha_q, \beta, \gamma$.

Coefficients of recalculation of distances between clusters are specified in Table 7.2 $\alpha_p, \alpha_q, \beta, \gamma$.

Table 7.2 Coefficients of distances recalculation between clusters

Name of a method	α_p	α_q	β	γ
Distance between the closest neighbours—the closest objects of clusters (Nearest neighbour)	1/2	1/2	0	−1/2
Distance between the farthest neighbours (Furthest neighbour)	1/2	1/2	0	1/2
The method of medians—the same centroid method, but the centre of the integrated cluster is calculated as an average of all objects (Median clustering)	1/2	1/2	−1/4	0
Average distance between clusters (Between—groups linkage)	1/2	1/2	0	0
Average distance between all objects of couple of clusters taking into account distances inside clusters (intra-groups linkage)	$\frac{k_p}{k_p+k_q}$	$\frac{k_q}{k_p+k_q}$	0	0
Distance between centres clusters (Centroid clustering), or centroid method. A lack of this method is that the centre of the integrated cluster is calculated as an average of the centres of the united clusters, without their volume	$\frac{k_p}{k_p+k_q}$	$\frac{k_p}{k_p+k_q}$	$\frac{-k_p k_q}{k_p+k_q}$	0
Ward's method. As distance between clusters the gain of the sum of squares of distances of objects to the centres of clusters received as a result of their association is calculated	$\frac{k_r+k_p}{k_r+k_q+k_p}$	$\frac{k_r+k_p}{k_r+k_q+k_p}$	$\frac{-k_r}{k_r+k_q+k_p}$	0

7.2.2 Divisional Algorithms

Divisional cluster algorithms, unlike agglomerative, on the first step represent all set of elements I as the only cluster. On each step of algorithm one of the existing clusters is recursively divided into two affiliated. Thus, clusters from top to down are iteratively formed. This approach isn't so in detail described in literature devoted to the cluster analysis, as agglomerative algorithms. It is applied when it is necessary to divide all set of objects on rather small amount clusters.

One of the first the divisional algorithms was offered by Smith Maknaoton in 1965 [1].

All elements are located on the first step in one cluster $C1 = I$.

Then the element, at which average value of distance from other elements in this cluster is the greatest is selected.

The chosen element is removed from a cluster of C1 and becomes the first member of the second cluster C2.

On each subsequent step an element in a cluster of C1 for which the difference between average distance to the elements which are in C2, and average distance to the elements remaining in C1 is the greatest is transferred to C2.. Transfer of elements from C1 in C2 proceed until the corresponding differences of averages become negative, i.e. so far there are elements located to elements of a cluster of C2 closer than to cluster elements of C1.

As a result one cluster is divided into two affiliated ones which will be split at the following level of hierarchy. Each subsequent level procedure of division is applied to one of the clusters received at the previous level. The choice of cluster to be split can be carried out differently.

In 1990 Kauffman and Rouzeuv suggested to choose at each level a cluster for splitting with the greatest diameter which is calculated on a formula [1]

$$D_C = \max d(i_p, i_q) \forall i_p, i_q \in C$$

Recursive division of clusters proceeds, so far all clusters or won't become sigleton (i.e. consisting of one object), or so far all members of one cluster won't have zero difference from each other.

7.2.3 Not Hierarchical Algorithms

The great popularity at the solution of clustering problems was acquired by the algorithms based on search of splitting a data set into clusters (groups). In many tasks algorithms of splitting are used owing to the advantages. These algorithms try to group data (in clusters) so that criterion function of splitting algorithm reaches an extremum (minimum). We'll consider three main algorithms of a clustering based on splitting methods. In these algorithms the following basic concepts are used:

- the training set (an input set of data) of M on which splitting is based;
- distance metrics:

$$d_A^2(m_j, c^{(i)}) = \left\| m_j - c^{(i)} \right\|^2 = (m_j - c^{(i)})^t A (m_j - c^{(i)}) \tag{7.6}$$

where the matrix A defines a way of distance calculation. For example, for a singular matrix distance according to Euclid metrics is used;

- vector of the centers of clusters C;
- splitting matrix on clusters U;
- goal function $J = J(M, d, C, U)$;
- set of restrictions.

Algorithm K-means (Hard C-means)

Consider in more detail algorithm for the example of clustering data on irises. For descriptive reasons the presentation is limited to only two parameters (instead of four)—length and width of a sepal. It allows to present this problem in two-dimensional space (Fig. 7.1). Points are noted by numbers of objects.

At the beginning k of any points in space of all objects are taken as possible centers. It isn't really critical which points will be initial centers, procedure of a choice of starting points will affect, mainly, on calculation time. For example, it may be the first k of objects of the set I. In this example it is points 1, 2 and 3.

Further iterative operation of two steps is carried out.

On the first step all objects were split into k = 3 groups, the closest to one of the centers. The proximity is defined by distance which is calculated by one of the ways

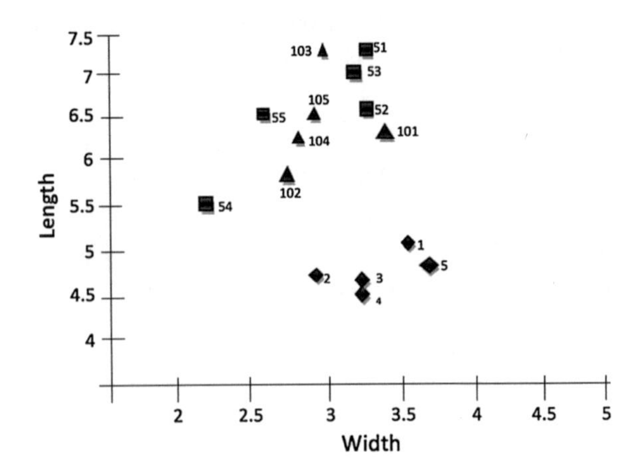

Fig. 7.1 Example of clustering data on irises by algorithm hard C-means

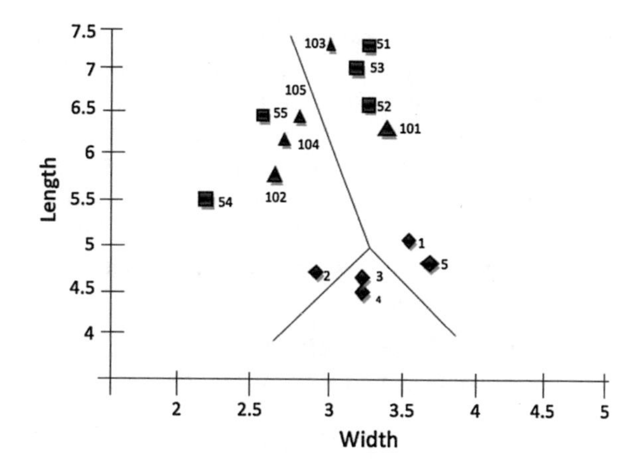

Fig. 7.2 Example of clustering data on irises by algorithm hard C-means

described earlier (for example, the Euclidean distance is used). Figure 7.2 illustrates splitting irises into three clusters.

On the second step the new centers of clusters are calculated. The centers can be calculated as average values of the variable objects referred to the created groups. The new centers, naturally, can differ from the previous ones.

In Fig. 7.3 the new centers and new division according to them are marked. It is natural that some points which earlier were related to one cluster, at new splitting are related to another (in this case such points are 1, 2, 5, etc.). The new centers in the Fig. 7.3 are marked with symbol "x", inside a circle.

The considered operation repeats recursively until the centers of clusters (respectively, and borders between them) cease to change. In our example the second splitting was the last. If to look at results of a clustering, it is possible to notice that the lower cluster completely coincides with the class Iris setosa. In other clusters there are representatives of both classes. In this case it was the result of a clustering with only two parameters, but not with all four.

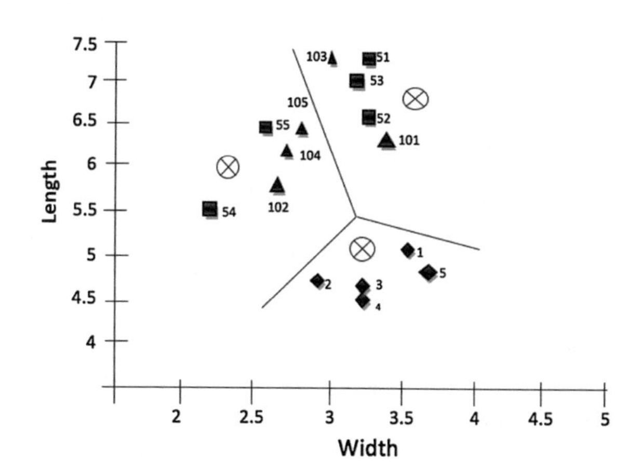

Fig. 7.3 Example of clustering data on irises by algorithm hard C-means

After objects are mapped in points of multidimensional space, procedure of automatic splitting into clusters becomes very simple. A problem is that initial objects can not always be presented in the form of points. Introduction of a metrics, distances between categorial variables or the relations is more difficult.

In summary it should be noted that the method K-means works well if data may be separated on compact, approximately spherical groups.

This algorithm is a prototype practically of all algorithms of fuzzy clustering, and its consideration will help the best understanding of the principles underlying in more complex algorithms.

Description of K-means Algorithm

Basic definitions and concepts within this algorithm are following:

- the training set $M = \{m_j\}_{j=1}^{d}$ d—number of points (vectors) of data;
- the distance metrics counted by a formula (7.6);
- vector of the centers of clusters $C = \{c^{(i)}\}_{i=1}^{c}$

  where

$$c^{(i)} = \frac{\sum_{j=1}^{d} u_{ij} m_j}{\sum_{j=1}^{d} u_{ij}}, \quad 1 \le i \le c \tag{7.7}$$

- splitting matrix $U = \{u_{ij}\}$

  where

$$u_{ij}^{(l)} = \begin{cases} 1, \; if \; d(m_j, c_i^{(l)}) = \min_{1 \le k \le c} d(m_j, c_k^{(l)}) \\ 0, \; otherwise \end{cases} \tag{7.8}$$

  Object function

$$J(M, U, C) = \sum_{i=1}^{c} \sum_{j=1}^{d} u_{ij} d_A^2(m_j, c^{(i)})$$

- set of restrictions

$$\{u_{ij}\} \in \{0, 1\}; \; \sum_{i=1}^{c} u_{ij} = 1; \; 0 < \sum_{j=1}^{d} u_{ij} < d \tag{7.10}$$

which defines that each vector of data can belong only to one cluster and doesn't belong to the rest. Each cluster contains not less than one point, but less than a total number of points.

Structurally the algorithm represents the following iterative procedure [1].

Step 1. To initialize initial splitting (for example, in a random way), to choose accuracy value δ (it is used in a condition of end of an algorithm), to initialize a number of iteration l = 0.

Step 2. To define the centers of clusters by the following formula:

$$c_i^{(l)} = \frac{\sum_{j=1}^{d} u_{ij}^{(l-1)} m_j}{\sum_{j=1}^{d} u_{ij}^{(l-1)}}, \quad 1 \leq i \leq c \tag{7.11}$$

Step 3. To update a splitting matrix to minimize squares of errors, using a formula

$$u_{ij}^{(l-1)} = \begin{cases} 1, \; if \; d(m_j, c_i^{(l)}) = \min_{1 \leq k \leq c} d(m_j, c_k^{(l)}) \\ 0, \; otherwise \end{cases} \tag{7.12}$$

Step 4. To check a condition $\left\| U^{(l)} - U^{(l-1)} \right\| < \delta$. If the condition is satisfied, finish process if it isn't true then pass to a step 2 with number of iteration $l = l + 1$. The main shortcoming inherent to this algorithm owing to discrete character of elements of a splitting matrix is the big size of splitting space. One way to overcome this shortcoming is the choice of elements of a splitting matrix by numbers from a unit interval. That is, belonging of a data element to a cluster has to be defined by membership function— the element of data can belong to several clusters with various degree of membership. In that case we come to a problem of fuzzy clustering. This approach found the embodiment in algorithm of fuzzy clustering—fuzzy method of K-means (Fuzzy C-Means).

7.3 Fuzzy C-Means Method

Consider a neural network with self-organization where training is performed without a teacher. The algorithm of self-organization refers a vector x to the corresponding cluster of data which is presented by its center, using a competitive training.

The basic form of algorithm of self-organization allows to find precisely position of the centers of the relevant groups (clusters) into which the output multidimensional space is split. These centers can be used further in hybrid algorithm of training of FNNs as initial values that considerably accelerates process of training and guarantees convergence to a global minimum.

Algorithm of Fuzzy C-means

Let's assume that in a network exists m fuzzy neurons with the centers in points $c_j, \; (j = 1, 2, \ldots, m)$. Initial values of these centers can be chosen randomly from areas of admissible values of the corresponding components of vectors $x_k, \; (k = 1, 2, \ldots, N)$ used for training. Let function of a fuzzification be set in the form of the generalized Gauss function expressed by a formula (3.45) in Chap. 3.

The vector entered in a network input x_k will belong to various groups represented by the centers c_j, with degree w_{kj}, and $0 < w_{kj} < 1$, and total degree of membership to all groups, is obviously, equal 1. Therefore

$$\sum_{j=1}^{m} w_{kj} = 1, \tag{7.13}$$

for all w_{kj} $(k = 1, 2, \ldots, N)$.

The function of an error corresponding to such representation can be defined as the sum of individual errors of membership to the centers c_i taking into account fuzziness degree β. Therefore,

$$E = \sum_{j=1}^{m} \sum_{k=1}^{N} w_{kj}^{\beta} \left\| c_j - x_k \right\|^2 \tag{7.14}$$

where β is a weight coefficient which accepts values from an interval $(1, \infty)$. The training goal of self-organization consists in such selection of the centers c_j, that for the whole set of the training vectors x_k—achievement of a minimum of function (7.14) at simultaneous fulfillment of conditions (7.13) is attained. Thus it is a problem of minimization of nonlinear function (7.14) with N constraints of type (7.13). The solution of this task can be transferred to minimization of Lagrange function defined by the form [3].

$$LE = \sum_{j=1}^{m} \sum_{k=1}^{N} w_{kj}^{\beta} \left\| c_j - x_k \right\|^2 + \sum_{k=1}^{N} \lambda_k \left(\sum_{j=1}^{m} w_{kj} - 1 \right) \tag{7.15}$$

where λ_k $(k = 1, 2, \ldots, N)$ are Lagrange's multipliers. In [4] it is proved that the solution of a task (7.15) can be presented in the form

$$c_j = \frac{\sum_{k=1}^{N} w_{kj}^{\beta} x_k}{\sum_{k=1}^{N} w_{kj}^{\beta}}, \tag{7.16}$$

$$w_{kj} = \frac{1}{\sum_{i=1}^{m} \left(\frac{d_{kj}^2}{d_{ij}^2} \right)^{\frac{1}{\beta - 1}}}, \tag{7.17}$$

where d_{kj}—is Euclidean distance between the center c_j and vector x_k, $d_{kj} = \left\| c_j - x_k \right\|$. As exact values of the centers c_j at the beginning of process aren't known, the training algorithm has to be iterative. It can be formulated in the following way:

1. To execute random initialization of coefficients w_{kj}, choosing their values from an interval [0, 1] so that the condition (7.13) be satisfied.

2. To define K centers c_j, in accordance with (7.16).
3. To calculate value of the error function according to expression (7.14). If its value appears below the established threshold or if reduction of this error of previous iteration is negligible, to finish calculations. The last values of the centers represent the required decision. Otherwise, go to step 4.
4. To calculate new values w_{kj} in a formula (7.17) and to pass to step 2.

We'll call such procedure the fuzzy self-organization algorithm C-means.

Repetition of iterative procedure leads to achievement of a minimum of function E which won't be a global minimum. The quality of the found centers estimated by value of an error function E essentially depends on preliminary selection of values w_{kj} and centers c_j. As the best will be such placement of the centers at which they settle down in the areas containing the greatest number of the shown vectors x_j. At such selection of the centers they will represent vectors of data x_j with the smallest total error.

Therefore the beginning of iterative procedure of calculation of optimum values of the centers has to be preceded by procedure of their initialization. Algorithms of peak and differential grouping of data belong to the most known algorithms of initialization.

7.4 Definition of Initial Location of the Centers of Clusters

7.4.1 Algorithm of Peak Grouping

The algorithm of peak grouping was offered by Jager and Filev [3, 5].

When using N input vectors the special grid which evenly covers space of these vectors is constructed. Nodes of this grid are considered as potential centers ϑ, for each of which peak function is calculated:

$$m(\vartheta) = \sum_{k=1}^{N} \exp\left\{\frac{-\|\vartheta - x_k\|^{2b}}{2\sigma^2}\right\} \tag{7.18}$$

where σ is some constant which is selected separately for each specific task.

Value $m(\vartheta)$ is considered as an assessment of height of peak function. It is proportional to quantity of vectors x_j, which get to the vicinity of the potential center ϑ. Great value $m(\vartheta)$ testifies to that the center ϑ locates in the area in which the greatest number of vectors is concentrated $\{x_k\}$.

The coefficient of σ influences final proportions between $m(\vartheta)$ and ϑ slightly.

After calculation of values $m(\vartheta)$ for all potential centers the first center is selected c_1, which has the greatest value $m(\vartheta)$. For a choice of the following centers it is necessary to exclude c_1 and nodes which are placed in close proximity to c_1.

It can be done by redefinition of peak function at the expense of separation of Gauss function from it with the center in a point c_1. Having designated this new function through $m_{new}(\vartheta)$, we receive:

$$m_{new}(\vartheta) = m(\vartheta) - m(c_1)\exp\left\{\frac{-\|\vartheta - c_1\|^{2b}}{2\sigma^2}\right\} \tag{7.19}$$

Note that this function has zero in a point c_1.

Then the same procedure repeats with the next center c_2, etc.

Process of finding of the following centers c_2, c_3 is realized consistently on the modified values $m_{new}(\vartheta)$, which turn out at an exception of the next neighbors of the center which was found at the previous stage. It comes to an end at the moment of localization of all the centers.

The method of peak grouping is effective at not really big dimension of a vector of X. Otherwise number of the potential centers increases as avalanche.

7.4.2 Algorithm of Differential Grouping

The algorithm of differential grouping is a modification of the previous algorithm, in which vectors x_j are considered as the potential centers ϑ. Peak function $D(x_i)$ in this case takes the form [2]:

$$D(x_i) = \sum_{j=1}^{N} \exp\left\{-\frac{\|x_i - x_j\|^{2b}}{(r_a/2)^2}\right\}, \tag{7.20}$$

where value of coefficient r_a defines the sphere of the neighborhood. On value $D(x_i)$ considerably influence only vectorsx_j, which are inside this sphere.

At the big density of points near x_i function value $D(x_i)$ is large. After calculation of values of peak function for each point x_i, the vector x is found, for which peak function $D(x)$ will appear to be the greatest. This point becomes the first center c_1.

Choice of the following center c_2 is performed after an exception of the previous center and all points which lie in its vicinity

As well as in the previous case peak function is redefined so

$$D_{new}(x_i) = D(x_i) - D(c_1)\exp\left\{-\frac{\|x_i - c_1\|^{2b}}{(r_b/2)^2}\right\}$$

At new definition of function D coefficients r_b designate new values of a constant which sets the sphere of the neighborhood of the following center. Usually a condition $r_b \geq r_a$ is used.

After modification of value of peak function a search of a new point x, for which $D_{new}(x_i) \rightarrow \max$ is performed. It becomes the new center.

Process of finding of the next center is resumed after the exception of all already selected points. Initialization comes to an end at the time of fixing of all centers which are provided by entry conditions.

7.5 Gustavson-Kessel's Fuzzy Cluster Analysis Algorithm

In classical algorithm fuzzy C = means elements of error function E are obtained by means of usual Euclid distance between a vector x and the center of a cluster c with:

$$d(x,c) = \|x - c\| = \sqrt{(x-c)^T(x-c)}$$

At such metrics of distance between two vectors the set of the points equidistant from the center represents a sphere with an identical scale on all axes. But if data form groups which form differs from spherical or if scales of separate coordinates of a vector strongly differ, such metrics becomes inadequate. In this case quality of a clustering can be increased considerably at the expense of the improved version of the self-organization algorithm which is called as Gustavson-Kessel's algorithm [2, 3].

The main changes of basic algorithm fuzzy C-means consist in introduction to a metrics calculation formula of the scaling matrix A. At such scaling the distance between the center c and vectors x is defined by a formula:

$$d(x,c) = \|x - c\| = \sqrt{(x-c)^T A(x-c)} \tag{7.21}$$

As scaling usually the positive-definite matrix is used, that is a matrix, at which all eigenvalues are real and positive.

Similar to the basic algorithm C-means the training goal of Gustavson-Kessel algorithm lies in such placement of the centers at which the criterion E is minimized:

$$E = \sum_k \sum_j w_{kj}^{\beta} d^2(x_k, c_j) \tag{7.22}$$

Description of Gustavson-Kessel algorithm:

1. To carry out initial placement of the centers in data space. To create an elementary form of the scaling matrix A.

2. To create a matrix of membership coefficients of all vectors x to the centers by a formula:

$$w_{kj} = \frac{1}{\sum_{i=1}^{m} \left(\frac{d_{kj}^2}{d_{ij}^2} \right)^{\frac{1}{\beta - 1}}} \tag{7.23}$$

3. To calculate new placement of the centers according to a formula:

$$c_j = \frac{\sum_{k=1}^{N} w_{kj}^{\beta} x_k}{\sum_{k=1}^{N} w_{kj}^{\beta}}, \tag{7.24}$$

4. To generate a covariance matrix for each vector:

$$S_j = \sum_{k=1}^{N} w_{kj}^{\beta} (x_k - c_j)(x_k - c_j)^T \tag{7.25}$$

5. To calculate a new scaling matrix for each *i*th centre by a formula:

$$A_j = \sqrt[n]{\det(S_j)} S_j^{-1} \tag{7.26}$$

6. If the last changes of centers and a covariance matrix are rather small in relation to the previous values (don't exceed the set values), finish iterative process, otherwise go to step 2.

7.6 Adaptive Robust Clustering Algorithms

Possibilistic Clustering Algorithms

Major drawbacks associated with a probabilistic approach (Fuzzy C-means algorithm)are connected with constraints (7.13). In the simplest case of two clusters ($m = 2$) is easy to see that the observation x_k, equally owned by both clusters and observation x_p, not belonging to any of them, may have the same levels of membership $w_{(k,1)} = w_{(k,2)} = w_{(p,1)} = w_{(p,2)} = 0.5$. Naturally, this fact decreasing the accuracy of classification, led to a **possibilistic approach** to the fuzzy classification [6]. In the possibilistic clustering algorithm goal function has the form

$$E(w_{k,j}, c_j) = \sum_{j=1}^{m} \sum_{k=1}^{N} w_{k,j}^{\beta} d^2(x_k, c_j) + \sum_{j=1}^{m} \mu_j \left(\sum_{k=1}^{N} (1 - w_{k,j})^{\beta} \right), \tag{7.27}$$

where scalar parameter $\mu_j > 0$ determines the distance on which membership level takes the value 0.5, that is if $d^2(x_k, c_j) = \mu_j$, then $w_{k,j} = 0.5$.

Minimization (7.27) by $w_{k,j}$, c_j, μ_j gives evident solution

$$w_{k,j} = \left(1 + \left(\frac{d^2(x_k, c_j)}{\mu_j} \right)^{\frac{1}{\beta - 1}} \right)^{-1}, \tag{7.28}$$

$$c_j = \frac{\sum_{k=1}^{N} w_{kj}^{\beta} x_k}{\sum_{k=1}^{N} w_{kj}^{\beta}}, \tag{7.29}$$

$$\mu_j = \frac{\sum_{k=1}^{N} w_{k,j}^{\beta} d^2(x_k, c_j)}{\sum_{k=1}^{N} w_{k,j}^{\beta}}, \tag{7.30}$$

It can be seen that the possibilistic and probabilistic algorithms are very similar and pass one into other by replacing the expression (7.27) to the formula (7.15), and vice versa. A common disadvantage of the considered algorithms is their computational complexity and the inability to work in real time. The algorithm (7.15) −(7.17) begins with the initial task of normal random partitions matrix W^0. On the basis of its values initial set of prototypes c_j^0 is calculated which then is used to calculate a new matrix W^1. Then this procedure is continued and sequence of solutions $c_j^1 W^2, \ldots, W^t, c_j^t W^{t+1}$ etc. is obtained., until the difference $\left\| W^{t+1} - W^t \right\|$ is less than a preassigned threshold ε.

Therefore, all available data sample is processed repeatedly.

The solution obtained using a probabilistic algorithm, is recommended as the initial conditions for possibilistic algorithm (7.28)−(7.30) [6, 7]. Parameter distance μ_j is initialized in accordance with (7.30) on the results of the probabilistic algorithm.

7.6.1 *Recurrent Fuzzy Clustering Algorithms*

Analysis of (7.15) shows that, for the calculation of membership levels $w_{k,j}$ instead of the Lagrangian (7.15) can be used its local modification:

$$L(w_{k,j}, c_j, \lambda_k) = \sum_{j=1}^{m} w_{k,j}^{\beta} d^2(x_k, c_j) + \lambda_k \left(\sum_{j=1}^{m} w_{k,j} - 1 \right) \tag{7.31}$$

Optimization of the expression (7.31) by the procedure of the Arrow-Hurwicz-Uzawa leads to an algorithm

$$w_{k,j} = \frac{d^2(x_k, c_{k,j})^{\frac{1}{\beta-1}}}{\sum_{l=1}^{m} \left(d^2(x_k, c_{k,l})\right)^{\frac{1}{\beta-1}}} \tag{7.32}$$

$$\begin{aligned} c_{k+1,j} &= c_{k,j} - \eta_k \nabla_{cj} L_k(w_{k,j}, c_{k,j}, \lambda_k) = \\ &= c_{k,j} - \eta_k w_{k,j}^{\beta} d(x_{k+1}, c_{k,j}) \nabla c_j \end{aligned} \tag{7.33}$$

Procedure (7.32), (7.33) is close to the learning algorithm Chang-Lee [8], and for $\beta = 2$ coincides with the gradient procedure clustering Park-Degger [9]:

$$w_{k,j} = \frac{\left\|x_k - c_{k,j}\right\|^{-2}}{\sum_{l=1}^{m} \left\|x_k - c_{k,l}\right\|^{-2}} \tag{7.34}$$

$$c_{k+1,j} = c_{k,j} - \eta_k w_{k,j}^2 (x_{k+1} - c_{k,j}) \tag{7.35}$$

Within the framework of possibilistic approach local criterion takes the form

$$E_k(w_{k,j}, c_j) = \sum_{j=1}^{m} w_{k,j}^{\beta} d^2(x_k, c_j) + \sum_{j=1}^{m} \mu_j (1 - w_{k,j})^{\beta} \tag{7.36}$$

and the result of its optimization has the form

$$w_{k,j} = \left(1 + \left(\frac{d^2(x_k, c_{k,j})}{\mu_j}\right)^{\frac{1}{\beta-1}}\right)^{-1} \tag{7.37}$$

$$c_{k+1,j} = c_{k,j} - \eta_k w_{k,j}^{\beta} d(x_{k+1}, c_{k,j}) \nabla c_j \tag{7.38}$$

where the distance parameter μ_j initialized according to (7.30).

In this case, N in Eq. (7.30) is a volume of data set used for initialization.

In the quadratic case, the algorithm (7.37), (7.38) is converted into a rather simple procedure and optimization result is of the form

$$w_{k,j} = \frac{\mu_j}{\mu_j + \left\|x_k - c_{k,j}\right\|^2} \tag{7.39}$$

wherein μ_j is the distance parameter which is initialized by the results of the probabilistic clustering (for example, using an algorithm Fuzzy C-means (7.15) −(7.17) according to the equation:

$$\mu_j = \frac{\sum_{k=1}^{N} w_{k,j}^2 \left\| x_k - c_j \right\|^2}{\sum_{k=1}^{N} w_{k,j}^2}, \tag{7.41}$$

7.6.2 *Robust Adaptive Algorithms of Fuzzy Clustering*

The considered above clustering methods can effectively solve the problem of classification with a substantial intersection of the clusters, however, it assumes that the data within each cluster are located compactly enough without sharp (abnormal) outliers.

However, it should be noted that the actual data are usually distorted by outliers, the share of which according to some estimates [10], is up to 20 % so that to speak of a compact placement of data is not always correct.

In this regard, recently, much attention was paid to problems of fuzzy cluster analysis of the data, the density distribution of which differs from the normal by presence of "heavy tails" [11, 12].

Robust Recursive Algorithm for Probabilistic Fuzzy Clustering

After standardization of feature vectors components so that all source vectors would belong to the unit hypercube $[0, 1]^n$, the objective function is constructed

$$E(w_{k,j}, c_j) = \sum_{j=1}^{m} \sum_{k=1}^{N} w_{k,j}^{\beta} D(x_k, c_j) \tag{7.42}$$

under constraints

$$\sum_{j=1}^{m} w_{k,j} = 1,\ k = 1, \ldots, N, \tag{7.43}$$

$$0 < \sum_{k=1}^{N} w_{k,j} \le N,\ j = 1, \ldots, m. \tag{7.44}$$

Here $D(x_k, c_j)$ is a distance between x_k and c_j in adopted metric. The result of clustering is assumed to be $N \times m$ matrix $W = \{w_{k,j}\}$, called "matrix of fuzzy decomposition." Typically, as the distance function $D(x_k, c_j)$ Minkowski metric L^p is applied [12]

$$D(x_k, c_j) = \left(\sum\nolimits_{i=1}^{n} \left|x_{k,i} - c_{j,i}\right|^p\right)^{\frac{1}{p}},\ p \geq 1, \tag{7.45}$$

where $x_{k,i}, c_{j,i}$ are the ith components of $(n \times 1)$—vectors x_k, c_j correspondingly.

Estimates relating to the quadratic objective functions are optimal when the data belong to the class of distributions with finite variance, the most famous member of which is a Gaussian.

Varying parameter p allows to improve the properties of the robustness of clustering procedures, however, the quality of assessment is determined by the type of data distribution. Thus, the estimates with $p = 1$ are optimal for the Laplacian data distribution, but their construction involves great computational expense. Quite realistic is the class of approximate normal distributions [12].

Approximately normal distributions are mixture of Gaussian density and distribution of some arbitrary density, which distorts with outliers the normal distribution. The optimal objective function in this case is the quadratic-linear, and tends to linear type as the distance from the minimum grows.

The most prominent representative of the approximate normal distribution density function is

$$p(x_i, c_i) = Se(c_i, s_i) = \frac{1}{2s_i} \sec h^2 \frac{x_i - c_i}{s_i}, \tag{7.46}$$

where c_i,s_i are parameters, determining a center and a width of the distribution.

This function resembles a Gaussian in the vicinity of the center, however, has a more heavy tails. With the distribution (7.46) is associated an objective function

$$f_i(x_i, c_i) = \beta_i \ln \cosh \frac{x_i - c_i}{\beta_i}, \tag{7.47}$$

where the parameter β_i defines steepness of this function, while in the vicinity of the minimum this function is very close to the quadratic, tending with the growth of X to a linear one.

Also interesting is the fact that the derivative of this function

$$f_i'(x_i) = \varphi(x_i) = \tanh \frac{x_i}{\beta_i}, \tag{7.48}$$

is a standard activation function of artificial neural networks (Back propagation, see Chap. 1). Using as a metric the following structure

$$D^R(x_k, c_j) = \sum\nolimits_{i=1}^{n} f_i(x_{k,i}, c_{j,i}) = \sum_{i=1}^{n} \beta_i \ln \cosh \frac{x_{k,i} - c_{j,i}}{\beta_i}, \tag{7.49}$$

is possible to introduce the objective function of robust classification [12]

$$\begin{aligned} E^R(w_{k,j}, c_j) &= \sum_{k=1}^{N}\sum_{j=1}^{m} w_{k,j}^{\beta} D^2(x_k, c_j) = \\ &= \sum_{k=1}^{N}\sum_{j=1}^{m} w_{k,j}^{\beta} \sum_{i=1}^{n} \beta_i \ln\cosh\frac{x_{k,i} - c_{j,i}}{\beta_i} \end{aligned} \tag{7.50}$$

and a corresponding Lagrangian

$$L = \sum_{k=1}^{N}\sum_{j=1}^{m} w_{k,j}^{\beta} \sum_{i=1}^{n} \beta_i \ln\cosh\frac{x_{k,i} - c_{j,i}}{\beta_i} + \sum_{k=1}^{N} \lambda_k \left(\sum_{j=1}^{m} w_{k,j} - 1\right) \tag{7.51}$$

where λ_k—is indefinite Lagrange multiplier, ensuring fulfillment of constraints (7.43), (7.44). The saddle point of the Lagrangian (7.51) can be found by solving the equations system of Kuhn-Tucker

$$\begin{cases} \frac{\partial L(w_{k,j}, c_j, \lambda_k)}{\partial w_{k,j}} = 0, \\ \frac{\partial L(w_{k,j}, c_j, \lambda_k)}{\partial \lambda_k} = 0, \\ \nabla c_j L(w_{k,j}, c_j, \lambda_k) = 0. \end{cases} \tag{7.52}$$

Solutions of the first and second equations lead to well-known results

$$\begin{cases} w_{k,j} = \frac{\left(D^R(x_k, c_j)\right)^{\frac{1}{1-\beta}}}{\sum_{l=1}^{m} \left(D^R(x_k, c_l)\right)^{\frac{1}{1-\beta}}} \\ \lambda_k = -\left(\sum_{l=1}^{m} \left(\beta D^R(x_k, c_l)\right)^{\frac{1}{1-\beta}}\right)^{1-\beta} \end{cases} \tag{7.53}$$

But the third equation

$$\nabla c_j L(w_{k,i}, c_j, \lambda_k) = \sum_{k=1}^{N} w_{k,j}^{\beta} \nabla c_j D^R(x_k, c_j) = 0, \tag{7.54}$$

evidently has no analytic solution. The solution of Eq. (7.54) can be obtained with the help of local modification of Lagrangian and recurrent fuzzy clustering algorithm [14].

Search of the Lagrangian local saddle point

$$L_k(w_{k,j}, c_j, \lambda_k) = \sum_{j=1}^{m} w_{k,j}^{\beta} D^R(x_k, c_j) + \lambda_k \left(\sum_{j=1}^{m} w_{k,j} - 1\right) \tag{7.55}$$

using procedures Arrow-Hurwitz-Udzawa leads to an algorithm

$$\begin{cases} w_{k,j}^{pr} = \frac{\left(D^R\left(x_k, c_j\right)\right)^{\frac{1}{1-\beta}}}{\sum_{l=1}^{m}\left(D^R\left(x_k, c_l\right)\right)^{\frac{1}{1-\beta}}} \\ c_{k+1,j,i} = c_{k,j,i} - \eta_k \frac{\partial L(w_{k,j}, c_j, \lambda_k)}{\partial c_{j,i}} = c_{k,j,i} + \eta_k w_{k,j}^{\beta} \tanh \frac{x_{k,i} - c_{k,j,i}}{\beta_i} \end{cases} \quad (7.56)$$

where $\eta_k$is a parameter of learning rate, $c_{k,j,i}$ is the ith component of the jth prototype calculated at the kth step.

But despite low computational complexity this algorithm (7.56) has the disadvantage inherent to all probabilistic clustering algorithm.

7.7 Robust Recursive Algorithm of Possibilistic Fuzzy Clustering

For possibilistic fuzzy clustering algorithms the criterion is the following expression [14]

$$E^R(w_{k,j}, c_j, \mu_i) = \sum_{k=1}^{N}\sum_{j=1}^{m} w_{k,j}^{\beta} D^2\left(x_k, c_j\right) + \sum_{j=1}^{m} \mu_i \sum_{k=1}^{N} (1 - w_{k,j})^{\beta} \quad (7.57)$$

Minimization of (7.57) by parameters $w_{k,j}$, c_j and μ_i leads to equations system

$$\begin{cases} \frac{\partial E^R(w_{k,j}, c_j, \mu_j)}{\partial w_{k,j}} = 0, \\ \frac{\partial E^R(w_{k,j}, c_j, \mu_j)}{\partial \lambda_k} = 0, \\ \nabla c_j E^R(w_{k,j}, c_j, \mu_j) = 0. \end{cases} \quad (7.58)$$

The solution of the first two equations of (7.58) leads to the well-known result

$$\begin{cases} w_{k,j}^{pos} = \left(1 + \left(\frac{D^R\left(x_k, c_j\right)}{\mu_j}\right)^{\frac{1}{\beta-1}}\right)^{-1} \\ \mu_j = \frac{\sum_{k=1}^{N} w_{k,j}^{\beta} D^R\left(x_k, c_j\right)}{\sum_{k=1}^{N} w_{k,j}^{\beta}} \end{cases} \quad (7.59)$$

while the third one

$$\nabla c_j E^R(w_{k,j}, c_j, \mu_j) = \sum_{k=1}^{N} w_{k,j}^{\beta} \nabla c_j D^R\left(x_k, c_j\right) = 0 \quad (7.60)$$

fully corresponds to (7.54).

Introducing the local modification of (7.57)

$$\begin{aligned} E_k^R &= \sum_{j=1}^{m} w_{k,j}^{\beta} D^R(x_k, c_j) + \sum_{j=1}^{m} \mu_j (1 - w_{k,j})^{\beta} = \\ &= \sum_{j=1}^{m} w_{k,j}^{\beta} \sum_{i=1}^{n} \beta_i \ln \cosh \frac{x_{k,i} - c_{j,i}}{\beta_i} + \sum_{j=1}^{m} \mu_j (1 - w_{k,j})^{\beta} \end{aligned} \tag{7.61}$$

and optimizing it we obtain:

$$\begin{cases} w_{k,j}^{pos} = \left(1 + \left(\frac{D^R(x_k, c_j)}{\mu_j}\right)^{\frac{1}{\beta - 1}}\right) \\ c_{k+1,j,i} = c_{k,j,i} - \eta_k \frac{\partial E_k^R(w_{k,j}, c_j, \mu_j)}{\partial c_{j,i}} = c_{k,j,i} + \eta_k w_{k,j}^{\beta} \tanh \frac{x_k - c_{k,j,i}}{\beta_i} \end{cases} \tag{7.62}$$

where the distance parameter μ_{kj} may be determined according to the second equation of the system (7.59) for k observations rather than the entire sample volume N.

It should be noted that the last equation of system (7.52) and (7.58) are identical and are determined only by choice of metrics. This makes possible to use any suitable metric for a particular case, which will determine only the setup procedure of prototypes if the equation for calculating the weights still remains the same.

Considered robust recursive methods may be used in a batch mode and in the on-line mode as well. In the last case the number of observation k represents a discrete time.

Experiments with a repository of data, distorted by abnormal outliers (emissions), have shown high efficiency of the proposed algorithms in the processing of the information given in the form of tables “object-property” [10, 11] and in the form of time series [14].

In particular, the problem of data classification of specially artificially generated sample containing three-dimensional cluster of data was considered, whose observations are marked the symbols “o”, “x” and “ + ” [12] (see. Fig. 7.4). Points in each cluster sampling are distributed according to the density of Laplace distribution having “heavy tails”

$$p(x_i) = \sigma \left(1 + (x_i - c)^2\right)^{-1} \tag{7.63}$$

where σ and c are width and center correspondingly.

The sample includes 9,000 observations (3,000 in each cluster) and is divided into training (7200 cases) and testing (1800 cases) subsamples [12, 14].

It should be noted that some observations are very far away from the centers of the clusters (Fig. 7.4a). Prototypes of the clusters are located in the central region of the data as shown in Fig. 7.4b. In order to find the correct prototypes clustering algorithm should be insensitive to outliers.

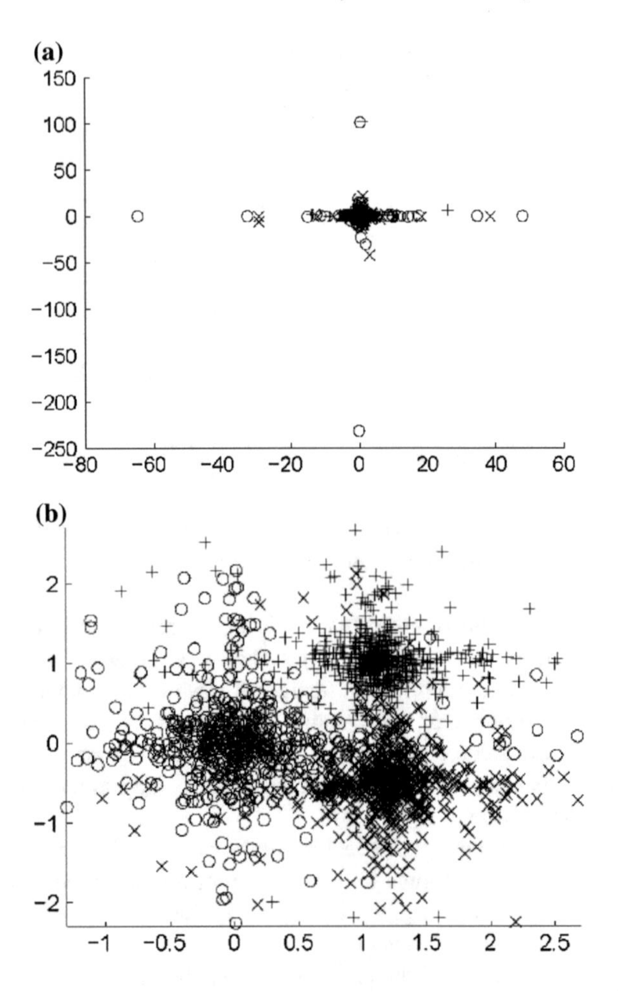

Fig. 7.4 Full sample (**a**) and its central part (**b**)

For all of the algorithms involved in the comparison, the procedure of the experiment was performed as follows. At the beginning of training a sample was clustered by appropriate algorithms and prototypes of clusters have been found. Then, training and testing samples were classified according to the results of clustering. Observations belonging to each cluster in the classification process are calculated in accordance with Eqs. (7.17), (7.56) or (7.62) depending on the type of clustering algorithm. The cluster, to which the observation belongs with a maximum membership degree, defines the class of this observation. Classification and training is performed in the on-line mode of receiving observations, where $\beta = 2$, $\beta_1 = \beta_2 = \beta_3 = 1$, $\eta(k) = 0.01$. The results are shown in Table 7.3 [12]

In the Fig. 7.5 it can be easily seen that the centers of the clusters (prototypes) produced by the algorithm « fuzzy C-means » by Bezdek, are shifted from the visual centers of the clusters, due to the presence of "heavy tails" of the data distribution density, in contrast to the robust methods with objective function (7.57)

Table 7.3 Results of classification. Classification error

Algorithm	Training sample	Testing sample
Fuzzy C-means	17.1 % (1229 obs.)	16.6 % (299 obs.)
Robust probabilistic	15.6 % (1127 obs.)	15.6 % (281 obs.)
Robust possibilistic	15.2 % (1099 obs.)	14.6 % (263 obs.)

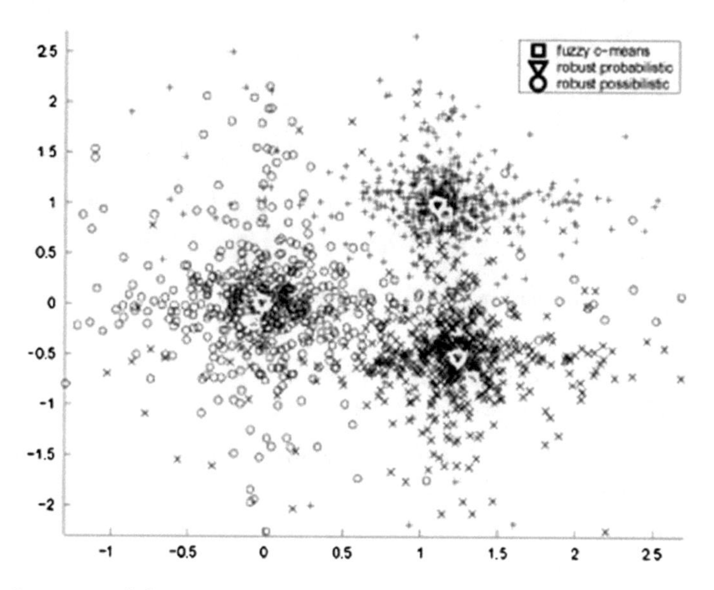

Fig. 7.5 Prototypes of clusters

and (7.61) in which prototypes are found more precisely, which is confirmed by the less classification error (see Table 7.3).

Continuous growth in the successful application of computational intelligence technologies in the areas of data analysis confirms the versatility of this approach. At the same time, real problems that arise in the processing of very large databases (Big Data), complicate the use of existing algorithms and tools and demand to be improved to meet the challenges of data mining in real time using the paradigms of CI and soft computing.

7.8 Application of Fuzzy Methods C-Means and Gustavson-Kessel's in the Problems of Automatic Classification

Consider application of fuzzy methods of cluster analysis C-means and Gustavson-Kessel's in the test and practical problems of automatic classification [3].

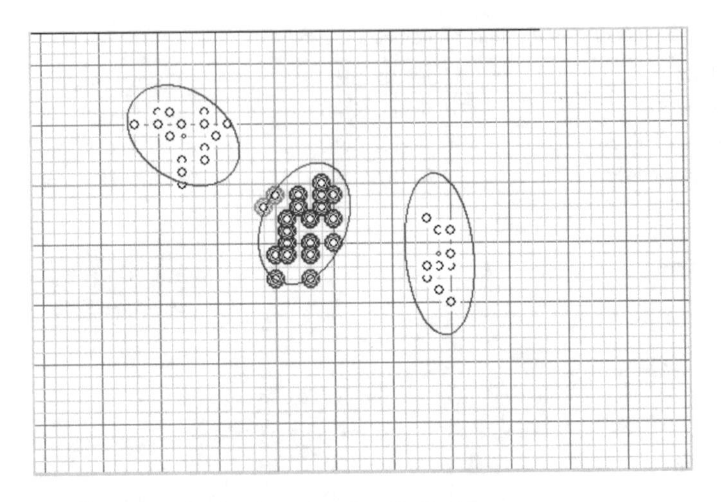

Fig. 7.6 Example of clustering by Gustafson-Kessel algorithm

Input data: set of points, number of clusters.
Output: Membership matrix U, cluster centers.

Example 7.1 Initial sampling points are shown in Fig. 7.6. The centers of the clusters are obtained using the methods:

1. peak grouping;
2. differential grouping.

Further, the method of finding clusters Gustafson-Kessel is used.

Placement of clusters in this example is shown in Fig. 7.6.

In the case of two clusters under the previous sample data, the results of clustering are given below (see. Fig. 7.7)

Example 7.2 Suppose that the original set of points is given in Fig. 7.8.

Consider first the case of two clusters. The corresponding results are presented in Fig. 7.8.

For three clusters results are shown in Fig. 7.9.

For four clusters: K = 4 corresponding solution is given below (Fig. 7.10).

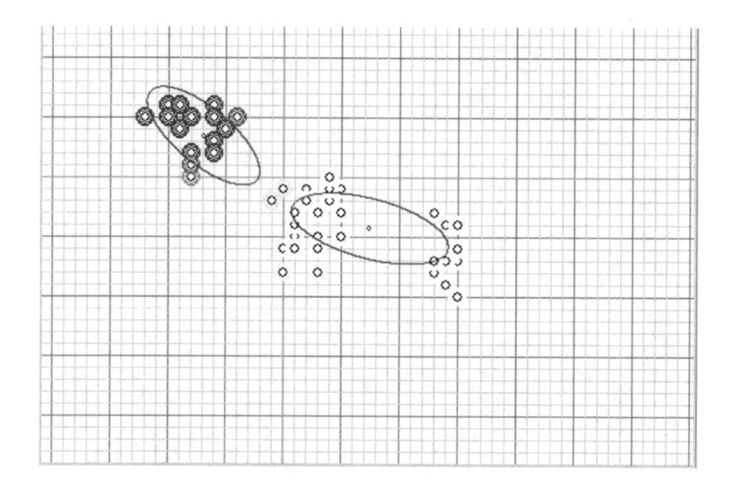

Fig. 7.7 Example of clustering by Gustafson-Kessel algorithm

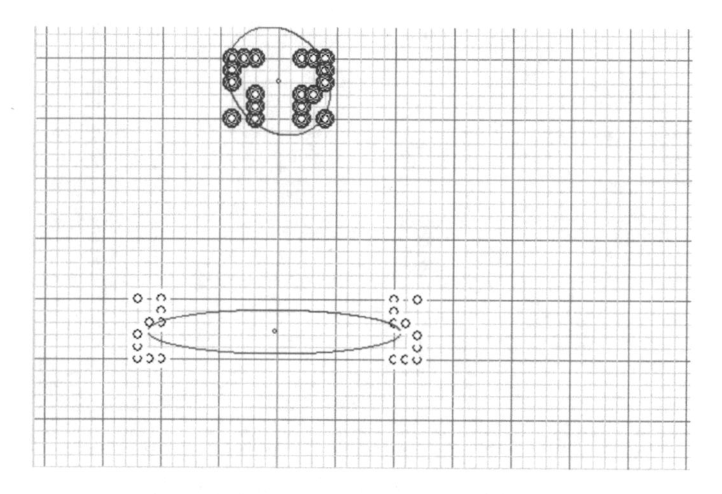

Fig. 7.8 Example of clustering by Gustafson-Kessel algorithm, k = 2

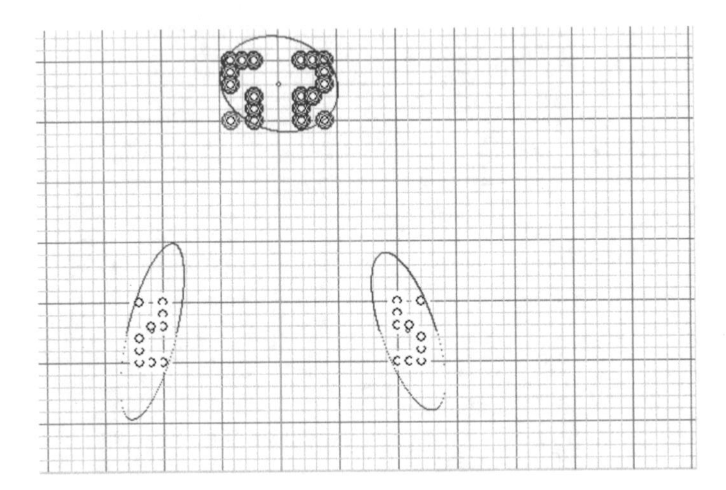

Fig. 7.9 Example of clustering by Gustafson-Kessel algorithm, k = 3

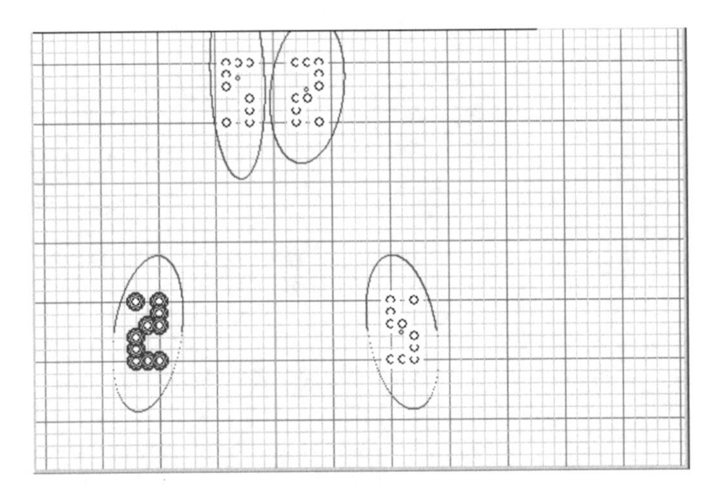

Fig. 7.10 Example of clustering by Gustafson-Kessel algorithm, k = 4

Example 7.3 Classification of the UN countries

These UN Millennium Indicators are presented in the Table 7.4.

In this experiment it was required to perform a clustering of the United Nations countries into 4 clusters by the above indicators. As a result of the clustering algorithm of Gustavson –Kessel application the following results were obtained (Tables 7.5, 7.6, 7.7, 7.8, and 7.9). As can be seen from the table in the first cluster are countries with relatively high rates of all indicators (compared to other countries in the sample). These are the countries of CIS, Eastern and western Europe, USA, Canada, the Balkans and Latin America countries.

Table 7.4 UN millenium indicators for countries

	Population percent below the poverty line	Percent of children for 5 years with an insufficient weight	Literacy	Gender equality. Percent of women among workers of the non-agricultural sphere
Afghanistan	70	48	50	17.8
Albania	25.4	14.3	99.4	40.3
Algeria	12.2	6	89.9	15.5
Angola	70	30.5	71.4	26.4
Argentina	15	5.4	98.6	47.6
Armenia	53.7	2.6	99.8	47
Azerbaijan	49.6	6.8	99.9	48.5
Bahrain	15	8.7	97	13.4
Bangladesh	49.8	47.7	49.7	24.2
Belize	40	6.2	84.2	44.4
Belarus	41.9	2	99.8	55.9
Butane	70	18.7	80	12
Benin	33	22.9	55.5	46
Bolivia	62.7	7.5	97.3	36.5
Bosnia Herzegovina	19.5	4.1	99.6	35.8
Botswana	70	12.5	89.1	47
Brazil	17.4	5.7	96.3	46.7
Bulgaria	12.8	2	99.7	52.2
Burundi	70	45.1	72.3	13.3
Burkina Faso	45.3	34.3	19.4	15.2
Cambodia	36.1	45.2	80.3	52.6
Cameroon	40.2	21	81.1	20.7
Verde's cap	40	13.5	89.1	39.1
It is central the African Republic	70	24.3	58.5	30.4
Fumes	64	28.1	37.3	5.5
Chile	17	0.7	99	37.3
China	4.6	10	98.9	39.5
Colombia	64	6.7	97.2	48.8
Congo	50	13.9	97.8	26.1
Costa Rica	22	5.1	98.4	39.5
Côte d'Ivoire	59	21.2	59.8	20.2
Croatia	20	0.6	99.6	46.3
Cuba	60	4.1	99.8	37.7

(continued)

Table 7.4 (continued)

	Population percent below the poverty line	Percent of children for 5 years with an insufficient weight	Literacy	Gender equality. Percent of women among workers of the non-agricultural sphere
Czech Republic	10	1	99.8	45.8
Democratic Republic of Congo	70	31.1	68.7	25.9
Djibouti	45.1	18.2	73.2	25
Dominican Republic	28.6	5.3	94	34.9
Ecuador	35	11.6	96.4	41.1
Egypt	16.7	8.6	73.2	21.6
El Salvador	48.3	10.3	88.9	31.1
Equatorial Guinea	50	18.6	92.7	10.5
Eritrea	53	39.6	60.9	35
Estonia	8.9	1	99.8	51.5
Ethiopia	44.2	47.2	57.4	39.9
Fiji	40	7.9	99.3	35.9
Gabon	40	11.9	59	37.7
Gambia	64	17	42.2	20.9
Georgia	11.1	3.1	99.8	45.2
Ghana	50	24.9	81.8	56.5
Guatemala	56.2	22.7	80.1	38.7
Guinea	40	23.2	50	30.3
Guinea-Bissau	48.7	25	44.1	10.8
Guyana	35	13.6	80	37.4
Haiti	45	17.3	66.2	39.5
Honduras	53	16.6	88.9	50.5
Hungary	17.3	3	99.5	47.1
India	28.6	47	64.3	17.5
Indonesia	27.1	26.1	98	30.8
Iran	30	10.9	86.3	17.2
Iraq	35	15.9	41	11.9
Jamaica	18.7	3.6	94.5	48
Jordan	11.7	4.4	99.4	24.9
Kazakhstan	34.6	4.2	99.8	48.7
Laos	38.6	40	78.5	42.1
Kenya	52	20.2	95.8	38.5
Democratic People's Republic of Korea	60	20.8	99.8	40.7

(continued)

Table 7.4 (continued)

	Population percent below the poverty line	Percent of children for 5 years with an insufficient weight	Literacy	Gender equality. Percent of women among workers of the non-agricultural sphere
Kuwait	20	9.8	93.1	24.1
Kyrgyzstan	64.1	11	99.7	44.1
Lebanon	20	3	92.1	25.9
Lesotho	50	17.9	87.2	24.7
Liberia	60	26.4	70.8	23.6
Libya	30	4.7	97	15
Madagascar	71.3	33.1	70.1	24.2
Malawi	65.3	21.9	63.2	12.5
Malaysia	40	12.4	97.2	38
Maldives	70	30.4	99.2	36.1
Mali	63.8	33.2	24.2	35.9
Mauritania	46.3	31.8	49.6	37
Mexico	30	7.5	96.6	37.4
Mozambique	69.4	23.7	62.8	11.4
Mongolia	36.3	12.7	97.7	49.4
Morocco	19	8.9	69.5	26.2
Nepal	42	48.3	70.1	11.8
Nicaragua	47.9	9.6	86.2	41.1
Niger	63	39.6	25.6	8.6
Nigeria	34.1	28.7	88.6	34
Pakistan	32.6	38	53.9	8.7
Panama	37.3	6.8	96.1	44
Papua New Guinea	37.5	7	68.6	35.4
Paraguay	21.8	4.6	96.3	42
Peru	49	7.1	96.6	37.2
Philippines	36.8	30.6	95.1	41.1
Poland	23.8	3	99.8	47.7
Moldova	23.3	3.2	98.7	54.6
Romania	21.5	5.7	97.8	45.3
Russian Federation	30.9	3	99.8	50.1
Rwanda	51.2	27.2	76.5	14.6
Senegal	33.4	22.7	51.5	25.7
Sri Lanka	25	29.4	97	43.2

(continued)

Table 7.4 (continued)

	Population percent below the poverty line	Percent of children for 5 years with an insufficient weight	Literacy	Gender equality. Percent of women among workers of the non-agricultural sphere
Serbia and Montenegro	30	1.9	99.8	44.9
Swaziland	40	10.3	88.1	31.3
Thailand	13.1	18.6	98	46.9
Trinidad and Tobago	21	5	99.8	41.3
Turkey	25	8.3	95.5	20.6
Turkmenistan	30	12	99.8	20
Tunisia	7.6	4	94.3	25.3
Uganda	44	22.8	80.2	35.6
Ukraine	31.7	3	99.9	53.6
United Arab Emirates	20	14.4	91.4	14.4
Tanzania	35.7	29.4	91.6	28.5
USA	5	1.4	99.1	48.8
Uzbekistan	27.5	7.9	99.7	41.5
Vietnam	50.9	33.1	94.1	51.8
Yemen	41.8	45.6	67.9	6.1
Zambia	72.9	28.1	81.2	29.4
Zimbabwe	34.9	13	97.6	21.8

Table 7.5 The centers of the clusters

28.25	5.96	97.86	43.77
29.85	12.51	82.14	23.67
59.72	26.00	65.40	24.48
40.39	34.25	74.51	31.35

In the second cluster are countries with smaller values of indicators, it's countries of North Africa and Middle East. In this cluster, is the lowest level of gender equality.

In the third cluster are the poorest countries with the lowest levels of literacy, as well as the low level of gender equality. Mainly it's African countries.

In the fourth cluster are poor countries with the most unfavorable conditions for the growth of children (See Table 7.6).

Example 7.4 Classification of the United Nations countries on sustainable development indicators.

Table 7.6 The matrix of belonging coefficients to different clusters (membership functions)

Afghanistan	0.055	0.162	0.495	0.288
Albania	0.599	0.114	0.073	0.214
Algeria	0.174	0.733	0.031	0.062
Angola	0.045	0.102	0.789	0.064
Argentina	0.939	0.024	0.016	0.021
Armenia	0.731	0.067	0.116	0.087
Azerbaijan	0.757	0.058	0.103	0.082
Bahrain	0.204	0.687	0.030	0.080
Bangladesh	0.031	0.071	0.187	0.712
Belize	0.251	0.252	0.266	0.231
Belarus	0.761	0.056	0.087	0.096
Butane	0.241	0.156	0.479	0.124
Benin	0.057	0.282	0.371	0.290
Bolivia	0.645	0.087	0.191	0.077
Bosnia Herzegovina	0.764	0.157	0.029	0.050
Botswana	0.530	0.130	0.215	0.124
Brazil	0.920	0.033	0.020	0.027
Bulgaria	0.914	0.033	0.025	0.028
Burundi	0.178	0.171	0.445	0.206
Burkina Faso	0.020	0.249	0.495	0.236
Cambodia	0.120	0.050	0.163	0.667
Cameroon	0.115	0.522	0.095	0.268
Verde's cap	0.428	0.197	0.160	0.215
It is central the African Republic	0.048	0.287	0.471	0.194
Fumes	0.021	0.117	0.681	0.180
Chile	0.758	0.158	0.034	0.049
China	0.800	0.091	0.040	0.070
Colombia	0.703	0.075	0.136	0.086
Congo	0.435	0.275	0.157	0.133
Costa Rica	0.930	0.039	0.012	0.019
Côte d'Ivoire	0.013	0.091	0.819	0.077
Croatia	0.928	0.032	0.018	0.022
Cuba	0.631	0.092	0.183	0.093
Czech Republic	0.919	0.040	0.018	0.022
Democratic Republic of Congo	0.038	0.108	0.784	0.069
Djibouti	0.036	0.604	0.203	0.158
Dominican respubl_ka	0.642	0.234	0.049	0.074
Ecuador	0.839	0.053	0.038	0.070
Egypt	0.065	0.791	0.068	0.076
El Salvador	0.389	0.243	0.219	0.148
Ecvatorial Guinea	0.347	0.331	0.140	0.182

(continued)

Table 7.6 (continued)

Eritrea	0.035	0.076	0.246	0.642
Estonia	0.909	0.037	0.026	0.028
Ethiopia	0.048	0.060	0.174	0.718
Fiji	0.716	0.130	0.067	0.087
Gabon	0.038	0.428	0.351	0.183
Gambia	0.035	0.200	0.509	0.256
Georgia	0.934	0.032	0.015	0.020
Ghana	0.180	0.107	0.264	0.448
Guatemala	0.100	0.143	0.577	0.181
Guinea	0.020	0.401	0.397	0.182
Guinea-Bissau	0.021	0.193	0.572	0.214
Guyana	0.126	0.435	0.206	0.233
Haiti	0.040	0.353	0.366	0.241
Honduras	0.368	0.122	0.275	0.235
Hungary	0.961	0.017	0.010	0.013
India	0.087	0.070	0.098	0.745
Indonesia	0.220	0.099	0.077	0.604
Iran	0.106	0.785	0.035	0.074
Iraq	0.034	0.306	0.464	0.196
Jamaica	0.815	0.077	0.051	0.057
Jordan	0.322	0.588	0.029	0.061
Kazakhstan	0.920	0.024	0.025	0.031
Laos	0.057	0.029	0.086	0.828
Kenya	0.375	0.131	0.318	0.177
Democratic People's Republic of Korea	0.293	0.128	0.451	0.127
Kuwait	0.232	0.689	0.023	0.055
Kyrgyzstan	0.609	0.093	0.213	0.085
Lebanon	0.240	0.656	0.040	0.063
Lesotho	0.276	0.360	0.204	0.160
Liberia	0.013	0.056	0.893	0.038
Libya	0.302	0.489	0.069	0.141
Madagascar	0.051	0.111	0.764	0.074
Malawi	0.044	0.120	0.715	0.120
Malaysia	0.725	0.100	0.069	0.106
Maldives	0.203	0.133	0.557	0.108
Mali	0.027	0.438	0.231	0.304
Mauritania	0.028	0.206	0.350	0.415
Mexico	0.900	0.054	0.017	0.029
Mozambique	0.050	0.117	0.721	0.112

(continued)

Table 7.6 (continued)

Mongolia	0.648	0.068	0.107	0.177
Morocco	0.055	0.757	0.101	0.086
Nepal	0.108	0.079	0.110	0.703
Nicaragua	0.299	0.206	0.290	0.204
Niger	0.020	0.170	0.533	0.276
Nigeria	0.096	0.040	0.052	0.813
Pakistan	0.065	0.172	0.163	0.600
Panama	0.930	0.023	0.021	0.025
Papua New Guinea	0.066	0.411	0.334	0.189
Paraguay	0.930	0.036	0.014	0.020
Peru	0.748	0.081	0.100	0.070
Philippines	0.119	0.045	0.096	0.740
Poland	0.966	0.013	0.009	0.012
Moldova	0.873	0.038	0.041	0.048
Romania	0.979	0.009	0.005	0.007
Russian Federation	0.924	0.023	0.024	0.029
Rwanda	0.138	0.315	0.240	0.308
Senegal	0.027	0.466	0.322	0.185
Sri Lanka	0.186	0.058	0.099	0.657
Serbia and Montenegro	0.921	0.031	0.021	0.027
Swaziland	0.293	0.433	0.125	0.150
Thailand	0.550	0.094	0.106	0.250
Trinidad and Tobago	0.931	0.035	0.013	0.021
Turkey	0.133	0.806	0.018	0.043
Turkmenistan	0.197	0.656	0.037	0.110
Tunisia	0.369	0.537	0.035	0.059
Uganda	0.089	0.145	0.241	0.525
Ukrain	0.867	0.036	0.044	0.053
United Arab Emirates	0.255	0.584	0.039	0.122
Tanzania	0.151	0.072	0.070	0.706
USA	0.905	0.043	0.025	0.027
Uzbekistan	0.889	0.046	0.023	0.042
Vietnam	0.190	0.077	0.308	0.426
Yemen	0.117	0.106	0.118	0.659
Zambia	0.147	0.125	0.659	0.069
Zimbabwe	0.249	0.585	0.046	0.120

Table 7.7 Indicators of the sustainable development

Albania	0.58387	0.54820	0.57383	0.57353
Algeria	0.56706	0.48474	0.26571	0.35047
Argentina	0.35296	0.63011	0.76377	0.49466
Armenia	0.65586	0.51200	0.33448	0.47481
Australia	0.56994	0.81968	0.78295	0.86663
Austria	0.68752	0.72971	0.80098	0.8154
Azerbaijan	0.60008	0.53918	0.35928	0.40572
Bangladesh	0.65282	0.21060	0.20669	0.17571
Belgium	0.614967	0.748405	0.817869	0.676777
Benin	0.498461	0.178393	0.334484	0.208119
Bolivia.	0.172206	0.433633	0.492253	0.298884
Bosnia and Herzegovina	0.548856	0.534668	0.492253	0.354646
Botswana	0.13233	0.418817	0.464936	0.347845
Brazil	0.216163	0.518054	0.626328	0.60656
Bulgaria	0.361631	0.583907	0.699515	0.518325
Cambodia	0.380226	0.23689	0.157964	0.204641
Cameroon	0.376477	0.201684	0.143939	0.2182
Canada	0.62295	0.77159	0.72201	0.84180
Chile	0.243463	0.64134	0.546863	0.731782
China	0.437748	0.463597	0.384858	0.332749
Colombia	0.157988	0.502915	0.546863	0.618515
Costa Rica	0.271513	0.557193	0.600368	0.724376
Croatia	0.601288	0.61875	0.676009	0.603607
Cyprus	0.962098	0.678039	0.600368	0.681226
Czech Republic	0.744175	0.717602	0.722017	0.702571
Denmark	0.761651	0.747263	0.743451	0.816827
Dominican Republic	0.303631	0.463597	0.464936	0.412767
Ecuador	0.294138	0.512001	0.600368	0.403071
Egypt	0.631237	0.399777	0.265716	0.419674
El Salvador	0.331064	0.457576	0.437827	0.504924
Estonia	0.553215	0.680678	0.699515	0.689583
Ethiopia	0.676386	0.101967	0.143939	0.201635
Finland	0.72688	0.752941	0.699515	0.85815
France	0.619443	0.754066	0.817869	0.794635
Gambia	0.324594	0.141861	0.265716	0.235642
Georgia	0.441675	0.516541	0.411085	0.47656
Germany	0.702424	0.768379	0.800988	0.806349

(continued)

Table 7.7 (continued)

Greece	0.588835	0.73447	0.743451	0.517882
Ґватемала	0.219163	0.316496	0.437827	0.325601
Honduras	0.167474	0.376757	0.359283	0.269843
Hungary	0.64978	0.67139	0.763774	0.663185
Iceland	0.962098	0.75068	0.743451	0.87829
India	0.536571	0.265352	0.225227	0.253416
Indonesia	0.537406	0.371083	0.359283	0.268663
Ireland	0.588616	0.778987	0.743451	0.752281
Israel	0.486303	0.754066	0.651581	0.644542
Italy	0.552527	0.733286	0.782957	0.668513
Jamaica	0.357959	0.5014	0.546863	0.405747
Japan	0.759359	0.767299	0.800988	0.793311
Yordaniya	0.517332	0.490799	0.359283	0.445905
Kazakhstan	0.655486	0.540692	0.265716	0.434827
Kenya	0.31728	0.211616	0.28762	0.234049
Korea, Republic	0.641912	0.759641	0.743451	0.610841
Kyrgyzstan	0.605705	0.36826	0.310566	0.317857
Latvia	0.558807	0.621604	0.699515	0.655443
Lithuania	0.520474	0.64134	0.676009	0.6509
Luxembourg	0.962098	0.73091	0.546863	0.802207
Madagascar	0.32533	0.178393	0.206696	0.19195
Malawi	0.490073	0.138213	0.28762	0.23739
Malaysia	0.344576	0.585379	0.437827	0.527639
Mexico	0.248442	0.594174	0.651581	0.552042
Moldova, Republic	0.510835	0.404146	0.437827	0.336335
Mongolia	0.542408	0.402688	0.334484	0.305619
Morocco	0.451213	0.325741	0.464936	0.427675
Mozambique	0.356033	0.080015	0.206696	0.241402
Namibia	0.047323	0.379606	0.464936	0.449997
Nepal	0.324226	0.172261	0.28762	0.3035
Niderlandi	0.655107	0.773727	0.782957	0.775775
Zealand is new	0.549644	0.79125	0.833616	0.854503
Nicaragua	0.239314	0.323085	0.334484	0.315336
Niger	0.40908	0.167985	0.157964	0.181214
Norway	0.744654	0.820582	0.763774	0.851428
Pakistan	0.619443	0.232537	0.130966	0.172761
Panama	0.239162	0.601457	0.573832	0.600787
Paraguay	0.245166	0.429174	0.546863	0.583011

(continued)

Table 7.7 (continued)

Peru	0.312213	0.554201	0.310566	0.522702
Filippini	0.386772	0.426208	0.310566	0.340265
Poland	0.589972	0.657887	0.722017	0.625403
Portugal	0.501937	0.657887	0.782957	0.667165
Rumuniya	0.650353	0.61875	0.600368	0.547301
Russian Federation	0.42253	0.548206	0.310566	0.515363
Senegal	0.486509	0.158067	0.28762	0.195048
Slovakia	0.744335	0.688527	0.651581	0.687639
Slovenia	0.650353	0.70137	0.763774	0.673112
South Africa	0.166191	0.366851	0.464936	0.29966
Spain	0.580784	0.743814	0.743451	0.724533
Sri Lanka	0.464124	0.456073	0.359283	0.450402
Sweden	0.757052	0.768379	0.699515	0.874694
Switzerland	0.600644	0.756307	0.722017	0.851068
Tajikistan	0.602294	0.343272	0.189316	0.253371
Tanzania, United Republic	0.520265	0.147865	0.225227	0.255087
Thailand	0.220891	0.450068	0.411085	0.400544
Trinidad i Tobago	0.463915	0.573568	0.464936	0.385968
Tunisia	0.452668	0.493827	0.464936	0.492711
Turkey	0.47499	0.487771	0.492253	0.455756
Uganda	0.381611	0.16714	0.244906	0.22529
Ukraine	0.716261	0.534668	0.464936	0.361609
United Arab Emirates	0.962098	0.684616	0.225227	0.51641
Great Britain	0.55376	0.72732	0.800988	0.780447
USA	0.452668	0.786202	0.937406	0.805513
Uruguay	0.419463	0.615887	0.743451	0.687183
Uzbekistan	0.53824	0.395423	0.225227	0.17794
Venezuela. Bolivar Republic	0.397572	0.513515	0.265716	0.30664
Vjetnam	0.520474	0.332429	0.265716	0.317299
Zambia	0.264436	0.145589	0.310566	0.254877
Zimbabwe	0.264436	0.035073	0.10799	0.1072

Table 7.8 Degrees of membership to clusters

3	Albania	0.40165	0.10648	0.49186
3	Algeria	0.08839	0.43869	0.47292
3	Argentina	0.38845	0.11707	0.49448
3	Armenia	0.17773	0.27465	0.54762
1	Australia	0.90018	0.02894	0.07086
1	Austria	0.95403	0.01378	0.03218
3	Azerbaijan	0.13099	0.24733	0.62168
2	Bangladesh	0.04774	0.8121	0.14015
1	Belgium	0.95258	0.01291	0.0345
2	Benin	0.017346	0.91288	0.069771
3	Bolivia.	0.086566	0.27213	0.6413
3	Bosnia and Herzegovina	0.1149	0.16889	0.71621
3	Botswana	0.090011	0.27373	0.63626
3	Brazil	0.21422	0.13172	0.65406
3	Bulgaria	0.29843	0.10605	0.59552
2	Cambodia	0.015168	0.92082	0.064017
2	Cameroon	0.017754	0.91106	0.071185
1	Canada	0.94729	0.01499	0.03771
3	Chile	0.34211	0.1277	0.53019
3	China	0.04406	0.22574	0.7302
3	Colombia	0.17747	0.15295	0.66959
3	Costa Rica	0.33409	0.12403	0.54188
1	Croatia	0.75759	0.052825	0.18958
1	Cyprus	0.68146	0.11573	0.2028
1	Czech Republic	0.94321	0.016686	0.0401
1	Denmark	0.92516	0.023355	0.051486
3	Dominican Republic	0.023839	0.05861	0.91755
3	Ecuador	0.08807	0.10437	0.80756
2	Egypt	0.10324	0.48662	0.41014
3	El Salvador	0.021848	0.041038	0.93711
1	Estonia	0.91918	0.018352	0.062473
2	Ethiopia	0.063802	0.77286	0.16334
1	Finland	0.91557	0.025895	0.058533
1	France	0.95921	0.011702	0.029093
2	Gambia	0.023284	0.87702	0.099697
3	Georgia	0.023663	0.038418	0.93792
1	Germany	0.94903	0.015455	0.03551
1	Greece	0.74098	0.062566	0.19646
3	Гватемала	0.067026	0.40502	0.52795
3	Honduras	0.071116	0.45306	0.47582
1	Hungary	0.95232	0.012626	0.03505
1	Iceland	0.76693	0.084196	0.14888
2	India	0.012701	0.93374	0.053564

(continued)

Table 7.8 (continued)

2	Indonesia	0.050291	0.64789	0.30182
1	Ireland	0.97523	0.006589	0.018182
1	Israel	0.75846	0.049725	0.19181
1	Italy	0.93061	0.01727	0.052117
3	Jamaica	0.037423	0.05336	0.90922
1	Japan	0.92541	0.023439	0.051152
3	Yordaniya	0.070166	0.17151	0.75832
3	Kazakhstan	0.16399	0.33124	0.50477
2	Kenya	0.020158	0.8778	0.10204
1	Korea, Republic	0.91524	0.02221	0.062547
2	Kyrgyzstan	0.066312	0.62744	0.30625
1	Latvia	0.82602	0.037702	0.13627
1	Lithuania	0.76748	0.046179	0.18634
1	Luxembourg	0.6903	0.11288	0.19682
2	Madagascar	0.021309	0.89052	0.088166
2	Malawi	0.014653	0.92942	0.055931
3	Malaysia	0.058977	0.05684	0.88418
3	Mexico	0.23705	0.117	0.64595
3	Moldova, Republic	0.064556	0.31673	0.61872
2	Mongolia	0.057925	0.56437	0.3777
3	Morocco	0.062452	0.26329	0.67426
2	Mozambique	0.029408	0.86369	0.1069
3	Namibia	0.1218	0.28352	0.59468
2	Nepal	0.025167	0.84944	0.12539
1	Niderlandi	0.97497	0.007208	0.017824
1	Zealand is new	0.8935	0.030837	0.075666
2	Nicaragua	0.051662	0.54775	0.40058
2	Niger	0.017697	0.91677	0.065531
1	Norway	0.90361	0.030465	0.065925
2	Pakistan	0.044592	0.82182	0.13359
3	Panama	0.21163	0.11324	0.67513
3	Paraguay	0.12109	0.13045	0.74846
3	Peru	0.0764	0.13887	0.78473
3	Filippini	0.043558	0.41156	0.54488
1	Poland	0.8745	0.028966	0.096534
1	Portugal	0.83716	0.037676	0.12517
1	Rumuniya	0.57831	0.094633	0.32706
3	Russian Federation	0.076257	0.13511	0.78863
2	Senegal	0.013131	0.93672	0.050145
1	Slovakia	0.89028	0.031444	0.078275
1	Slovenia	0.97281	0.007269	0.019919
3	South Africa	0.08185	0.34815	0.57
1	Spain	0.98221	0.004552	0.013235

(continued)

Table 7.8 (continued)

3	Sri Lanka	0.043578	0.13993	0.8165
1	Sweden	0.89026	0.034693	0.075048
1	Switzerland	0.93911	0.017102	0.043793
2	Tajikistan	0.045867	0.78327	0.17087
2	Tanzania, United Republic	0.015506	0.92838	0.056119
3	Thailand	0.055589	0.17785	0.76656
3	Trinidad i Tobago	0.059986	0.083263	0.85675
3	Tunisia	0.026507	0.033837	0.93966
3	Turkey	0.039097	0.051998	0.90891
2	Uganda	0.01028	0.94557	0.044149
3	Ukraine	0.23303	0.27103	0.49594
1	United Arab Emirates	0.36305	0.28152	0.35543
1	Great Britain	0.94643	0.014397	0.039177
1	USA	0.79335	0.060098	0.14655
1	Uruguay	0.666	0.069556	0.26445
2	Uzbekistan	0.043994	0.77022	0.18578
3	Venezuela. Bolivar Republic	0.06372	0.38883	0.54745
2	Vjetnam	0.028441	0.81113	0.16043
2	Zambia	0.03963	0.78231	0.17806
2	Zimbabwe	0.061394	0.75431	0.18429
Number of a cluster		1	2	3
	GINI	Ihd	Iql	Isd
	0.63499	0.724	0.73749	0.73457
	0.44656	0.236	0.25218	0.24782
	0.38459	0.493	0.45717	0.45412
	Criterion	9.8738	Hi-Beni	0.438

Table 7.9 Degrees of membership of the countries to clusters

4	Albania	0.28424	0.3116	0.072582	0.33157
4	Algeria	0.027483	0.071321	0.10306	0.79814
2	Argentina	0.22755	0.53851	0.067084	0.16686
4	Armenia	0.070525	0.11933	0.091308	0.71884
1	Australia	0.86029	0.070031	0.022093	0.047588
1	Austria	0.94636	0.0251	0.008865	0.019672
4	Azerbaijan	0.033017	0.073697	0.051201	0.84209
3	Bangladesh	0.042206	0.084422	0.63857	0.2348
1	Belgium	0.92169	0.039019	0.011551	0.027739
3	Benin	0.017031	0.046297	0.82568	0.111
2	Bolivia	0.061197	0.50498	0.18867	0.24515
4	Bosnia and Herzegovina	0.057183	0.1661	0.071758	0.70496
2	Botswana	0.060793	0.52973	0.18418	0.22529
2	Brazil	0.068824	0.80042	0.041411	0.089343

(continued)

2	Bulgaria	0.14689	0.65894	0.051013	0.14315
3	Cambodia	0.009822	0.028878	0.90035	0.060951
3	Cameroon	0.009908	0.028354	0.90497	0.056772
1	Canada	0.92961	0.034214	0.010931	0.025244
2	Chile	0.18624	0.58983	0.067466	0.15646
4	China	0.019842	0.10456	0.078945	0.79666
2	Colombia	0.066882	0.7671	0.056491	0.10953
2	Costa Rica	0.16698	0.6328	0.060572	0.13965
1	Croatia	0.61489	0.18965	0.045698	0.14976
1	Cyprus	0.57037	0.14719	0.085695	0.19675
1	Czech Republic	0.91329	0.037491	0.013955	0.03526
1	Denmark	0.90361	0.042127	0.016682	0.037581
2	Dominican Republic	0.032441	0.67082	0.072181	0.22456
2	Ecuador	0.040697	0.79243	0.045605	0.12127
4	Egypt	0.04537	0.097315	0.16448	0.69284
2	El Salvador	0.03225	0.70707	0.054953	0.20573
1	Estonia	0.8311	0.091463	0.020293	0.057147
3	Ethiopia	0.050831	0.094557	0.63814	0.21647
1	Finland	0.88749	0.050062	0.019109	0.043343
1	France	0.94945	0.024991	0.007902	0.017653
3	Gambia	0.011694	0.037745	0.88941	0.061153
4	Georgia	0.038094	0.26718	0.053444	0.64128
1	Germany	0.94112	0.027184	0.009862	0.021834
1	Greece	0.60917	0.19229	0.051299	0.14724
2	Guatemala	0.052387	0.36066	0.31644	0.27051
3	Honduras	0.054833	0.31568	0.35575	0.27374
1	Hungary	0.90577	0.04526	0.013551	0.035419
1	Iceland	0.68922	0.12112	0.063855	0.12581
3	India	0.019765	0.051379	0.75142	0.17743
4	Indonesia	0.030765	0.093396	0.26095	0.61489
1	Ireland	0.96099	0.019763	0.005589	0.013659
1	Israel	0.61194	0.22343	0.041591	0.12304
1	Italy	0.86916	0.06992	0.017282	0.043638
2	Jamaica	0.038649	0.71492	0.050335	0.19609
1	Japan	0.90616	0.041501	0.016386	0.035957
4	Jordan	0.009514	0.034653	0.018891	0.93694
4	Kazakhstan	0.069722	0.12096	0.11663	0.69268
3	Kenya	0.013179	0.047963	0.85807	0.080792
1	Korea. Republic	0.85087	0.070721	0.021213	0.057195
4	Kyrgyzstan	0.03518	0.085876	0.22919	0.64975
1	Latvia	0.69365	0.16621	0.035339	0.1048
1	Lithuania	0.61283	0.2234	0.040208	0.12357
1	Luxembourg	0.58655	0.1471	0.084004	0.18234

(continued)

3	Madagascar	0.010267	0.031539	0.90351	0.054684
3	Malawi	0.011209	0.029699	0.89214	0.066953
2	Malaysia	0.049112	0.71574	0.043663	0.19148
2	Mexico	0.079655	0.78924	0.03832	0.092786
4	Moldova, Republic	0.024628	0.095535	0.092677	0.78716
4	Mongolia	0.022514	0.068667	0.14794	0.76088
4	Morocco	0.048106	0.2518	0.17008	0.53001
3	Mozambique	0.014971	0.041838	0.87376	0.069428
2	Namibia	0.075055	0.54377	0.17888	0.20229
3	Nepal	0.017185	0.062621	0.81948	0.10071
1	Netherlands	0.9755	0.011725	0.003857	0.008913
1	New Zealand	0.84821	0.07761	0.023867	0.050316
3	Nicaragua	0.042242	0.23973	0.4443	0.27372
3	Niger	0.008523	0.022791	0.92278	0.045911
1	Norway	0.87596	0.055127	0.021828	0.047085
3	Pakistan	0.039064	0.080179	0.66428	0.21648
2	Panama	0.067097	0.81058	0.034832	0.087487
2	Paraguay	0.039361	0.83131	0.041076	0.088257
2	Peru	0.065334	0.46554	0.10782	0.3613
4	Philippine	0.029716	0.15589	0.21217	0.60223
1	Poland	0.76646	0.12003	0.028969	0.084542
1	Portugal	0.71572	0.16448	0.033888	0.085911
1	Romania	0.42991	0.22585	0.068687	0.27555
4	Russian Federation	0.058523	0.27694	0.090317	0.57422
3	Senegal	0.00987	0.026109	0.90456	0.059459
1	Slovakia	0.82228	0.073597	0.027661	0.076458
1	Slovenia	0.94376	0.027013	0.008162	0.02106
2	South Africa	0.060062	0.42837	0.25723	0.25433
1	Spain	0.96064	0.020311	0.005394	0.013656
4	Sri Lanka	0.015491	0.078825	0.040186	0.8655
1	Sweden	0.85291	0.064036	0.025878	0.057175
1	Switzerland	0.91403	0.042608	0.013135	0.03023
3	Tajikistan	0.040625	0.091683	0.48922	0.37848
3	Tanzania	0.013224	0.032817	0.87388	0.080083
2	Thailand	0.045151	0.57189	0.13572	0.24724
4	Trinidad and Tobago	0.052559	0.26387	0.06346	0.62011
4	Tunisia	0.049968	0.38186	0.056285	0.51189
4	Turkey	0.050238	0.3103	0.058481	0.58098
3	Uganda	0.002341	0.00716	0.97664	0.013863
4	Ukraine	0.11937	0.15783	0.11911	0.60369
4	United Arab Emirates	0.25172	0.18882	0.17458	0.38488
1	Great Britain	0.91502	0.044743	0.012211	0.028022
1	USA	0.69788	0.16177	0.047239	0.093105

(continued)

1	Uruguay	0.48376	0.33651	0.051159	0.12857
3	Uzbekistan	0.038792	0.098392	0.4853	0.37751
4	Venezuela	0.039499	0.16547	0.19342	0.6016
4	Vietnam	0.027619	0.082803	0.4324	0.45718
3	Zambia	0.026225	0.093996	0.75376	0.12602
3	Zimbabwe	0.040152	0.099806	0.71609	0.14395
Number of a cluster		1	2	3	4
Centers of clusters					
		0.64429	0.73332	0.74677	0.74678
		0.29515	0.51657	0.53844	0.51317
		0.419	0.19409	0.23834	0.22998
		0.51726	0.46924	0.3707	0.39482
		Criterion 1	9.4268	Hi-Beni	0.39492

Investigations of fuzzy clustering method C-means by indicators of sustainable development for the countries of the United Nations were carried out. For this, the data of the World Data Center in Ukraine (WDC) were used (See Table 7.7).

As sustainable development indicators the following indices were taken:

- Index GINI-GINI
- Ihd—index of health status
- Iql—standard of living index
- Isd—index of sustainable development.

As algorithm of initial centers placement the algorithm of differential grouping was applied. Clustering was carried out for a different number of clusters K = 3, 4, 5. Besides the value of optimized criterion the quality of splitting will be evaluated by the indicator of Hi-Beni:

$$\chi = \frac{d_{av}}{D_{av}},$$

where d_{av} is the average intra-cluster distance, D_{av}—average inter-cluster distance. This indicator should be minimized.

Experiment 1. K = 3

Let us analyze the results (See Table 7.8). The *first cluster* contains countries with the highest values of all parameters. These are the countries of Western Europe, as well as some other. Namely, Australia, Austria, Belgium, Great Britain, Hungary, Denmark, Iceland, Ireland, Israel, Italy, Cyprus, Latvia, Lithuania, Luxembourg, Netherlands, New Zealand, Norway, Poland, Portugal. USA, Slovakia, Slovenia, Croatia, Czech Republic, Sweden, Switzerland, Uruguay.

The *second cluster* contains countries with an average value of the index GINI, and minimum values of all other indicators. These are the countries of Africa and

South-East Asia. These include: Bangladesh, Egypt, Zambia, Zimbabwe, India, Indonesia, Cambodia, Cameroon, Kyrgyzstan, Nicaragua, Niger, Pakistan, Uganda, Senegal, Tajikistan, Tanzania, and others.

The *third cluster* contains countries with average values of all the indicators and the small value of the index GINI. It includes the CIS countries, Latin America and some of the most developed countries of Asia and Africa. Namely, Armenia, Albania, Algeria, Argentina, Brazil, Bolivia, Bulgaria, Bosnia and Herzegovina, Venezuela, Honduras, Guatemala, Georgia, Jordan, Kazakhstan, China, Costa Rica, Colombia, Mexico, Moldova, Peru, Paraguay, Russian Federation, Trinidad and Tobago, Tunisia, Turkey, Ukraine, Chile, South Africa, Jamaica.

Experiment 2. K = 4

It is interesting to analyse dynamics of changes of clusters after transition from K = 3 to K = 4 (See Table 7.9).

The countries with the greatest values of all indicators fall to the first cluster. The structure of this cluster practically didn't change. In the second cluster there are countries with the minimum value of an index GINI and average values of all other indicators. Here are the countries of Latin America: Argentina, Brazil, Panama, Paraguay, Peru, Uruguay, etc.

The countries with the minimum values of all indicators except GINI index fall to the third cluster. Here the countries from the second cluster of the previous clustering at K = 3 fall. Namely, Bangladesh, Benin, Zambia, Zimbabwe, India, Cambodia, Cameroon, Kenya, Mozambique, Nepal, Pakistan, Senegal, Tadzhikistan, Tanzania, Uzbekistan.

The countries with average values of all indicators fall to the fourth cluster. Here the countries from the third cluster of the previous clustering fall, namely: Venezuela, Vietnam, Ukraine, the Russian Federation, Azerbaijan, Georgia, Indonesia, Jordan, Kyrgyzstan, Sri-Lanka. Value of an indicator of Chi- Beni decreased from 0.438 to 0.39492.

Experiment 3. K = 5

For this experiment we present only the average data for cluster centers (see below)

Centers of clusters:			
0.52645	0.45648	0.35255	0.37995
0.56042	0.67078	0.70818	0.64717
0.41643	0.18501	0.23361	0.22605
0.68519	0.75586	0.75276	0.80007
0.27997	0.47653	0.48824	0.45777
Criterion 1	9.0011	Hi Beni	0.38816

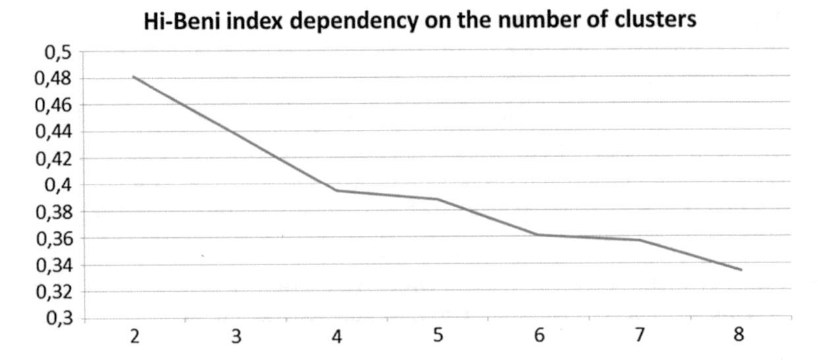

Fig. 7.11 Hi-Beni index versus number of clusters

Consider the dependence of the index Hi-Beni on the number of clusters K (Fig. 7.11).

As the chart above shows, the value of Hi-Beni index significantly decreases when $K = 2 - 4$, then its value is changing slightly. Therefore, the optimal number of clusters lies in vicinity of $K = 4$.

7.9 Conclusions

Cluster analysis includes a set of different classification algorithms. In general, whenever it is necessary to classify the "mountains" of information to suitable for further processing groups, cluster analysis is very useful and effective. Cluster analysis is needed for the classification of information, it can be used in a certain way to structure the variables and to find out which variables should be combined in the first place, and which should be considered separately.

A great advantage of the cluster analysis is that it allows to split the objects not only by one parameter but by a set of attributes as well. In addition, cluster analysis unlike most mathematical and statistical methods do not impose any restrictions on the form of these objects, and allows to treat a variety of raw data of almost arbitrary nature. This is important, for example, in the situation when indicators are diverse views, and it's impossible to use traditional econometric approaches.

As any other method, cluster analysis has certain disadvantages and limitations: in particular, the content and the number of clusters depend on the criteria selected for partition. For the reduction of the original data set to a more compact form there may be some distortion, and characteristics of individual objects may be lost by replacing them with the characteristics of parameters of the cluster center.

The main disadvantage of the considered methods of fuzzy clustering C-means and Gustavson-Kessel is that they can only be used when the number of clusters K is known. But usually, the number of clusters is unknown, and visual observations in the multidimensional case simply don't lead to a success.

References

1. Durant, B., Smith, G.: Cluster analysis.- M.Statistica 289 p. (rus)
2. Zaychenko, Y.P.: Fundamentals of Intellectual Systems Design, 352 p. Kiev Publ. house, Slovo (2004). (in Russ.)
3. Zaychenko, Y.P.: Fuzzy models and methods in intellectual systems, 354 p. Publ. House "Slovo", Kiev (2008). Zaychenko, Y.: Fuzzy Group Method of Data Handling under fuzzy input data. In: System research and information technologies, №3, pp. 100–112 (2007). (in Russ.)
4. Osovsky, S.: Neural networks for information processing, transl. from pol. M Publ. house Finance and Statistics, 344 p. (2002). (in Russ.)
5. Yager, R.R., Filev, D.P.: Approximate clustering via the mountain method. IEEE Trans. Syst. Man Cybern. **24,** 1279–1284 (1994)
6. Krishnapuram, R., Keller, J.: A possibilistic approach to clustering. IEEE Trans. Fuzzy Syst. **1**, 98–110 (1993)
7. Krishnapuram, R., Keller, J.: Fuzzy and possibilistic clustering methods for computer vision. IEEE Trans. Fuzzy Syst. **1**, 98–110 (1993)
8. Chung, F.L., Lee, T.: Fuzzy competitive learning. Neural Netw. **7**, 539–552 (1994)
9. Park, D.C., Dagher, I.: Gradient based fuzzy C-means (GBFCM) algorithm. In: Proceedings of IEEE Inernational Conference on Neural Networks, pp. 1626−1631 (1984)
10. Bodyanskiy, Y., Gorshkov Y., Kokshenev, I., Kolodyazhniy, V.: Robust recursive fuzzy clustering algorithms. In: Proceedings of East West Fuzzy Colloquim 2005, pp. 301–308, HS, Zittau/Goerlitz (2005)
11. Bodyanskiy, Y., Gorshkov, Y., Kokshenev, I., Kolodyazhniy, V.: Outlier resistant recursive fuzzy clustering algorithm. In: Reusch, B. (ed.) Computational Intelligence: Theory and Applications. Advances in Soft Computing, vol. 38, pp. 647–652. Springer, Berlin (2006)
12. Bodyanskiy, Y.: Computational intelligence techniques for data analysis. In: Lecture Notes in Informatics, vol. P-72, pp. 15–36. GI, Bonn (2005)
13. Vasiliyev, V.I.: Pattern-recognition systems. Naukova Dumka, Kiev (1988). (in Russ)
14. Bodyanskiy, Y., Gorshkov, Y., Kokshenev, I., Kolodyazhniy, V., Shilo, O.: Robust recursive fuzzy clustering-based segmentation of biomedical time series. In: Proceedings of 2006 International Symposium on Evolving Fuzzy Systems, pp. 101–105, Lancaster, UK (2006)

Chapter 8
Genetic Algorithms and Evolutionary Programing

8.1 Introduction

The chapter is devoted to widely used genetic algorithms (GA) and deeply connected with them evolutionary programing. In Sect. 8.2 genetic algorithms are considered and their properties analyzed. The different variants of main GA operators: cross-over are considered, various presentations of individuals- binary and floating—point are considered and different cross-over operators for them are described.

In the Sect. 8.3 various implementations of mutation operator are presented and their properties are discussed. Parameters of GA are considered, deterministic, adaptive and self-adaptive parameters modifications are described and discussed. In the Sect. 8.5 different selection strategies are considered and analyzed. In the Sect. 8.6 the application of GA for solution of practical problem of computer network structural synthesis is described. The Sect. 8.7 is devoted to detail description and analysis of evolutionary programing (EP). Main operators of EP: mutations and selection are considered and their properties analyzed. Different variants of algorithms EP are described. In the Sect. 8.8 differential evolution is considered, its properties discussed and the algorithm of DE is presented.

8.2 Genetic Algorithms

Genetic algorithms (GAs) are the first algorithmic models developed to simulate genetic systems. First proposed by Fraser [1], it was the extensive work done by Holland [2] that popularized GAs. It is also due to his work that Holland was generally recognized as the *father of genetic alorithms*.

GAs model genetic evolution, where the characteristics of individuals are expressed using genotypes. The main driving operators of GA are selection (to

M.Z. Zgurovsky and Y.P. Zaychenko, *The Fundamentals of Computational Intelligence: System Approach*, Studies in Computational Intelligence 652,
DOI 10.1007/978-3-319-35162-9_8

model survival of the fittest) and recombination through application of a crossover operator (to model reproduction).

Consider the general description of genetic algorithm.

Preliminary stage. Create and initialize the initial population $C(0)$.

Choose the stop condition, assign populations counter $t = 0$.

Iteration t

1. Estimate fitness index $f(X_i(t))$ of every individual $X_i(t)$.
2. Execute reproduction to create offspring.
3. Choose new population $C(t+1)$.
4. Check the stop condition, if it fulfills then stop otherwise $t = t+1$ and go to the next iteration.

Canonical Genetic Algorithm

The canonical GA (CGA) proposed by Holland [2] follows the general algorithm descripted above, with the following implementation specificity:

- A bit string representation of individuals was used.
- Proportional selection was used to select parents for recombination.
- One-point crossover (see Sect. 8.2) was used as the primary method to
- produce offspring.
- Uniform mutation (refer to Sect. 8.3) was proposed as a background operator of small importance.

It is worth to note that mutation was not considered as an important operator in the original GA implementations. Only in later implementations the explorative power of mutation was used to improve the search capabilities of GAs.

Since CGA, several variations of the GA have been developed that differ in representation scheme, selection operator, crossover operator and mutation operator. Consider the main approaches to implementation of GA illustrating all its diversity.

Crossover

Crossover operators can be divided into three main categories based on the arity (i.e. the number of parents used) of the operator. This results in three main classes of crossover operators []:

- **asexual**, where an offspring is generated from one parent.
- **sexual**, where two parents are used to produce one or two offspring.
- **multi-recombination**, where more than two parents are used to produce one or more offspring.

Each pair (or group) of parents have a probability, p_c of producing offspring. Usually, a high crossover probability is used.

In selection of parents, the following questions need to be considered:

- Due to probabilistic selection of parents, it may happen that the same individuals will be selected as both parents, in which case the generated offspring will be a copy of the parent. The parent selection process should therefore incorporate a test to prevent such non-desirable case.

- It is also possible that the same individual takes part in more than one application of the crossover operator. This becomes a problem when fitness-proportional selection schemes are used. In addition to parent selection and the recombination process, the crossover operator considers a replacement policy. If one offspring is generated, the offspring may replace the worst parent. Such replacement can be based on the restriction that the offspring must be more fit than the worst parent, or it may be rejected. Crossover operators have also been implemented where the offspring replaces the worst individual of the population. In the case of two offspring, similar replacement strategies can be used.

Binary Representations

Most of the crossover operators for binary representations are sexual, being applied to two selected parents. If $x_1(t)$ and $x_2(t)$ denote the two selected parents, then the recombination process is described in Algorithm 8.1. In this algorithm, $m(t)$ is a mask that specifies which bits of the parents should be swapped to generate the offsprings, $\tilde{x}_1(t)$ and $\tilde{x}_2(t)$. Several crossover operators have been developed to compute the mask:

- **One-point crossover**: Holland [2] suggestedthat segments of genes be swapped between the parents to create their offspring, and not single genes. A one-point crossover operator was developed that randomly selects a crossover point, and the bitstrings after that point are swapped between the two parents. One-point crossover is illustrated in Fig. 8.1b. The mask is computed using Algorithm 8.1.
- **Two-point crossover**: In this case two bit positions are randomly selected, and the bitstrings between these points are swapped as illustrated in Fig. 8.1c. The mask is calculated using Algorithm 8.2. This operator can be generalized to an n-point crossover [3].
- **Uniform crossover**: The n_x-dimensional mask is created randomly [3] as summarized in Algorithm 8.3. Here, p_x is the bit-swapping probability. If $p_x = 0.5$, then each bit has an equal chance to be swapped. Uniform crossover is illustrated in Fig. 8.1a.

Algorithm 8.1. One-Point Crossover Mask Calculation
Select the crossover point $\xi \sim U(1, n_x - 1)$;
Initialize the mask: $m_j(t) = 0$, for all $j = 1, \ldots, n_x$;
for $j = \xi + 1$ **to** n_x **do**
$m_j(t) = 1$;
end

Algorithm 8.2. Two-Point Crossover Mask Calculation
Select the two crossover points, $\xi_1, \xi_2 \sim U(1, n_x)$;
Initialize the mask: $m_j(t) = 0$ for all $j = 1, \ldots, n_x$
for $j = \xi_1 + 1$ **to** ξ_2 **do**
$m_j(t) = 1$
end

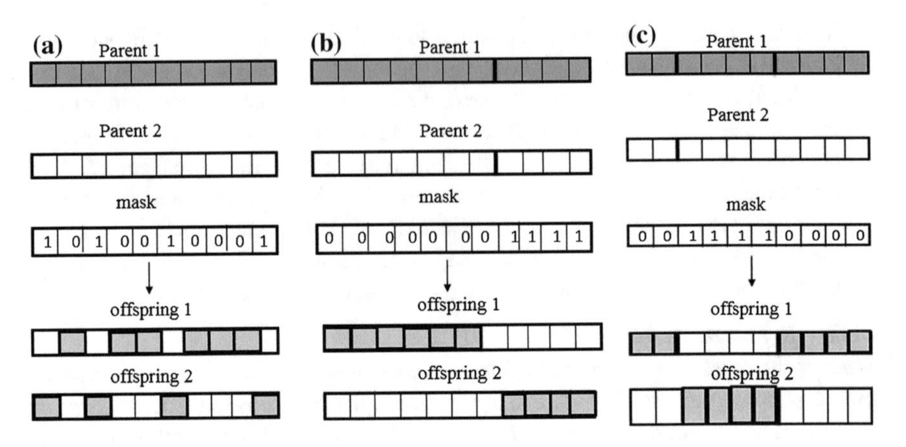

Fig. 8.1 Crossover operators for binary representations. **a** Uniform crossover, **b** one-point crossover, **c** two-point crossover

Algorithm 8.3. Uniform Crossover Mask Calculation
Initialize the mask: $m_j(t) = 0$ for all $j = 1, \ldots, n_x$
for $j = 1$ **to** n_x **do**
if $U(0,1) \leq p_x$ **then**
$m_j(t) = 1$ **end**
end

Bremermann [4] proposed the first multi-parent crossover operators for binary representations. Given n_μ parent vectors, $x_1(t), \ldots, x_{n\mu}(t)$, majority mating generates one offspring using

$$\tilde{x}_{l_j}(t) = \begin{cases} 0, \; if \; n'_\mu > \dfrac{n_\mu}{2}, \; l = 1, \ldots, n_x \\ 1, otherwise \end{cases} \tag{8.1}$$

where n'_μ is the number of parents with $x_{l_j}(t) = 0$.

A multi-parent version of n-point crossover was also proposed, where $n_\mu - 1$ identical crossover points are selected in the n_μ parents. One offspring is generated by selecting one segment from each parent.

8.2.1 *Floating-Point Representations*

The crossover operators discussed above (excluding majority mating) can also be applied to floating-point representations as discrete recombination strategies. In contrast to these discrete operators where information is swapped between parents, intermediate recombination operators, developed specifically for floating-point representations, blend components across the selected parents.

One of the first floating-point crossover operators is the linear operator proposed by Wright [3]. From the parents, $x_1(t)$ and $x_2(t)$, three candidate offspring are generated as $(x_1(t)+x_2(t))$, $(1.5x_1(t)-0.5x_2(t))$ and $(-0.5x_1(t)+1.5x_2(t))$.

The two best solutions are selected as the offspring. Wright also proposed a directional heuristic crossover operator where one offspring is created from two parents using

$$\tilde{x}_{ij}(t) = U(0,1)(x_{2j}(t) - x_{1j}(t)) + x_{2j}(t) \tag{8.2}$$

subject to the constraint that parent $x_2(t)$ cannot be worse than parent $x_1(t)$.

The arithmetic crossover was suggested in [5], which is a multi-parent recombination strategy that takes a weighted average over two or more parents. One offspring is generated using

$$\tilde{x}_{ij}(t) = \sum_{l=1}^{n_\mu} \gamma_{lj} x_{lj}(t), \tag{8.3}$$

where $\sum_{l=1}^{n_\mu} \gamma_{lj} = 1$.

A special case of the arithmetic crossover operator is obtained for $n_\mu = 2$, in which

$$\tilde{x}_{ij}(t) = (1-\gamma)x_{1j}(t) + \gamma x_{2j}(t) \tag{8.4}$$

with $\gamma \in [0,1]$. If $\gamma = 0.5$, the effect is that each component of the offspring is simply the average of the corresponding components of the parents.

Eshelman and Schaffer [6] developed a variation of the weighted average given in Eq. (8.4), referred to as the blend crossover (BLX-α), where

$$\tilde{x}_{ij}(t) = (1-\gamma_j)x_{1j}(t) + \gamma_j x_{2j}(t) \tag{8.5}$$

with $\gamma_j = (1+2\alpha)U(0,1) - \alpha$.

The BLX-α operator randomly picks, for each component a random value in the range

$$\left[x_{1j}(t) - \alpha\left(x_{2j}(t) - x_{1j}(t)\right), x_{2j}(t) + \alpha\left(x_{2j}(t) - x_{1j}(t)\right)\right] \tag{8.6}$$

BLX-α assumes that $x_{1j}(t) < x_{2j}(t)$. It was found that $\alpha = 0.5$ works well.

The BLX-α has the property that the location of the offspring depends on the distance between the parents. If this distance is large, then the distance between the offspring and its parents will be also large. The BLX-α allows a bit more exploration than the weighted average of Eq. (8.3), due to the stochastic component in producing the offspring.

Michalewicz et al. [5] developed the two-parent geometrical crossover to produce a single offspring as follows:

$$\tilde{x}_{ij}(t) = \left(x_{1j}x_{2j}\right)^{0.5} \tag{8.7}$$

The geometrical crossover can be generalized to multi-parent recombination as follows:

$$\tilde{x}_{ij}(t) = (x_{ij}^{\alpha_1}, x_{2j}^{\alpha_2}, \ldots, x_{n_{\mu j}}^{\alpha_n}), \tag{8.8}$$

where n_μ is the number of parents, and

$$\sum_{l=1}^{n_\mu} \alpha_l = 1,\ \alpha_l > 0 \tag{8.9}$$

Deb and Agrawal [7] developed the binary crossover (SBX) to simulate the behavior of the one-point crossover operator for binary representations. Two parents, $x_1(t)$ and $x_2(t)$ are used to produce two offspring, where for $j = 1, \ldots, n_x$.

$$\tilde{x}_{1j}(t) = 0.5\left[(1+\gamma_j)x_{1j}(t) + (1-\gamma_j)x_{2j}(t)\right] \tag{8.10}$$

$$\tilde{x}_{2j}(t) = 0.5\left[(1-\gamma_j)x_{1j}(t) + (1+\gamma_j)x_{2j}(t)\right] \tag{8.11}$$

where

$$\gamma_j = \begin{cases} (2r_j)^{\frac{1}{\eta+1}}, & \text{if } \mathrm{r_j} \le 0,5 \\ \left(\frac{1}{2(1-r_j)}\right)^{\frac{1}{\eta+1}}, & \textit{otherwise} \end{cases} \tag{8.12}$$

$r_j \sim U(0,1)$ и $\eta > 0$—is the distribution index.

It was suggested to take $\eta = 1$ [7].

The SBX operator generates offspring symmetrically about the parents, which prevents bias towards any of the parents. For large values of η there is a higher probability that offspring will be created near the parents. For small η values, offspring will be more far from the parents.

Now consider multi-parent crossover operators. The main objective of these multi-parent operators is to intensify the explorative capabilities compared to two-parent operators.

The unimodal distributed (UNDX) operator was developed where two or more offspring are generated using three parents. The offspring are created by an ellipsoidal probability distribution, with one of the axes formed along the line that connects two of the parents. The extent of the orthogonal direction is determined from the perpendicular distance of the third parent from the axis. The UNDX operator can be generalized to work with any number of parents, with $3 \le n_\mu \le n_s$. For the generalization, $n_\mu - 1$ parents are randomly selected and their center of mass (mean), $x(t)$, is calculated, where $x_j(t) = \sum_{l=1}^{n_\mu - 1} x_{lj}(t)$.

From the mean, $n_\mu - 1$ direction vectors, $d_l(t) = x_l(t) - x(t)$ are computed, for $l = 1, \ldots, n_\mu - 1$.

Using the direction vectors, the direction cosines are computed as

$$e_l(t) = d_l(t)/|d_l(t)|, \tag{8.13}$$

where $|d_l(t)|$ is the length of vector $d_l(t)$. A random parent, with index n_μ is selected. Let $x_{n\mu}(t) - x(t)$ be the vector orthogonal to all $e_l(t)$, and $\delta = |x_{n\mu}(t) - x(t)|$.

Let $e_l(t)$, $l = n_\mu, \ldots, n_s$ be the orthonormal basis of the subspace orthogonal to the subspace spanned by the direction cosines, $e_l(t)$, $l = 1, \ldots, n_\mu - 1$.

Offspring are then generated using

$$\tilde{x}_i(t) = \bar{x}(t) + \sum_{l=1}^{n_\mu - 1} N(0, \sigma_1^2)||d_l||e_l + \sum_{l=n_\mu}^{n_s} N(0, \sigma_2^2)\delta e_l(t) \tag{8.14}$$

where
$$\sigma_1 = \frac{1}{\sqrt{n_\mu - 2}} \text{ and } \sigma_2 = \frac{0,35}{\sqrt{n_s - n_\mu - 2}}. \tag{8.15}$$

Using Eq. (8.14) any number of offsprings can be created, sampled around the center of mass of the selected parents. A higher probability is assigned to create offspring near the center rather than near the parents. The effect of the UNDX operator is illustrated in Fig. 8.2a for $n_\mu = 4$.

Tsutsui and Goldberg [8] proposed the simplex crossover (SPX) operator as another center of mass approach to recombination. $n_\mu = n_x + 1$ Parents are chosen independently for n_x—dimensional search space. These n_μ parents form simplex. Simplex is extended in each of n_μ directions as shown in Fig. 8.2. For $n_x = 2, n_\mu = 3$ and $\bar{x}(t) = \sum_{l=1}^{n_\mu} x_l(t)$ extended simplex is determined by points

$$(1 + \gamma)(x_l(t) - \bar{x}(t)) \tag{8.16}$$

for $l = \overline{1, n_\mu} = 3$; и $\gamma \geq 0$. The offsprings are generated uniformly in extended simplex.

In [9] a variation of the UNDX operator was proposed, which was called as parent-centric crossover (PCX). Instead of generating offspring around the center of mass of the selected parents, offsprings are generated around selected parents. PCX selects n_μ parents and computes their center of mass, $x(t)$. For each offspring to be generated one parent is selected uniformly from the n_μ parents. A direction vector is calculated for each offspring as

$$d_l(t) = x_l(t) - x(t) \tag{8.17}$$

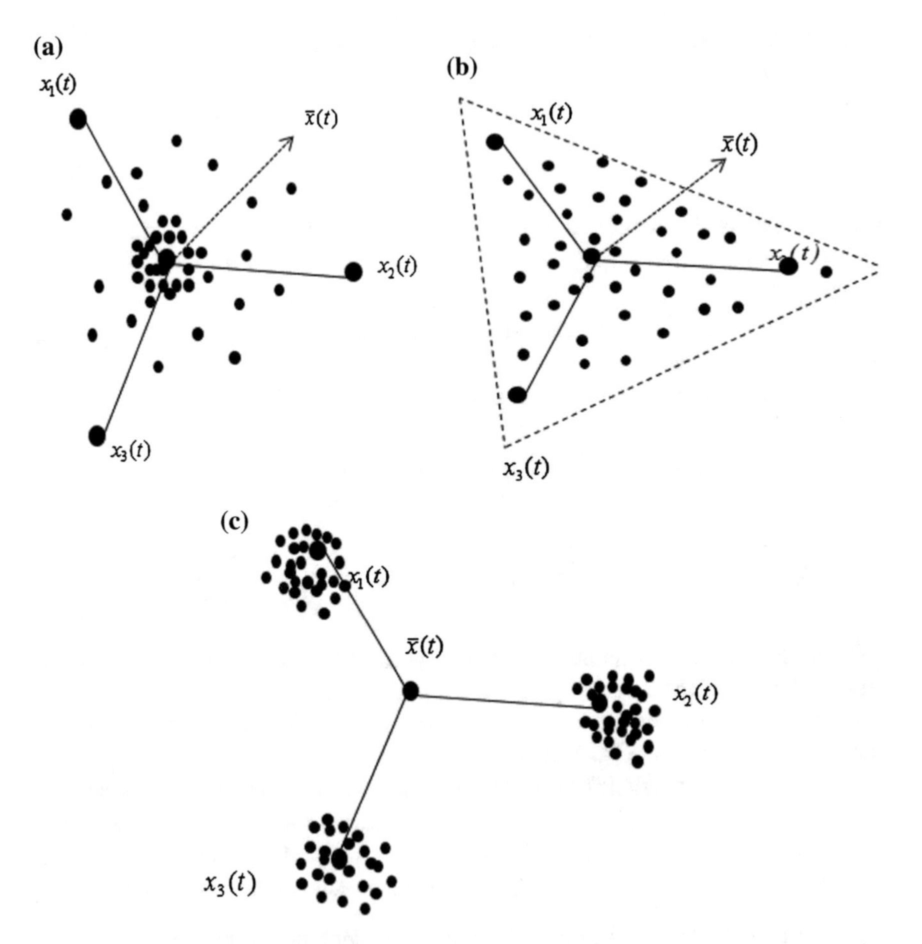

Fig. 8.2 Illustration of crossover operators based on center of mass for several parents (points represents potential offsprings)

where $x_l(t)$ is the randomly selected parent. From all the other $n_\mu - 1$ parents perpendicular distances, δ_l, for $i = l = 1, \ldots, n_\mu$, are calculated to the line $d_i(t)$. The average over these distances is calculated, i.e.

$$\bar{\delta} = \sum_{\substack{l=1 \\ l \neq 1}}^{n_\mu} \frac{\delta_l}{n_\mu - 1} \tag{8.18}$$

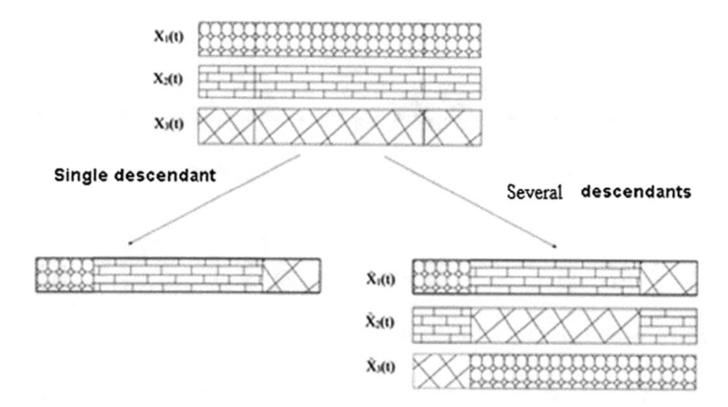

Fig. 8.3 Diagonal crossover

Then offspring is generated using

$$\tilde{x}_i(t) = \bar{x}(t) + \sum_{l=1}^{n_\mu - 1} N(0, \sigma_1^2)||d_l(t)||e_l + \sum_{l=n_\mu}^{n_s} N(0, \sigma_2^2)\delta e_l(t) \tag{8.19}$$

where $x_i(t)$ is the randomly selected parent of offspring $\tilde{x}_i(t)$, and $\{e_l(t)\}$ are the orthonormal basis of the subspace orthogonal $d_l(t)$. The effect of the PCX operator is illustrated in Fig. 8.2c.

The diagonal crossover operator is a generalization of n-point crossover for more than two parents: $n \geq 1$ crossover points are selected and applied to all of the $n_\mu = n + 1$ parents. One or $n + 1$ offspring can be generated by selecting segments from the parents along the diagonals as illustrated in Fig. 8.3, for $n = 2$, $n_\mu = 3$.

8.3 Mutations

The aim of mutation is to introduce new genetic material into an existing individual; that is, to add diversity to the genetic characteristics of the population. Mutation is used in support of crossover to ensure that the full range of allele is accessible for each gene. Mutation is applied at a certain probability, p_m, to each gene of the offspring, $\tilde{x}_i(t)$, to produce the mutated offspring, $x_i(t)$. The mutation probability, also referred to as the mutation rate, is usually a small value, $p_m \in [0, 1]$, to ensure that good solutions would not be distorted too much. Given that each gene is mutated at probability p_m, the probability that an individual will be mutated is given by $\Pr ob(\text{mutation } \tilde{x}_i(t)) = 1 - (1 - P_m)^{n_x}$.

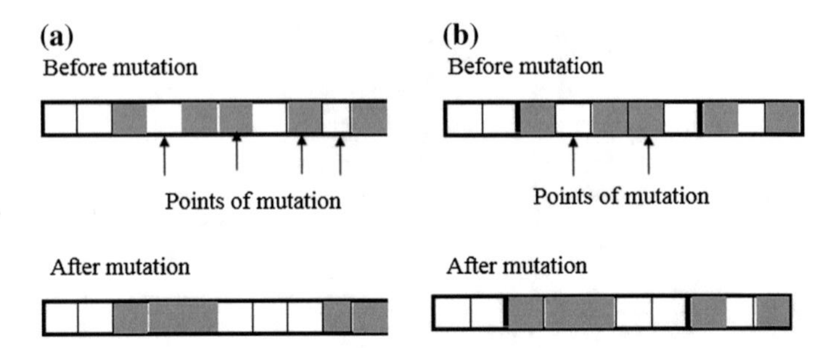

Fig. 8.4 Mutation operators for binary presentation. **a** Random mutation, **b** Inorder mutation

Assuming binary representations, if $H(\tilde{x}_i(t), x(t))$ is the Hamming distance between offspring $\tilde{x}_i(t)$ and its mutated version $x_i(t)$, then the probability that the mutated version resembles the original offspring is given by $\Pr ob(x_i'(t) \approx \tilde{x}_i(t)) = P_m^{H(\tilde{x}_i(t);x_i'(t))}(1 - P_m)^{n_x - H(\tilde{x}_i(t);x_i'(t))}$.

Binary Representations

For binary representations, the following mutation operators have been developed:

- **Uniform (random) mutation**, where bit positions are chosen randomly and the corresponding bit values are negated as illustrated in Fig. 8.4a
- **Inorder mutation**, where two mutation points are randomly selected and only the bits between these mutation points undergo random mutation. Inorder mutation is illustrated in Fig. 8.4b
- **Gaussian mutation**: For binary representations of floating-point decision variables, it was proposed that the bit-string that represents a decision variable be converted back to a floating-point value and mutated with Gaussian noise. For each chromosome random numbers are obtained from a Poisson distribution to determine the genes to be mutated. The bit-strings representing these genes are then converted. To each of the floating-point values is added the stepsize $N(0, \sigma_j)$, where σ_j is 0.1 of the range of that decision variable. The mutated floating-point value is then converted back to a bitstring. Hinterding showed [10] that Gaussian mutation on the floating-point representation of decision variables provided superior results to bit flipping.

For large dimensional bit-strings, mutation may significantly increase the computational cost of the GA. In order to reduce computational complexity, was proposed to divide the bit-string of each individual into a number of bins [3]. The mutation probability is applied and if a bin is to be mutated, one of its bits is randomly selected and flipped.

Floating-Point Representations

As indicated by Hinterding [10] and Michalewicz [5], better performance is obtained by using a floating-point representation when decision variables are floating-point values and by applying appropriate operators to these representations, than to convert to a binary representation. This resulted in the development of mutation operators for floating-point representations. One of the first proposals was a uniform mutation, where

$$x'_{ij}(t) = \begin{cases} \tilde{x}_{ij}(t) + \Delta\left(t, x_{\max} - x'_{ij}(t)\right), & \textit{if a random digit is } 0; \\ \tilde{x}_{ij}(t) + \Delta\left(t, \tilde{x}_{ij}(t) - x_{\min j}(t)\right); & \textit{if a random digit is } 1 \end{cases} \quad (8.20)$$

where $\Delta(t, x)$ returns random values from the range $[0, x]$.

8.4 Parameters Control

In addition to the population size, the performance of a GA is influenced by the mutation rate, p_m, and the crossover rate, p_c. In early GA studies very low values for p_m and relatively high values for p_c were proposed. Usually, the values for p_m and p_c are kept static. It is, however, widely accepted that these parameters have a significant influence on performance, and optimal settings for p_m and p_c can significantly improve performance. To obtain such optimal settings through empirical parameter tuning is a time consuming process. A solution to the problem of finding best values for these control parameters is to use dynamically changing parameters.

Although dynamic, and self-adjusting parameters have been used for Evolutionary programing (EP) and ES (see Sects. 8.5) already in the 1960s, Fogarty [11] provided one of the first studies of dynamically changing mutation rates for GAs. In this study, Fogarty concluded that performance can be significantly improved using dynamic mutation rates. Fogarty used the following schedules where the mutation rate exponentially decreases with generation number.

Dynamic genetic algorithms were a continuation of the idea of generational algorithms that are changing the size of the population. In contrast to genetic algorithms with fixed parameters, dynamic genetic algorithms are able to modify and adjust their parameters, such as the size of the population, the essence of genetic operators, the probability of their use, and even the genotype of the algorithm in the process. The following main strategic parameters are used that can be changed during operation of the genetic algorithm [11]:

- The operator and mutation probability of mutation;
- The operator crossover and the probability of a crossover;
- The mechanism of selection and its pressure (selective pressure);
- Population size.

The researcher is always able to determine the sub-optimum values of the algorithm parameters empirically, but as genetic search process is inherently dynamic and adaptive process, it makes sense to assume that the optimal parameters of the algorithm may change during operation. Often, to improve the application of genetic algorithm it uses dynamic mutation and crossover, as well as changes the size of the population. There are two basic approaches to the determination of the

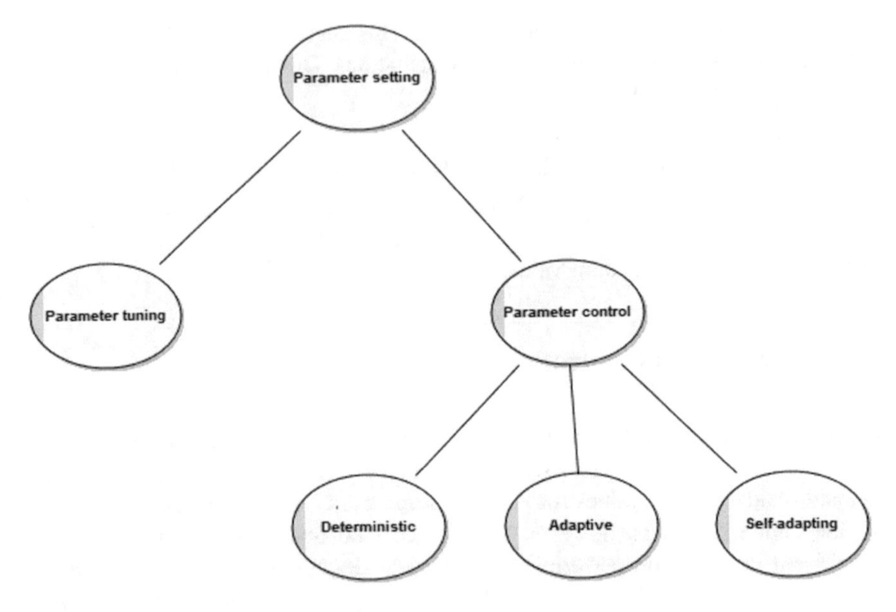

Fig. 8.5 Classification of methods of GA parameters setting (control)

parameters of the genetic algorithm: adjustment of parameters (parameter tuning) and control parameters (Fig. 8.5) [12].

Parameters adjusting consists in running a number of tests to determine at which parameter values genetic algorithm behaves well. Thus, the optimal probabilities of crossover and mutation, population size, and other parameters are experimentally investigated.

De Jong (de Jong) [13] conducted a large-scale study (for a number of test functions) to identify the optimal parameters of the genetic algorithm in the traditional formulation and recommended to use: population size 50, single-point crossover probability of 0.6 and mutation probability: 0.001. But, of course, these values are not universal. Control of the parameters of the algorithm involves starting with some initial values of parameters may defined by the first method. Further, during operation of the algorithm, these parameters are modified in various ways.

Depending on the method of modifying parameters such genetic algorithms are divided into several categories [13]:

1. deterministic (deterministic)
2. adaptive
3. self-adapting (self-adaptive).

The deterministic genetic algorithms parameters are changed during operation of the algorithm on some predetermined law, which often depends on the time (the number of generations, or the amount of computation of the objective function).

The main problem here is to find the form of the aforementioned law for maximum performance of the genetic algorithm as part of the task.

Consider how the concept of deterministic genetic algorithms can be applied in practice. For example, in the genetic algorithm with non-fixed population size each individual is attributed to the maximum age, i.e. the number of generations, after which the individual dies [14]. The introduction of the new algorithm parameter—the age enables to include operator selection in a new population. Age of each individual is different and depends on its adaptability. In every generation at step crossover in the usual way additional population of descendants is created. The size of the additional population is proportional to the size of the general population $V\,A(t) = size\,A(t) * p$, where p—the probability of reproduction.

For reproducing the individuals are selected from the general population with equal probability, regardless of their fitness. After applying the mutation and crossover to offspring is attributed age according to the value of their fitness. Then from the general population are removed those individuals whose lifespan has expired, and added the descendants of the intermediate population. Thus, the size of the population after one iteration of the algorithm is calculated according to the formula:

$$size\,A(t+1) = size\,A(t) + size\,B(t) - D(t), \tag{8.21}$$

where $D(t)$—the number of individuals who die in the generation t.

To organize deterministic dynamic mutations Back (Back) and Schutz (Schuts) [3] used a decreasing function of time of the following form:

$$p\;(t) = (2 + \frac{l-2}{T-1}t)^{-1}. \tag{8.22}$$

where T—the number of iterations of the genetic algorithm, and l—length of the chromosome. The probability of mutation obtained by this formula is bounded in the interval $(0, 1/2]$.

One of the first dynamically varying intensity parameters mutations was proposed by Fogarty [11]. Fogarty proposed the following scheme in which the mutation rate decreases exponentially with the number of generation t:

$$P_m(t) = \frac{1}{240} + \frac{0,11375}{2^t} \tag{8.23}$$

Alternatively, Fogarty also proposed for the binary representation the intensity of the mutations on the 1-bit:

$$P_m(j) = \frac{0,3528}{2^{j-1}},\;\; j = 1, 2, \ldots, n_b,\;\; n_b, \tag{8.24}$$

where n_b—indicates how the least significant bit.

These two schemes were combined and the following dependence for P_m was obtained:

$$P_m(j,t) = \frac{28}{1905 \times 2^{j-1}} + \frac{0,4026}{2^{t+j-1}} \tag{8.25}$$

Various schemes may be used to reduce the intensity of mutations. In the above scheme is suggested exponential version.

An alternative may be a linear scheme, which provides a slow change, allowing a large number of experiments. However, a slow reduction in the intensity of mutations can also be devastating for found good solutions.

Virtually all such formulas include the condition $1/t$, where t—number of generations. In this case, the initial probability of mutation is quite high and decreases as the algorithm runs. Thus, in the early stages of a genetic algorithm mutations play a key role in identifying the most promising areas of the search, but later for their study a crossover is used.

Such formulas don't eliminate the need to set parameters, but significantly increase the efficiency of the genetic algorithm by researchers estimates.

Note that the probability of mutations in these studies deterministically changes in time and does not depend on the actual search results. The question of the optimality of this approach is controversial. Alternatively, the current search results can be used to adjust the parameters of the genetic algorithm.

In adaptive genetic algorithms information about the algorithm results through feedback effect on the values of its parameters. However, unlike the self-adapting algorithms, the values of these parameters are common to all individuals of a population. But since the essence of the mechanism is determined by the feedback researcher, the question how to make it the optimal remains open.

This method of control algorithm was first used by Rechenberg (Rechenberg) [12]. He postulated that one of the five mutations should lead to the emergence of a better individual. Further, he linked the likelihood of mutations with a frequency of good mutations. Thus, if the frequency of good mutations is above $1/5$, the probability of mutation increases and vice versa. In the formalized form Rechenberg rule usually is such:

$$\text{if } \varphi(k) < 1/5 \to (p_m/\lambda)$$

$$\text{if } \varphi(k) = 1/5 \to (p_m)$$

$$\text{if } \varphi(k) > 1/5 \to (\lambda p_m),$$

where $\varphi(k)$—the success rate of mutations in k iterations, and -λ is training coefficient. (usually $\lambda = 1.1$).

To improve efficiency of the genetic algorithm Terence (Thierens) proposed using two adaptive mutations change the scheme. (Constant Gain scheme) has the form [12]:

1. The individual (x, p_m) is mutated in 3 different ways

$$M(x, p_m/\omega) \rightarrow (x_1)$$

$$M(x, p_m) \rightarrow (x_2)$$

$$M(x, \omega p_m) \rightarrow (x_3)$$

2. Selection of the fittest individual and the probabilities of new mutations

$$MAX\{(x, p_m), (x_1, p_m/\lambda), (x_2, p_m), (x_3, \lambda p_m)\}$$

In expressions above λ is a training coefficient, and ω- factor of study. Usually λ and ω have different values $(1 < \lambda < \omega)$. In order to avoid fluctuations in the learning process, it is limited as $\omega > \lambda$. Typical values of coefficients are: $\lambda = 1.1$ and $\omega = 1.5$.

Declining scheme promotes the use of more aggressive step size, at the same time suppressing the random fluctuations that may occur at a high learning rate.

1. The individual is mutated in 3 different ways

$$M(x, \omega p_m) \rightarrow (x_1); M(x, p_m) \rightarrow (x_2); M(x, p_m/\omega) \rightarrow (x_3)$$

2. The decrease of parent mutation probability $(x, p_m) \rightarrow (x, \gamma p_m)$
3. Selection of the fittest individual and new mutation probability

$$MAX\{(x, \gamma p_m), (x_1, \lambda p_m), (x_2, p_m), (x_3, \lambda p_m)\}$$

The difference between the two schemes is that the second scheme has no ability to increase the probability of mutations and current mutation probability will decrease, until a next successful descendant. Used coefficients lie within the following bounds: $\lambda > 1, \omega > 1$ and $0.9 < \gamma < 1$.

Typical coefficients values are $\lambda = 2.0$, $\omega = 2.0$ and $\gamma = 0.95$. To adapt the probabilities of crossover and mutation Shi (Shi) used a set of fuzzy logic rules. Table 8.1 describes a set of fuzzy rules relating to the probability of mutations (MP) [12].

In the expressions above, BF—best fitness value, UN—number of iterations since the last change BF, MP—the probability of a mutation. This set of rules, of course, provides the giving of membership functions for different metrics and mutation probabilities. For example, Sugeno fuzzy inference may be used. In order

Table 8.1 Set of fuzzy adaptive rules (Shi) for mutations

1. If BF is low or middle MP is low
2. If BF is middle and UN is low then, MP is low
3. If BF is middle and UN is middle then, then MP is middle
4. If BF is high and UN is low then MP is low
5. If BF is high and UN is middle then, MP is high
6. If UN is high then, MP is chosen randomly (high, middle or low)

to organize the dynamic changes of the crossover or mutation operator different probabilities, constant or linear functions are selected for each fuzzy inference.

In the original formulation of a set of rules by Shi (Shi) as the metrics was used the variance of fitness with high, medium and low fuzzy values.

For optimization problems with constraints penalty functions are frequently used. This idea can be applied to the organization of the adaptive crossover and mutation. So, for solving graph coloring problem Eiben and van der Hauw suggested the use of an evaluation function with stepwise adaptation of weights: when a certain constraint is violated, the value of penalty for the corresponding variable increases $eval(x) = f(x) + W * penalty(x)$, where $f(x)$ is the objective function, $penalty(x)$—the penalty function, and w—weighting factor that determines weight of penalty to be appointed in violation of restrictions. If all constraints are satisfied, then $penalty(x) = 0$.

For the organization of the mechanism of self-adaptive population size Greffenstet used a meta-approach based on the use of other genetic algorithm. Thus, each iteration of meta-genetic algorithm has initiated the launch of the target genetic algorithm to adapt the parameters. The result of the meta-algorithm was optimal population size for the target algorithm.

The idea of organizing self-adapting mutation operator was first implemented by Scheffel (Shaffel) in evolutionary strategy, where he tried to control the step mutations. Back (Back) disseminated his ideas on genetic algorithms. He added to the individual extra bits that coded the probability of mutation. Each coded probability of mutation applies both to itself and to the corresponding gene (Table 8.2):

Experimental studies have shown that by using this approach, a significant increase in performance of the genetic algorithm is achieved.

Of course, it is also possible to combine the scheme adapting the parameters of the genetic algorithm in a frame of one task. For example, the organization of the mechanism of adaptation both a crossover and mutation can significantly increase the performance of the genetic algorithm.

Thus, control of parameters allows to use a variety of the most suitable parameter values in different stages of genetic research to maximize the performance of the algorithm. Adaptive and self-adapting genetic algorithms complicate

Table 8.2 Extended chromosome with coded mutations probabilities

$x_1\ x_2\ x_3 \ldots x_{n-1}\ x_n\ \ldots\ x_m\ p_1\ p_2\ p_3\ \ldots\ p_{n-1}\ p_n\ \ldots\ p_m$
$x_1\ x_2\ x_3 \ldots x_{n-1}\ x_n\ \ldots\ x_m\ p'_1\ p'_2\ p'_3\ \ldots\ p'_{n-1}\ p'_n\ \ldots\ p'_m$

the algorithm itself, delegating him the task of determining the value of strategic options, but at the same time, they eliminate the need for researchers to adjust parameters manually.

The next step in the development of dynamic algorithms is the emergence of multi-level dynamic genetic algorithms (nGA) [12]. The lower level of the algorithm itself solves the task of improving the decisions population. The upper levels are genetic algorithms that solve the optimization problem to improve the parameters of the algorithm of the lower level as the objective function is usually used the speed of the algorithm of the lower level and the rate of improvement of their population from generation to generation.

A good strategy is to link the intensity of mutations with the individual fitness index: the higher the index of fitness of the individual, the less must be the probability of mutations of its gene, and conversely the lower is fitness index the greater is the probability of mutations.

For floating-point representations, the performance of the GA also affects the step size of mutations. The ideal strategy is to start with the large size mutations to provide more random jumps in the search space. The step size of mutations decreases with time, so that a very small change is produced at the end of the search process. The step size may also be proportional to the individual fitness index, so that less suitable individuals had a greater step size than the more suitable.

The intensity of the crossover P_c also has a significant impact on the performance of GA. Since the optimum value of P_c depends on the task for dynamic adjustment may be used the same alternative strategy that for mutations.

In addition to crossover choice of the best operator of evolution is also a problem-oriented. In general, the definition of the best set of operators and control parameters values are essentially the problems of multi-criteria optimization.

8.5 Strategy Selection

The above GA greatly differ from biological evolution in that the population size is fixed. It allows to describe the selection process consisting of two steps:

- Selection of the parents;
- Replacement strategy, which decides whether to replace a descendant of parents and what kind of parents should be replaced.

There are two main classes of GA, based on the strategy used for replacement parents: generational GA and steady state GA [3]. In the generational GA replacement policy replaces all parents to their descendants, after the descendants were generated and subjected to mutations. As a result, there is no overlay (intersection) between the current population and new population (assuming that elitism is not used).

In a steady state GA after generation and mutation of descendant it's immediately decided: to keep a parent or offspring for the next generation. Thus, there is overlap between the current and next generations.

The value of the intersection between the current and the new generation is called "generational gap". Generational GA have zero gap, while the steady-state GA have large gaps between generations.

A large number of strategies for GA steady state were developed:

- To replace the worst, when the descendant replaces the worst individual of the current generation;
- Random replacement when the descendant replaces a randomly selected individual from the current generation;
- "Replacement of the oldest" strategy (first-in-first-out), which should replace the oldest of the individual from the current population. This strategy has a high probability of replacing one of the best individuals.
- Tournament strategy that is randomly selected group of individuals, and the worst individual in the group is replaced by a descendant. Alternatively, the size of the tournament is two, and the worst individual is replaced by a descendant with a probability $0,5 \leq p_r \leq 1$.
- Conservative strategy, combines the strategy to replace the oldest one with the upgraded deterministic tournament strategy. The size of the tournament consists of two individuals, one of which is the oldest individual in the current population. The worst of the two is replaced by a descendant. This approach ensures that the older individual is not lost, if he has the best fitness index.
- A strategy of "elitism" in which the best parents are excluded from the selection and automatically transferred to the next generation.
- Competition "parent-child", which uses a strategy of selection to decide whether to replace by a descendant of one of their parents.

Island Genetic Algorithms

GA is well suited for a parallel implementation. Three main categories are distinguished in parallel GA:

- Uniform GA type (master-slave), in which the evaluation of the fitness index is distributed across multiple processors;
- uniform cell GA, where each individual is assigned to one processor and each processor is assigned to only one individual. For each of the individual is determined the scope of a small neighborhood, breeding and reproduction are limited only by the scope of the neighborhood.
- Multi-population or island GA, which use a lot of populations, each on a separate processor. Information exchange between populations is performed through migration policy. Although island GA have been developed for a parallel implementation of many processors, they can also be implemented in a uniprocessor system.

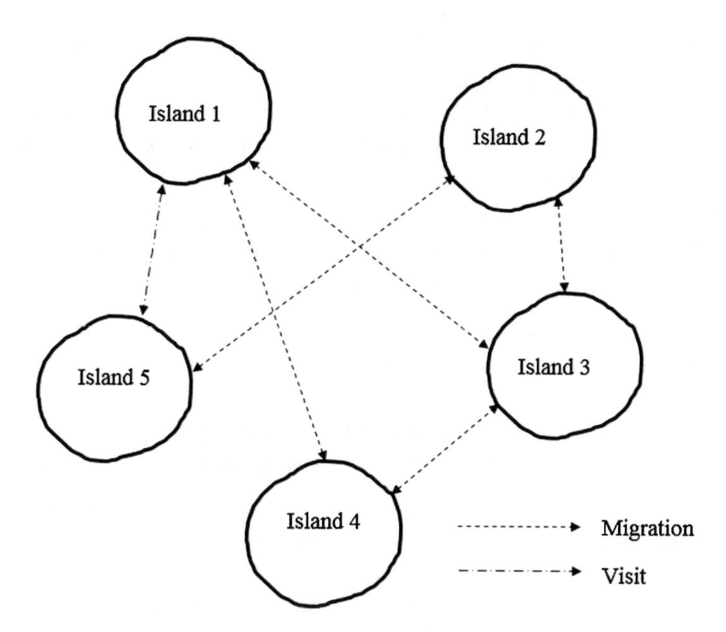

Fig. 8.6 Routes of migration and visits in island genetic algorithms

Let us consider in more detail the island GA. In the island GA, several populations are developing in parallel, in a cooperative structure. In such a model, the GA has lots of islands, each island has one population. Selection, crossover and mutation occurs in each population independently of the other. In addition, individuals are allowed to migrate between islands (or populations) as shown in Fig. 8.6.

An integral part of the island GA is migration policy governing the exchange of information between the islands. Migration policy determines [3]:

- The communication topology, which defines the way of individuals migration between the islands. For example, a ring topology (as shown in Fig. 8.6) allows the exchange of information between the neighboring islands. The communication topology determines how fast (or slow) good decisions are propagating to other populations. For loosely coupled structure (such as a ring topology) islands are more isolated from one another, and dissemination of good decisions is carried slower. Loosely topology also contributes to the appearance of the solution set. Tightly structure have a greater speed of information dissemination, which can lead to premature convergence.
- The intensity of migration, which determines the frequency of migrations. Closely related to the intensity of the migration is the question when the migration should take place. If migration occurs too early, the number of good building blocks for migrants may be too small to have any effect on the final results. Typically, migration occurs when each population converged to some

decision. After an exchange of individuals, the development process of the population is restarted again.
- Selection mechanism used to decide exactly which individuals will migrate.
- Replacement strategy is used to decide which individual destination node will be replaced by a migrant.

On the basis of breeding strategies and replacement island GAs can be grouped into 2 classes of algorithms: namely, static and dynamic island GAs. For statistical island GAs deterministic strategy of selection and replacement is used, for example, [3]

- Good migrant replaces the bad individual;
- A good migrant replaces a randomly selected individual;
- Randomly selected migrant replaces a bad individual;
- Randomly selected migrant replaces a randomly selected individual.

For a good selection of the migrant any of the fitness-proportional selection operators can be used. For example, elitism strategy means the best individual of one population moves to another population.

Gordon [10] used the tournament selection, considering the two randomly selected individuals. The best of the two migrates, while the worst is replaced by an individual which won the tournament from the neighboring population.

Dynamic models do not use the topology to determine the migration routes. Instead, decisions about migration are performed in a probabilistic way. Migration takes place with a certain probability. If migration from the island takes place, the island destination is also determined probabilistically. It may be used tournament selection using the average values of the population fitness index. Additionally, we can use the strategy of decision-making to decide whether immigrants to be accepted. For example, immigrant can be accepted with probability if his fitness index (FI) is higher than the values of the other FI in this island.

Another important aspect to be considered in the island GA, it must be initialized as a separate population. Of course, it is possible to use a random approach which may lead to that different populations will cover the same part of the search space. The best approach is to initialize different populations so that they would cover different parts of the search space, and contribute to the formation of a sort of niche in the individual islands. Next, you step on the meta-level to combine solutions from different islands.

Another type of “island” is a cooperative co-evolutionary GA (CCGA) [3]. In this case, instead of allocating all individuals for several (sub) populations, each population is given one or more genes (one decision variable) to optimize. Subpopulations are mutually exclusive, each is intended to develop one or more genes. Each population therefore optimizes one (or several) genes, i.e. no sub-population does have the necessary information to solve the problem independently. Conversely, information from all populations is to be combined to construct a solution.

Stop conditions. GA may use different conditions stop Among the most common should be mentioned:

- The number of iterations of the algorithm;
- Achieved fitness value of the index of the best individual;
- If the value of FI of the best individual does not change for a predetermined number of iterations

Multi-criteria Optimization (MCO)
Much attention is paid to the use of GA for solving multi-criteria optimization problems:

$$\min_{x} f(x; c_i), i = \overline{1, k} \tag{8.26}$$

Algorithm VEGA. One of the first GA algorithms for solving problems of the MCO is the vector genetic algorithm (VEGA—Vector Evaluated Genetic Algorithm) [3].

Each sub-population is associated with one objective function. Further the selection is applied to each subpopulation to create a pool of applicants. As a result of selection process the best individuals in each of the objective function are included in the pool of pretendents. Next crossover is performed by selecting parents from a pool of pretenders.

8.6 The Application of Genetic Algorithm in the Problem of Networks Structural Synthesis

The important application of GA are combinatorial optimization problems, in particularly computer `networks structural synthesis problems. Consider the formulation of computer networks with technology MPLS synthesis problem [12].

There are given a set of network nodes $X = \{x_j\}$ $j = \overline{1, n}$—MPLS routers (so-called LRS—Label Switching Routers), their locations over regional area, channels capacities $D = \{d_1, d_2, \ldots, d_k\}$ and their costs per unit length $C = \{c_1, c_2, \ldots, c_k\}$, classes of service (CoS) are determined, matrices of input demands for the kth CoS are known $H(k) = \|h_{ij}(k)\|$ $i, j = \overline{1, n}$; $k = 1, 2, \ldots, K$, where $h_{ij}(k)$ is the intensity of flow of the kth CoS which is to transfer from node i to node j in sec (*Mbits/s*).

Additionally the constraints are introduced on the Quality of Service (QoS) for each class k as constraint on mean PTD (packets transfer delay) $T_{inp,k}$, $k = \overline{1, K}$ and packets loss ratio (PLR).

It's demanded to find network structure as a channels set $E = \{(r, s)\}$, choose channels capacities $\{\mu_{rs}\}$ and find flows distributions of all classes so that ensure the transmission demands of all classes $H(k)$ in full volume with mean delays T_{av}, not exceeding the given values $T_{inp,k}$ and by this the constraint on packets loss ratio (PLR) should be fulfilled and total network cost be minimal [15].

Let's construct the mathematical model of this problem.

It's demanded to find such network structure E for which

$$\min_{\{\mu_{rs}\}} C_{\Sigma}(M) = \sum_{(r,s)\in E} C_{rs}(\{\mu_{rs}\}) \tag{8.27}$$

under constraints

$$T_{av}(\{\mu_{rs}\};\{f_{rs}\}) \leq T_{inp,k} \; k = \overline{1,K} \tag{8.28}$$

$$f_{rs} < \mu_{rs} \text{ for all } \quad (r,s) \tag{8.29}$$

$$\mu_{rs} \in D \tag{8.30}$$

$$PLR_k(\{\mu_{rs}\};\{f_{rs}\}) \leq PLR_{k\,inp} \tag{8.31}$$

where $PLR_k(\{\mu_{rs}\};\{f_{rs}\})$—is packets loss ratio for the kth flow, $PLR_{k\,inp}$—given constraint on its value.

This problem belongs to so-called NP-complete optimization problems. For its solution general genetic algorithm was developed using two operators: crossover and mutation [12].

Define matrix of channels $K = \|k_{ij}\|$, where

$$k_{ij} = \begin{cases} 1, \; if\, \exists(i-j) \\ 0, \; otherwise, \; i.e. \neg\exists(i-j) \end{cases}, \tag{8.32}$$

for each network structure. Then initial population of different structures in given class of multi-connected structures is generated with connectivity coefficient 2. For synthesis we'll use semi-uniform crossover which is grounded for small population size.

Parents (structures $E_i(k)$ are chosen randomly with probability inverse proportional to cost $C_{\Sigma}(E_i(k))$, each parent is determined by matrix $K^i, i = 1,2$. In the process of semi-uniform crossover each descendant receives exactly a half of quantity of parents genes. The crossover mask is presented as the following matrix $M = \|m_{ij}\|$

$$\text{where } m_{ij} = \begin{cases} 0, & if\, p \geq p0 \\ 1, & if\, p < p0 \end{cases},$$

where parameter $p0 = 0.5$ and $p \in [0,1]$—is a random value.

This process of crossover may be written so:

$$E(k)' = \left\| e(k)'_{ij} \right\| = \begin{cases} (i-j)^1, \ k^1_{ij} = 1, if \ m_{ij} = 0; \\ (i-j)^2, \ k^2_{ij} = 1, if \ m_{ij} = 1. \end{cases}$$

In this crossover only one descendant is generated that to maximize the algorithm productivity. In case if after crossover isolated sub-graphs were generated then with direct channels they are connected to a root. Further for generated descendant-structure $E(k)'$ capacities assignment and flows distribution problems are solved [15] and network total cost is calculated. Then using cost value C_Σ decision is made whether to include new structure $E(k)'$ in the set of local optimal structures (new population) Π or not.

After crossover operation the mutations are performed. Note that basic GA algorithms use unconditional mutations. Mutations consist in deleting or adding new channels into network structure. During the improvement of basic GA algorithm in tandem with semi-uniform crossover the following schemes mutations probability change were investigated [12]: deterministic and adaptive.

In deterministic case mutation probability is determined using function depending on time. Thus we 'll change mutation probability according to the law: $\sigma(t) = 1 - ct/T$, where $0 \le t \le T$, a $c \in (0,1)$—is a coefficient. Note that as time increases the mutation probability will decrease.

Note that main properties of this approach are:

1. mutation probability change doesn't depend on success of its application in the process of genetic search;
2. the researcher fully controls the mutation probability change by definite formula;
3. mutation probability change is fully predictable.

For implementation of adaptive version of mutation probability change Rechenberg rule was used [12]:

$$\sigma(t) = \begin{cases} \sigma(t-1)/\lambda, \varphi(t-1) < 1/5 \\ \sigma(t-1)\lambda, \varphi(t-1) > 1/5 \\ \sigma(t-1), \varphi(t-1) = 1/5 \end{cases}, \tag{8.33}$$

where $\varphi(t)$ is good mutation percentage, and $\lambda = 1.1$ is a learning coefficient.

Note that main properties of such approach are the following:

1. mutation probability change depends on success of its application in process of genetic search;
2. the researcher has information how mutation probability will be changed:
3. mutation probability change is non-predictable.

Self-adaptive mutation may be organized at the level of chromosomes (network structures) and at the level of genes(channels). For this should be given the law of mutation probability change. But implementation of this approach leads to significant cut in algorithm productivity and in this problem it isn't used [12]. As alternative to the scheme of unconditional crossover and dynamic mutation the scheme with unconditional mutation and dynamic crossover was explored in this work.

In the process of algorithm perfection in pair with unconditional mutation the following schemes of crossover probability change were applied and investigated:

- *Deterministic*. Implementation of deterministic scheme of crossover probability change is based on hypothesis that at different stages of genetic search crossover may be more or less significant, therefore as a function of crossover probability change should be used non-monotonic function. That's why in this problem we used the function $\sigma(t) = |\sin(t)|$, where $0 \le t \le T$.
- *Adaptive*. Define adaptive crossover as an operator whose probability decreases if population is sufficiently homogeneous and increases in case if the population is heterogeneous. As the measure of homogeneity/heterogeneity we chose the value

$$C_\Delta = \max(C_\Sigma(E_i(k)) - C_\Sigma(E_j(k)), \;\; i \in [1,\ldots,n], j \in [1,\ldots,n], i \neq j,$$

where n is a population size (in this case n = 3). It's reasonable to suggest that in case of similar individuals in population the crossover operation will be inefficient and vice versa. Therefore in adaptive crossover the law of crossover probability change takes the form:

$$\sigma(t) = \begin{cases} \sigma(t-1)\lambda, C_\Delta > C^* \\ \sigma(t-1)/\lambda, C_\Delta < C^* \\ \sigma(t-1), C_\Delta = C^* \end{cases}, \tag{8.34}$$

where C^* is threshold value, $\lambda = 1.1$ is a learning coefficient.

- *Self-adaptive*. The implementation of self-adaptive crossover is unreasonable from algorithm productivity criterion as well as the implementation of self-adaptive mutation.

The above-considered versions of GA were applied and investigated in the project of Ukrainian global MPLS network design [12]. After series of experiments with different initial input data and parameters, such as volumes of traffic, given sets of channels capacities and their costs per unit length the following results were obtained:

- a *combination of unconditional crossover and dynamic deterministic mutation*: with $c = 0.3$ this realization proved to be the one of the most successful (performance increase by 15 %) that confirms the hypothesis about distribution of participation between crossover and mutations at the different stages of genetic search.

Thus, mutations play key role at the initial stage of search when exploring the search space while at the final stage it's more efficient to use crossover for finding optimal solution.

- *a combination of unconditional crossover and dynamic adaptive mutation*: this realization didn't allow to achieve stable decrease of algorithm run time, that confirms idea about inefficiency of this approach for the considered problem;
- *a combination of unconditional crossover and dynamic self-adaptive mutation*: this realization is unreasonable as it significantly complicates the genetic search and leads to algorithm performance decrease (productivity cut);
- a *combination of dynamic deterministic crossover and unconditional mutation*: this realization didn't ensure the stable performance increase, perhaps it's needed more fine adjustment of crossover probability change;
- a *combination of dynamic adaptive crossover and unconditional mutation: such realization proved to be the most successful* and ensured the performance increase by 20–22 %. Thus, the hypothesis about some positive properties of crossover which mutation operator doesn't have was confirmed. Note that application of crossover is grounded only if a population is heterogeneous, i.e. the individuals differ from each other sufficiently.
- a *combination of dynamic self-adaptive crossover and unconditional mutation*: such realization proved to be inefficient due to complication of search process.

As illustration of the above described experimental results in the Fig. 8.7 the initial structure of Ukrainian global network is presented, and in the Fig. 8.8 the optimized structure obtained with application of modified genetic algorithm which realizes a combination of dynamic adaptive crossover and unconditional mutation.

Note that cut in total network cost for optimized structure in comparison with initial structure equals to 14,250 thousand $ − 10,023 thousand $ = 4227 thousand

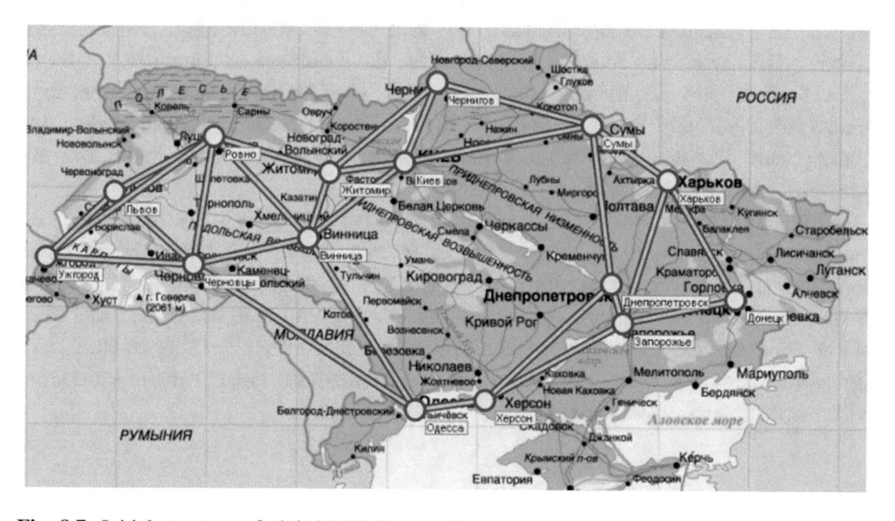

Fig. 8.7 Initial structure of global MPLS network

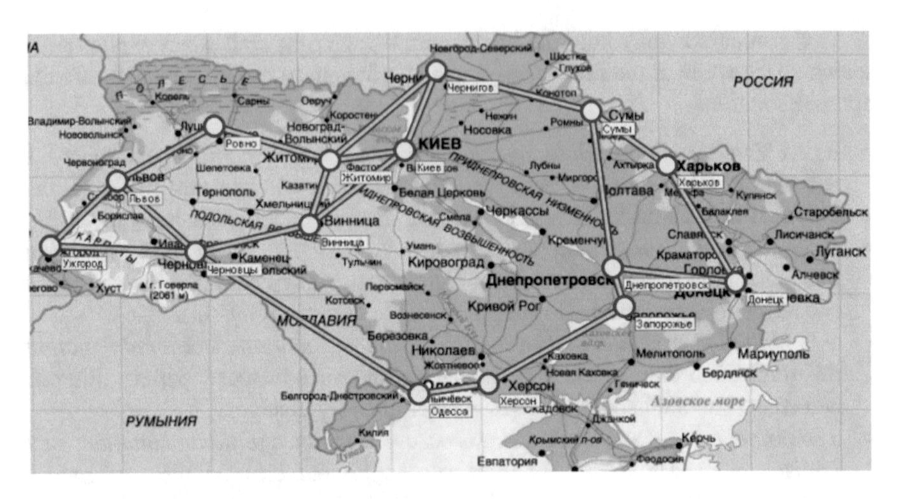

Fig. 8.8 Optimized structure of global MPLS network

$, i.e. the cost after optimization is by 30 % less than the cost of initial network structure. Additionally the suggested algorithm productivity increased by 22 % in comparison with basic genetic algorithm [12].

8.7 Evolutionary Programing

Evolutionary programing (EP) appeared owing to the pioneer work of Vogel, Owen and Walsh [16],who used the modeling of evolution for the development of artificial intelligence (AI). Though the EP has the same overall objective of the natural process of evolution simulation with GA and genetic programing, it differs significantly from them by that EP focuses on the development of behavioral patterns, rather than genetic, i.e. EP considers so- called "phenotypic" evolution rather than genotypic. Iteratively EA uses two of three evolution operators, namely mutations that provide the necessary variety and selection. Recombination operators are generally not used within the EP.

EP was invented in the early 60s as an alternative approach to AI that focused on models of human intelligence. Because of its conclusion that intelligence can be regarded as "property which allows the system to adapt its behavior to achieve the desired goal in a wide range of environments", a model was developed that simulates the evolution of behavior traits. Therefore, in contrast to other evolutionary algorithms, fitness—index (FI) measures the "behavioral noise" of the individual with respect to its environment.

EP uses 4 main components of evolutionary algorithms (EA) [16]:

(1) Initialization: like other paradigms initialized starting position of individuals so as to cover uniformly the area of optimization.
(2) Mutations: mutation operator aims to introduce variation into the population, i.e. generate new candidates. Each parent produces one or more offsprings by mutation operator. A large number of mutation operators have been developed.
(3) Rating: fitness function is used to quantify the "behavioral noise" of individuals. While the fitness function provides an absolute measure of fitness, to assess how well the individual solves the problem of optimization, survival in the EP is normally based on the relative measure fitness index. This measure allows you to compare a group of individuals between themselves and the individuals that survive and move to the next generation, are selected on the basis of this measure.
(4) Selection: The main aim of the operator—is to select individuals who are moving into the next generation. Selection—is a competitive process in which the parents and their offsprings are competing among themselves for survival.

Operators of mutation and selection are used to iteratively until no stop condition is satisfied.

They can be used in any of the stopping conditions considered for GA.

Comparing GA and EP are the *following differences* should be noted:

- EP focuses on phenotypic rather than genotypic evolution;
- In accordance with the above EP uses no operators of recombination; there is no exchange of genetic material.
- EP uses relative fitness function, to quantify with respect to the functioning of a randomly selected group of individuals;
- Selection based on competition. Those individuals who are functioning better than group of rivals receive greater probability of inclusion into the next generation;
- Parents and offspring are struggling for survival;
- The behavior of individuals influences strategy parameters that determine the amount of variations between parents and descendants.

Operators EP

Operators of mutations

Since the mutations is the only means of introducing variability into the population EP, it is very important to design the mutation operator so that to provide a compromise between the phases the study—use.

After the initial phase of the study, individuals should be allowed to use the information about the search space for fine-tuning decisions.

For this discussion, assume that the objective is to minimize the continuous function without limitation: $f : R^n x \rightarrow R$. If $x_i(t)$ denotes the pretendent for a decision presented by the ith individual, where t is time, every $x_i(t) \in R$, $j = \overline{1, n_x}$.

In general, the mutation is described as $x'_{ij}(t) = x_{ij}(t) + \Delta x_{ij}(t)$, wherein $x'_{ij}(t)$—a descendant of the parent $x_{ij}(t)$ generated by adding the step size $\Delta x_{ij}(t)$ to the parent.

On the step size acts noise with some probability distribution, where the deviation of the noise is determined by the strategic option σ_{ij}. In general, the step size is computed as

$$\Delta x_{ij}(t) = \Phi(\sigma_{ij})\xi_{ij}(t) \tag{8.35}$$

where $\Phi : R \rightarrow R$ is a function, which scales the effect of noise $\xi_{ij}(t)$.

On the basis of the characteristics of the scaling function Φ, EP algorithms can be divided into the following three categories [3]:

- Non-adaptive EP, in this case $\Phi(\sigma) = \sigma$. Otherwise, variations in step size—static.
- Dynamic EP, in which the deviation in the step value changes over time, using some deterministic function Φ.
- Self-learning EP in which variations in step size change dynamically. The system is trained to find the best values σ_{ij} in parallel with the decision variables. Since deviations σ_{ij} affect the behavior of individuals in the case of dynamic and self-learning EP, these deviations are considered as strategic parameters. As in the case with all of evolutionary algorithms (EA), EP utilizes a stochastic search process.

Stochastic is introduced by calculating the step size as a function of the noise $\xi_{ij}(t)$, generated using a probability distribution. The following distributions are used for EP:

1. *Uniform*. Noise is sampled from a uniform distribution

$$\xi_{ij}(t) \sim u(x_{\min j}, x_{\max j}) \tag{8.36}$$

It should be noted that the $E\{\xi_{ij}\} = 0$ in order to prevent any displacement by noise where $E\{*\}$ operator of mathematical expectation. Wong and Yurevich offered uniform mutation operator

$$\Delta x_{ij}(t) = U(0,1)\left(\hat{y}_j(t) - x_{ij}(t)\right)$$

This operator causes all individuals to do casual steps toward better individual that is very similar to the social component used in the swarm optimization algorithms [17].

2. *Gaussian*. For Gaussian mutation operators, noise is generated from a normal distribution with zero mean: c.

In comparison with other distributions, Gaussian distribution density is defined as

$$f_G(x) = \frac{1}{\sqrt{2\pi}} e^{-\frac{x^2}{2\sigma^2}} \tag{8.37}$$

3. *Cauchy* distribution. For operators mutation Cauchy distribution

$$\xi_{ij}(t) \sim C(0, \upsilon) \tag{8.38}$$

where υ—a parameter of the scale, the probability density function of Cauchy is given

$$f_c(x) = \frac{1}{\alpha} \frac{\upsilon}{\upsilon + x^2} \tag{8.39}$$

where $\upsilon > 0$. The corresponding distribution function is given by

$$F_c(x) = \frac{1}{2} + \frac{1}{\pi} arctg\left(\frac{x}{\upsilon}\right) \tag{8.40}$$

Cauchy distribution has wider tails than a Gaussian distribution and therefore produces more mutations.

4. *Levi distribution*. For Levi distribution [3] $\zeta_{ij}(t) \sim L(v)$. Levi distribution with center in coordinates beginning is given so

$$F_{L,U,\gamma}(x) = \frac{1}{\pi} \int_0^\infty e^{-\gamma q^v} \cos(qx) dq, \tag{8.41}$$

where $\gamma > 0$ is a scale coefficient and $0 < v < 2$ regulates distribution form. If $v = 1$, then obtain Cauchy distribution, if $v = 2$, then obtain Gaussian distribution. For $|x| \gg 1$ Levi distribution may be approximated so: $f_L(x) \sim \alpha x^{-(\upsilon + 1)}$.

5. *Exponential*. In this case $\xi_{ij}(t) \sim E(0, \xi)$, and the density function of the exponential distribution is given as

$$f_{E,\xi}(x) = \frac{\xi}{2} e^{-\xi|x|} \tag{8.42}$$

where $\xi > 0$ regulates variance (which is equal $\frac{2}{\xi^2}$).

Random numbers can be calculated so

$$x = \begin{cases} \frac{1}{\xi}\ln(2y), & \textit{if } y \le 0.5 \\ -\frac{1}{\xi}\ln 2(1-y), & \textit{if } y > 0.5 \end{cases}$$

where $y \sim U(0,1)$.

It's easy to note that $E(0,\xi) = \frac{1}{\xi}E(0,1)$.

6. *Chaotic*. Chaotic distribution is used for noise generation according to. $\xi_{ij}(t) \sim R(0,1)$,

where $R(0,1)$ presents a chaotic sequence in the space *(1;1)* generated according to recurrence [3]

$$x_{t+1} = \sin\left(\frac{2}{x_t}\right)x_t, \quad t = 0,1,2,\ldots \tag{8.43}$$

7. *Combined distribution*. The average mutation operator, was proposed which is a linear combination of Gaussian and Cauchy distributions [3]. In this case

$$\xi_{ij}(t) = \xi_{N,ij}(t) + \xi_{C,ij}(t) \tag{8.44}$$

where $\xi_{N,ij}(t) \sim N(0,1)$ and $\xi_{C,ij}(t) \sim C(0,1)$.

The resulting distribution generates more very small and large mutations compared to a Gaussian distribution. It has been proposed also an adaptive averaged mutations operator

$$\Delta x_{ij}(t) = \gamma_{ij}(t)\left(c_{ij}(0,1) + \upsilon_{ij}(t)N(0,1)\right) \tag{8.45}$$

where $\gamma_{ij}(t) = \sigma ij(t)$—general scaling, $\upsilon_{ij}(t) = \frac{\sigma_{1ij}}{\sigma_{2ij}}$—determines the form of distribution probability function.

Thus the Cauchy distribution, thanks to the large tail, generates more large mutations than the Gaussian distribution.

Levy distribution has tails between Gauss and Cauchy distributions. Thus, the Cauchy distribution should be used when solutions-pretendents are far from the optimum, and the Gauss—near the optimum for fine-tuning. Distribution of Levi as a combination provides a good balance between exploration and optimization.

Another factor that plays an important role in the balance between the study area and the optimization of this process is a way strategic parameters calculated and adjusted as these parameters directly affect the step size

Selection Operators

Selection operators are used for determination of those offsprings who had survived and go to the next generation. In the initial work on EP and the majority of their variations (develop) a new generation was selected from all the parents and their offsprings, i.e. parents and offsprings are competing for survival. Unlike GA competition is based on the relative fitness index, but not on the absolute one.

Absolute measures FIs are real fitness function, which estimate how closely is a candidate to the optimal solution. On the other hand, the relative performance index of fitness determines how much better the individual is relatively (as compared with a group of randomly selected applicants from parents and children).

In this presentation we use μ to denote the number of parents and λ—number of offsprings.

The first step in the process is to count index FI relative to each parent $x_i(t)$ and descendant $x_i(t)$.

Let $P(t) = C(t) \cup (C''(t))$ is competing pool and let $u_i(t) \in P(t),\ i = 1,\ldots,\mu+\lambda$ be an individual in a competitive game. Then for each $u_i(t) \in P(t)$ a group of competing n_p individuals randomly selected from $P(t) \backslash u_i(t)$. For each $u_i(t)$ is calculated index

$$S_i(t) = \sum\nolimits_{l=1}^{n_p} s_{ie}(t) \tag{8.46}$$

$$\text{where } s_{ie}(t) = \begin{cases} 1, \ \mathit{if}\, f(u_i(t)) < f(u_e((t)) \\ 0, \ \mathit{otherwise} \end{cases} \tag{8.47}$$

The alternative strategy was suggested in which [3]

$$s_{ie}(t) = \begin{cases} 1, \ \mathit{if}\ r_1 < \dfrac{f(u_l(t))}{f(u_i(t)) + f(u_l((t))} \\ 0, \ \mathit{otherwise} \end{cases} \tag{8.48}$$

where n_p is chosen as $n_p = [2\mu r_2 + 1],\ r_1, r_2 \sim U(0,1)$.

In this case $f(u_i(t) << f(u_l(t)))$, this will be if index FI for $u_i(t)$ is much better than for $u_l(t)$ then for individual $u_i(t)$ is assigned greater probability of getting winning score 1.

On the basis of the index S_i assigned to each individual $u_i(t)$ any of the methods of selection can be used.

Elitism: μ best individuals from the combined population $P(t)$ move to the next population $C(t+1)$.

Selection: M best individuals are stochastically selected from the tournament.

The proportional selection: Each individual is attributed to the likelihood of being selected

$$P_s(u_i(t)) = \frac{s_i(t)}{\sum_{l-1}^{\mu+\lambda} s_l(t)} \tag{8.49}$$

Roulette method can also be used to select μ individuals for the next generation.

Various methods have been proposed for the solution which of the parents or descendants will move to the next generation. In particular, it was suggested that the number of offsprings generated by each parent is determined by its FI. The higher is the FI of parent, more offsprings it produces. The best of the descendants generated by parent is chosen and competes with the parent for survival.

Competition between parent and descendant is based on the *annealing method* [3]. The descendant $x_l'(t)$ survives and proceeds to the next generation,

$$\text{if } f\left(x_l'(t)\right) < f(x_l(t)), \ \text{ or if } e^{-\left(f\left(x_l'(t)\right) - f(x_{il}(t))\right)/\tau(t)} > U(0,1) \tag{8.50}$$

otherwise parent survives, where $\tau(t)$—so-called temperature coefficient, wherein $\tau(t) = \gamma\tau(t-1), \ 0 < \gamma < 1.$

Strategic Parameters

As mentioned above, the step sizes are dependent on strategic parameters that form part of EP mutation operators.

For further discussion it is assumed that each component has its own strategic option and individuals are presented as

$$x_i(t) = (x_i(t), \sigma_i(t))$$

Static Strategic Options

The simplest way to manage the strategic options is to use a record of their value. In this case, the function of strategic options $\Phi(\sigma, X)$- is linear, i.e.

$$\Phi\left(\sigma_{ij}(t)\right) = \sigma_{ij} \tag{8.51}$$

The offsprings then are calculated as

$$x_{ij}'(t) = x_{ij}(t) + N_{ij}(0, \sigma_{ij}) \tag{8.52}$$

where $\Delta x_{ij}(t) = N_{ij}(0, \sigma_{ij})$.

The disadvantage of this approach is that too low σ_{ij} limits search ranges and slows convergence. On the other hand, too large a value for σ_{ij} limits possibility of fine tuning solutions.

Dynamic Strategy

One of the first approaches to change the values of the strategic options with the time—is to link them with the values of the FI

$$\gamma_{ij}(t) = \gamma_i(t) = \gamma f(x_i(t)) \tag{8.53}$$

In this case offsprings are generated according to

$$x_{ij}'(t) = X_{ij}(t) + N(0, \sigma_i(t)) = x_{ij}(t) + \sigma_i(t)N(0,1), \tag{8.54}$$

where $\gamma \in N(0,1)$.

If you have information about the global optimum, individual error can be used instead of the absolute error of the individual index FI. However, such information is generally unavailable. As an alternative, it may be used phenotypic distance from the best individual in accordance

$$\sigma'_{ij}(t) = \sigma_i(t) = |f(\hat{y}) - f(x_i)| \tag{8.55}$$

where $\hat{y}$—is the most proper individual.

A distance in solutions space may be also used $\sigma'_{ij}(t) = \sigma_i(t) = d(\hat{y}, x_i) = \|\hat{y} - x_i\|$, where $d(*, *)$—determines Euclid norm between vectors.

The advantage of this approach is that the worse is an individual, the more it should be mutated. On the other hand, the greater (better) is the individual, the less the child must be differed from the parent for fine adjustment of good solutions.

This approach has several drawbacks:

1. If the value of FI is very large, the size of the steps will also be too large, which may lead to the fact that the individual will jump over a good minimum.
2. The problem is further complicated if the optimal value of F—large non-negative quantity. If the value FI from "good" individuals is large, it will lead to large steps and the resulting offspring is removed from the good decisions. In such cases, if the information about the optimal value of the function is available, using the error value is more adequate.

Many proposals have been made to control the step size depending on the value of the fitness function. Fogel suggested an additive approach that

$$x'_{ij}(t) = x_{ij}(t) + \sqrt{\beta_{ij}(t) f(x_i) + \gamma_{ij}} + N_{ij}(0, 1) \tag{8.56}$$

where β_{ij} and γ_{ij} are proportionality constant and displacement parameter.

For training of recurrent neural networks it has been proposed [3]

$$x'_{ij}(t) = x_{ij}(t) + \beta \sigma_{ij} N_{ij}(0, 1), \tag{8.57}$$

where β is proportionality coefficient and

$$\sigma_{ij} = U(0, 1) \left[1 - \frac{f(x_i(t))}{f(x_{\max}(t))} \right], \tag{8.58}$$

$f_{\max}$—is maximal FI value for current population.

In this case, the problem of maximizing f is solved and $f(x_i(t))$ takes on positive values. It was suggested that the deviation be proportional to the normalized values of the FI

$$x'_{ij}(t) = x_{ij}(t) + \beta_{ij} \sigma_i N_{ij}(0, 1), \tag{8.59}$$

where β is proportionality coefficient and deviation is calculated as

$$\sigma_i(t) = \frac{f(x_i(t))}{\sum_{l=1}^{\mu} f(x_i(t))}, \tag{8.60}$$

μ is population size. In this approach it's assumed that function f is minimized.

It was suggested that deviation value was proportional to distance from the best individual [3]

$$\sigma_{ij}(t) = \beta_{ij}\left|\hat{y}_j - x_{ij}(t)\right| + \gamma, \tag{8.61}$$

where $\gamma > 0$—is small value, and coefficient β_{ij} is determined from condition $\beta_{ij} = \frac{\beta\sqrt{d(x_{\max}, x_{\min})}}{\pi}$, $\beta \in [0, 2]$, $d(x_{\max}, x_{\min})$ determines search space width as Euclidian distance between $x_{\min}$ and $x_{\max}$.

Adaptation (Self-organization) of EP

Two major problems concerning strategic parameters are the amount of noise mutation to be added, and a step size of mutations. In order to provide a truly self-organizing behavior, strategic parameters should be developed (or study) in parallel with variables of EP, which use such a mechanism called self-learning For the first time the idea of self-learning and self-organization in relation to neural networks was formulated by Ivakhnenko [18] and Rozenblatt [19]. As for the EP, the first proposals for the creation of a self- learning EP were made by Vogel et al. authors in [20]. These methods can be divided into the following three categories:

1. *Additive methods*. The first self-learning algorithm EP was proposed by Vogel [16] is additive, in which

$$\sigma_{ij}(t+1) = \sigma_{ij}(t) + \eta\sigma_{ij}(t)N_{ij}(0, 1) \tag{8.62}$$

where η is called a learning speed (rate).

In the first application of this approach $\eta = 1/6$. If $\sigma_{ij}(t) \leq 0$, then $\sigma_{ij} = \gamma$,where γ is small positive constant (usually $\gamma = 0.001$), that ensure non-zero deviations.

As an alternative Fogel suggested

$$\sigma_{ij}(t+1) = \sigma_{ij}(t) + \sqrt{f_\sigma\left(\sigma_{ij}(t)\right)} + N_{ij}(0, 1) \tag{8.63}$$

where $f_\sigma\left(\sigma_{ij}(t)\right) = \begin{cases} \sigma_{ij}(t), & \mathit{if}\ \sigma_{ij}(t) > 0, \\ \gamma, & \mathit{if}\ \sigma_{ij}(t) < 0. \end{cases}$.

2. The *multiplicative method.* Janda and Wang proposed a multiplicative method in which

$$\sigma_{ij}(t+1) = \sigma(0) \cdot \left(\lambda_1 e^{-\lambda_2 \frac{t}{n_t}} + \lambda_3\right) \tag{8.64}$$

where $\lambda_1, \lambda_2, \lambda_3$—adjustable parameters and nt—maximum number of iterations.

1. *Lognormal methods.* They are taken from the literature on evolutionary strategies [3]

$$\sigma_{ij}(t+1) = \sigma_{ij}(t)\, e^{\tau N_i(0,1) + \tau' N'_{ij}(0,1)} \tag{8.65}$$

$$\text{where } \tau = \frac{1}{\sqrt{2n_x}}; \quad \tau' = \frac{1}{\sqrt{2\sqrt{n_x}}},$$

n_x—is vector x dimension.

Self-learning EP exhibit undesirable behavior of stagnation due to the tendency that the strategy parameters converge quickly. Because of this deviation are quickly becoming too small, thus limiting the search. Search fades for a while, until the parameters of the strategy will not grow strongly enough, thanks to random variations.

Consider the most famous implementation of algorithms EP.

Classic EP

According to [3] Classic EP uses Gaussian mutations More specifically, the classical EP uses log-normal self-learning by the Eq. (8.65) and generates offspring, using Eq. (8.5). For the construction of a new population from the current population of the parent and the children the strategy of elitism is used.

Fast EP

A number of authors [21] adapted the classical EPO to create a fast EP (fast EP) by replacing the Gaussian noise with mutations by Cauchy distribution, according to the expression (8.37) where $v = 1$. A descendant is generated using

$$x'_{ij}(t) = x_{ij}(t) + \sigma_{ij}(t) C_{ij}(0,1) \tag{8.66}$$

which uses self-lognormal self-learning.

Wider tails of Cauchy distribution produce larger step sizes, and as a result accelerate the convergence. Analysis of fast EP showed that the size of the steps may be too large for the appropriate functioning, whereas mutation of the Gaussian distribution are the best opportunities for fine-tuning decisions. This has led to the development of a fast algorithm EP. In this algorithm, each parent produces two children, one using a Gaussian mutation and the other—the Cauchy mutation. The best descendant is chosen, who competes with their parents for survival.

An alternative approach is to start the search using mutation by Cauchy distribution, and after approaching the optimum to switch to a Gaussian distribution. However, this strategy creates the problem of determining the moment of switching.

Exponential EP

EP algorithm was proposed with a double exponential probability distribution [3]. Descendants are generated using the expression

$$x'_{ij}(t) = x_{ij}(t) + \sigma_{ij}\frac{1}{\xi}E_{ij}(0,1), \tag{8.67}$$

where σ_{ij} is learning parameter, and the variation of the distribution is controlled by parameter ξ. The less ξ, the greater the variation, and vice versa. To provide good properties search at the beginning ξ must be initialized by small value, which constantly increases over time.

Fast EP

In order to further improve the rate of convergence of the EP it was suggested to accelerate EP, which uses variation operators [3]:

- Direction operator to determine the direction of the search based on the values of fitness-function;
- Gaussian mutation operator specified in (8. 5).

Individuals are presented as

$$x_i(t) = (x_i(t), \rho_i(t), a_i(t)), \tag{8.68}$$

where $\rho_i(t) \in \{-1, 1\}$ $j = 1, \ldots, n_x$ determines search direction for each component

*j*th individual, and a_i contains age of individual.

Age is used to boost the broader research (search) if the descendants are worse than their parents.

Generation of descendants occurs in 2 steps. At the first step s the age of each individual is adapted and the direction of the search is determined(assuming minimization).

If the individual's FI improved, the search continues in the direction of improvement. If the FI has not improved, the age is incremented, which results in an large step size, according to the following expression

$$a_i(t) = \begin{cases} 1, & if\ f(x_i(t)) < f(x_i(t-1)) \\ a_i(t-1)+1, & otherwise \end{cases}, \tag{8.69}$$

and

$$\rho_{ij}(t) = \begin{cases} sign\,(x_{ij}(t) - x_{ij}(t-1)), & if\ f(x_i(t)) < f(x_i(t-1)) \\ \rho_{ij}(t-1), & otherwise \end{cases} \tag{8.70}$$

If $a_i(t) = 1$, then

$$\begin{aligned} \sigma_i(t) &= \gamma_1 f(x_i(t)) \\ x'_{ij}(t) &= x_{ij}(t) + \rho_{ij}(t) N(0, \sigma_i(t)) \end{aligned} \tag{8.71}$$

Otherwise, if $a_i(t) > 1$, then

$$\begin{aligned} \sigma_i(t) &= \gamma_2 f(x_i(t)) \\ x'_{ij}(t) &= x_{ij}(t) + N(0, \sigma_i(t)) \end{aligned}$$

where γ_1,γ_2 are positive constants.

Selection procedure is done by a descendant of the competition directly from its parent using the absolute index FI.

8.8 Differential Evolution

The positions of individuals in the search space provide valuable information about the fitness landscape. Under condition of the use of good random initialization method with uniform distribution to generate a population, individuals give a good presentation of the whole search space with a relatively large distance between individuals. Over time, as search process proceeds the distance between the individuals becomes less and all individuals converge to the same decision. If distances between individuals are large, the individuals have to make large steps to explore as much as possible the search space.

On the other hand, if the distances between individuals are small, the step sizes must also be small for study local areas. This behavior is achieved by so- called **differential evolution** [22] when calculating the size of the step mutations as the weighted distances between randomly selected individuals. Using the vector difference has many advantages.

Firstly, information about the fitness landscape features presented by the current population may be used to control the search.

Secondly, according to the central limit theorem, the step sizes mutation approaches a Gaussian distribution, assuming that the population is sufficiently large to provide a large number of difference vector.

Mutations

DE mutation operator makes a trial vector for each individual from the current population by mutations in the target vector with a weighted difference. This trial vector then will be used by the crossover operator to generate offspring. For each parent $x_i(t)$ test vector $u_i(t)$ is generated as follows: [22].

Select target vector $x_{i1}(t)$ from the population such that $i \neq i_1$. Next, randomly select two individuals x_{i2} and x_{i3} of the population, such that $i \neq i_1 \neq i_2$ where

$i_2, i_3 \sim U(1, u_\Delta)$. Using these individuals, then the trial vector is calculated by the perturbation of the target vector:

$$u_i(t) = x_{i1}(t) + \beta(x_{i2}(t) - x_{i3}(t)) \tag{8.72}$$

where $\beta \in [0, \infty]$—a scaling factor, which regulates the value of differential variation.

Crossover

The operator implements discrete crossover recombination of the trial vector $u_i(t)$ and the parental vector $x_i(t)$ to generate a descendant $x_i(t)$. Crossover is implemented as follows

$$x'_{ij}(t) = \begin{cases} u_{ij}(t), & \text{if } j \in I \\ x_{ij}(t), & \text{otherwise} \end{cases} \tag{8.73}$$

where $x_{ij}(t)$ denotes the *j*th element of the vector $x_i(t)$ and I—set of indices of elements that should be subjected to changes (in other words, multipoint crossover).

To determine the set I may be used a variety of methods, of which the following two approaches are used most frequently:

The binomial crossover. Crossover points are randomly selected from a set of possible crossover points $\{1, 2, \ldots, n_x\}$, where n_x—dimension of the problem. In final form, this process is described in DE algorithm.

In this algorithm p_r—the likelihood that considered crossover point will be included. The larger the p_r, the more crossover points will be included. This means that more elements of the test vector are used to generate a descendant and the less of the parent vector. To provide a difference of at least one element from the parent multipoint crossover or initialization includes randomly selected point j^x.

Exponential crossover. Starting from randomly selected index exponential crossover operator selects a sequence of adjacent crossover points, considering the list of potential crossover points as a circular array. By this at least one crossover point is selected, and since this index the next is selected until such condition: $U(0,1) \geq p_r$ or $|I| = n_x$ will be true.

Selection

Selection is used to determine which individuals take part in the mutation operation to generate trial vector and which of the parents or the children go into the next generation. Regarding the mutation operator a variety of methods of selection may be used.

Random selection is generally used to select parents from which difference vector calculated. For most algorithms DE target vector is selected or randomly, or the best individual s selected. To construct a new generation of the population a deterministic selection is used: descendant replaces parent if its FI is better than the parent's, otherwise parent goes to the next generation. This ensures that the average FI of generation is not deteriorated.

Describe a general DE algorithm [3]
Set the counter of generation $t = 0$.
Initialize control parameters β and p_r;
Create and initialize a population of $C(0)$ of x_s individuals
until another stop condition for each individual $x_i(t) \in C(t)$ do
estimate FI $f(x_i(t))$;
Create a test vector $u_i(t)$ using a mutation operator;
create a descendant $x_i'(t)$, using a crossover operator;
if $f(x_i'(t))$ is better than $f(x_i(t))$, then,
add a descendant $x_i'(t) \in C(t+1)$;
end;
otherwise $x_i(t) \in C(t+1)$;
end; end; end.
Choose the best individual.

The Control Parameters

In addition to a population size, the performance of DE is affected by two control parameters: a scale factor β and the recombination probability p_r. Consider the effect of these parameters. The size of the population as indicated in the Eq. (8.64) has a direct impact on the properties of DE algorithms. The more individuals in the population, the more available the differential vectors and more directions can be investigated. However, it should be born in mind that the more individuals are in the population, the greater is the computational cost of one generation.

Empirical studies provide the following relation $n_s = 10n_x$ [3].

The nature of the mutations provides a lower bound on the number of individuals: $n_s > n_\upsilon + 1$, where n_υ—the number of used differentials (differences). For n_υ differences it's required $2n_\upsilon$ different individuals, two for each difference.

The scaling factor. The scaling factor $\beta \in (0, \infty)$ adjusts the gain variation of the difference $(x_{i2} - x_{i1})$. Less is β, the smaller is the mutation and the longer time is required for convergence of the algorithm. Large values of β improve search capabilities of the algorithm, but can cause skipping over a good optimum point. Therefore the value of β must be small enough to allow to explore narrow valleys using differences and on the other hand, large enough to provide the necessary diversity.

The probability of recombination. The probability of recombination p_r directly affects the diversity of DE. This parameter controls the number of elements of the parent $x_i(t)$, which vary. The higher the probability of the recombination, the more variability is introduced into a new population, thereby increasing the diversity of properties and research. Increase in p_r often leads to faster convergence, while decreasing p_r increases robustness search. Most implementations strategies of DE accept control parameters constant. To find the optimal parameters self-learning strategies DE were developed [3].

References

1. Fraser, A.S.: Simulation of genetic systems by automatic digital computers I: introduction. Aust. J. Biol. Sci. **10**, 484–491 (1957)
2. Holland, J.H.: Adaptation in Natural and Artificial Systems. University of Michigan Press, Ann Arbor (1975)
3. Engelbrecht, A.: Computational Intelligence. An Introduction, 2nd edn. John Wiley & Sons, Ltd., pp. 630 (2007)
4. Bremermann, H.J.: Optimization through Evolution and Recombination. In: Yovits, M.C., Jacobi, G.T., Goldstine, G.D. (eds.) Self-organization Systems, pp. 93–106. Spartan Books (1962)
5. Michalewicz, Z.: Genetic Algorithms + Data Structures = Evolutionary Programs, 3rd edn. Springer, Berlin (1996)
6. Eshelman, L.J., Schaffer, J.D.: Real-Coded Genetic Algorithms and Interval Schemata. In: Whitley, D. (ed.) Foundations of Genetic Algorithms, vol. 2, pp. 187–202. Morgan Kaufmann, San Mateo (1993)
7. Deb, K., Agrawal, R.B.: Simulated binary crossover for continuous space. Complex Syst. **9**, 115–148 (1995)
8. Yager, R.R., Filev, D.P.: Approximate clustering via the mountain method. IEEE Trans. Syst. Man Cybern. **24**, 1279–1284 (1994)
9. Deb, K., Joshi, D., Anand, A.: Real-coded evolutionary algorithms with parent-centric recombination. In: Proceedings of the IEEE Congress on Evolutionary Computation, pp. 61–66 (2002)
10. Hinterding, R.: Gaussian mutation and self-adaption for numeric genetic algorithms. In: Proceedings of the International Conference on Evolutionary Computation, vol. 1, pp. 384 (1995)
11. Fogarty, T.C.: Varying the Probability of Mutation in the Genetic Algorithm. In: Schaffer, J.D. (ed.) Proceedings of the Third International Conference on Genetic Algorithms, pp. 104–109. Morgan Kaufmann, San Mateo, C.A. (1989)
12. Zaychenko, H.Y., Anikiev, A.S.: Efficiency of genetic algorithm application for MPLS network structure synthesis. Vestnik of ChNTU №1.- Chercassy, pp. 176–182 (2008). (rus)
13. de Jong, K.A., Morrison, R.W.: A test problem generator for non-stationary environments. In: Proceedings of the IEEE Congress on Evolutionary Computation, pp. 2047–2053 (1999)
14. Michalewicz, Z.: Genetic algorithms, numerical optimization, and constraints. In: Proceedings of the 6th International Conference on Genetic Algorithms, pp. 1–158 (1995)
15. Zaychenko, H.Y., Zaychenko, Y.P.: MPLS Networks: Modeling, analysis and optimization. Kiev. NTUU KPI, pp. 240 (2008). (rus)
16. Fogel, L.J., Owens, A., Walsh, M.: Artificial intelligence through simulated evolution. John Wiley & Sons (1966)
17. Eberhart, R.C., Shi, Y.: Particle swarm optimization: developments, applications and resources. In: Proceedings of the IEEE Congress on Evolutionary Computation, vol. 1, pp. 27–30 (2001)
18. Ivakhnenko, A.G.: Self-learning systems of pattern recognition and automatic control. Publishing House Technika, Kiev (1969). (rus)
19. Rosenblatt, F.: Principles of Neurodynamics. Perceptrons and Theory of Brain Mechanisms. M.: Publishing house, Mir (1965). (rus)
20. Fogel, D.: Review of Computational intelligence: imitating life. IEEE Trans. Neural Netw. **6**, 1562–1565 (1995)
21. Yuryevich, J., Wong, K.P.: Evolutionary programming based optimal power flow algorithm. IEEE Trans. Power Syst. **14**(4), 1245–1250 (1999)
22. Storn, R., Price, K.: Differential evolution—a simple and efficient heuristic for global optimization over continuous spaces. J. Global Optim. **11**(4), 341–359 (1997)

Chapter 9
Problem of Fuzzy Portfolio Optimization Under Uncertainty and Its Solution with Application of Computational Intelligence Methods

9.1 Introduction

The problem of constructing an optimal portfolio of securities under uncertainty is considered in this chapter. The main objective of portfolio investment is to improve the investment environment, giving securities such investment characteristics that are only possible in their combination. The global market crisis of recent years has shown that the existing theory of investment portfolio optimization and forecasting stock indices exhausted and revision of the basic theory of portfolio management is needed. Therefore in this work the novel theory of investment portfolio optimization under uncertainty is presented based on fuzzy set theory and efficient forecasting methods.

The direct problem of fuzzy portfolio optimization and dual problem are considered. In the direct problem structure of a portfolio is determined which will provide the maximum profitableness at the given risk level (Sect. 9.3). In dual problem structure of a portfolio is determined which will provide the minimum risk level at the set level of critical profitableness (Sect. 9.7). In the Sect. 9.3 fuzzy portfolio model was constructed for triangular membership functions and algorithm of its solution was considered. In the Sect. 9.4 fuzzy portfolio model was constructed for bell-shaped membership functions and in the Sect. 9.5 for Gaussian membership functions.

In the Sect. 9.6 the experimental investigations of the suggested fuzzy portfolio model were considered and comparison with classic Markovitz model was presented.

That to increase groundness of portfolio optimization and to decrease the possible risk it was suggested to use forecasting share prices for portfolio model. The input data for the optimization system are predicted by Fuzzy Group Method of Data Handling (FGMDH) (Sect. 9.8). The comparative analysis of optimal portfolio obtained by using different membership functions was fulfilled (Sect. 9.8). The experimental investigations of the suggested theory were carried out and comparison with classical portfolio model was performed (Sects. 9.7 and 9.8).

M.Z. Zgurovsky and Y.P. Zaychenko, *The Fundamentals of Computational Intelligence: System Approach*, Studies in Computational Intelligence 652,
DOI 10.1007/978-3-319-35162-9_9

The suggested portfolio optimization system is an effective tool for the operational management of portfolio investments.

The problem of investment in securities had arisen with appearance of the first stock markets. The main objective of portfolio investment is to improve the investment environment, giving securities such investment characteristics that are only possible in their combination. Careful processing and accounting of investment risks have become an integral and important part of the success of each company. However, more and more companies have to make decisions under uncertainty, which may lead to undesirable results. Particularly serious consequences may have the wrong decisions at long-term investments. Therefore, early detection, adequate and the most accurate assessment of risk is one of the biggest problems of modern investment analysis.

Historically, the first and the most common way to take account of uncertainty was the use of probability theory. The beginning of modern investment theory was put in the article H. Markowitz, “Portfolio Selection”, which was released in 1952. In this article mathematical model of optimal portfolio of securities was first proposed. Methods of constructing such portfolios under certain conditions are based on theoretical and probabilistic formalization of the concept profitability and risk. For many years the classical theory of Markowitz was the main theoretical tool for optimal investment portfolio construction, after which most of the novel theories were only modifications of the basic theory.

New approach in the problem of investment portfolio construction under uncertainty is connected with fuzzy sets theory. Fuzzy sets theory was created about half a century ago in the fundamental work of Lotfi Zadeh [1, 2]. This theory came into use in the economy in the late 70s. By using fuzzy numbers in the forecast parameters decision-making person was not required to form probability estimates.

The application of fuzzy sets technique enabled to create a novel theory of fuzzy portfolio optimization under uncertainty and risk deprived of drawbacks of classical portfolio theory by Markovitz. The main source of uncertainty is changing stock prices of securities at the stock market as the decision on portfolio is based on current stock prices while the implementation of portfolio is performed in future and portfolio profitablenes depends on future prices which are unknown at the moment of decision making. Therefore that to raise the reliability of decision concerning portfolio and cut possible risk it’s needed to forecast future prices of stocks. For this the application of inductive modelling method, so-called Fuzzy Froup Method of Data Handling (FGMDH) seems to be very perspective. And thanks to the results obtained using FGMDH system is devoid of subjective expert risk. The main goals of this chapter are to review the main results in fuzzy portfolio optimization theory, to consider and analyze so-called direct and dual problem of optimization, to estimate the application of FGMDH for stock prices forecasting while constructing investment portfolio and to carry out experimental investigations for estimation of the efficiency of the elaborated theory.

9.2 Direct Problem of Fuzzy Portfolio Optimization

Problem Statement

The purpose of the analysis and optimization of an investment portfolio is a research in the area of portfolio optimization, and also the comparative analysis of structure of the effective portfolios received with the use of classical model of Markovitz and fuzzy-set model of a investment portfolio optimization.

Let us consider a share portfolio of N components and its expected behavior at time interval $[0, T]$. Each of a portfolio component $i = \overline{1, N}$ at the moment T is characterized by it's financial profitableness r_i (evaluated at a point T as a relative increase in the price of the asset for this period) [3]. The holder of a share portfolio —the private investor, or an investment company—operates the investments, being guided by certain reasons. On one hand, the investor tries to maximize the profitableness. On the other hand, he fixes maximum permissible risk of an inefficiency of the investments.

Assume the capital of the investor be equal 1. The problem of share portfolio optimization consists in a finding of a vector of share prices distribution in a portfolio $x = \{x_i\} i = \overline{1, N}$ maximizing the expected income at the set risk level.

In process of practical application of Markovitz model its drawbacks were detected:

1. The hypothesis about normality of profitableness distributions in practice does not prove to be true.
2. Stationarity of price processes is not always confirmed in practice.
3. At last, the risk of stocks is considered as a dispersion (standard deviation of the prices from expected value) or its volatility, i.e. the decrease in profitableness of securities in relation to the expected value, and profitableness increase are estimated in this model absolutely the same. While for the proprietor of securities these events are absolutely different. These weaknesses of Markovitz theory caused necessity of development essentially new approach for definition of an optimum investment portfolio.

Let's consider the main principles and ideas of a fuzzy method for portfolio optimization [3–6].

The risk of a portfolio is not its volatility, but possibility that expected profitableness of a portfolio will appear below some pre-established planned value.

- Correlation of stock prices in a portfolio is not considered and not taken into account.
- Profitableness of each security is not random, but fuzzy number. Similarly, restriction on extremely low level of profitableness can be both usual scalar and fuzzy number of any kind. Therefore, to optimize a portfolio in such statement may mean, in that specific case, the requirement to maximize expected profitableness of a portfolio in time moment T at the fixed risk level of a portfolio.

- Profitableness of a security on termination of ownership term is expected to be equal r and lies in a settlement range.

For ith security denote:

$\bar{r}_i$—the expected profitableness of the ith security;
r_{1i}—the lower border of profitableness of the ith security;
r_{2i}—the upper border of profitableness of the ith security.
$r_i = (r_{1i}, \bar{r}_i, r_{2i})$—profitableness of the ith security is a triangular or Gaussian fuzzy number.

Then profitableness of a portfolio:

$$r = (r_{\min} = \sum_{i=1}^{N} x_i r_1; \bar{r} = \sum_{i=1}^{N} x_i \bar{r}_i; r_{\max} = \sum_{i=1}^{N} x_i r_{2i}) \tag{9.1}$$

where x_i is the weight of the ith security in a portfolio (its portion), and

$$\sum_{i=1}^{N} x_i = 1, \quad 0 \le x_i \le 1 \tag{9.2}$$

Critical level of profitableness of a portfolio at the moment of T given by an investor may be fuzzy triangular type number $r^* = (r_1^*, \bar{r}^* \, r_2^*)$ or may be crisp value.

9.3 Fuzzy-Set Approach with Triangular Membership Functions

To define structure of a portfolio which provides the maximum profitableness at the set risk level, it is required to solve the following problem [4]:

$$\{x_{opt}\} = \{x\} | r \to \max, \beta = const \tag{9.3}$$

where r is a portfolio profitableness, β is a desired risk, vector x satisfies (9.2).

Let us consider a risk estimation of portfolio investments. On Fig. 9.1 membership function r and criterion value r^* are shown.

Point with ordinate α_1—the crossings point of two membership functions. Let us choose any level of membership α and define corresponding intervals $[r_1, r_2]$ and $[r_1^*, r_2^*]$. At $\alpha > \alpha_1$, $r_1 > r_2^*$, intervals are not crossed, and the risk level equals to zero. Level α_1 is a top border of risk zone. At $0 \le \alpha \le \alpha_1$ intervals are crossed.

Calculate the risk value β. For this calculate shaded area of the phase space S_α (see Fig. 9.2).

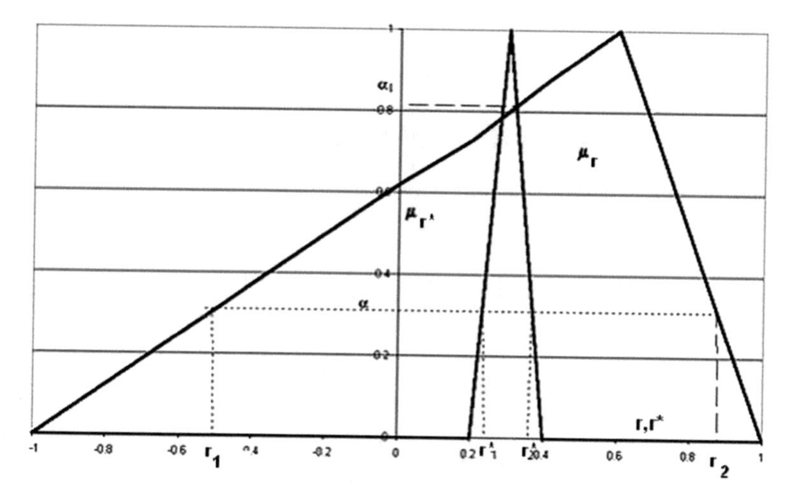

Fig. 9.1 Membership functions of r and r^*

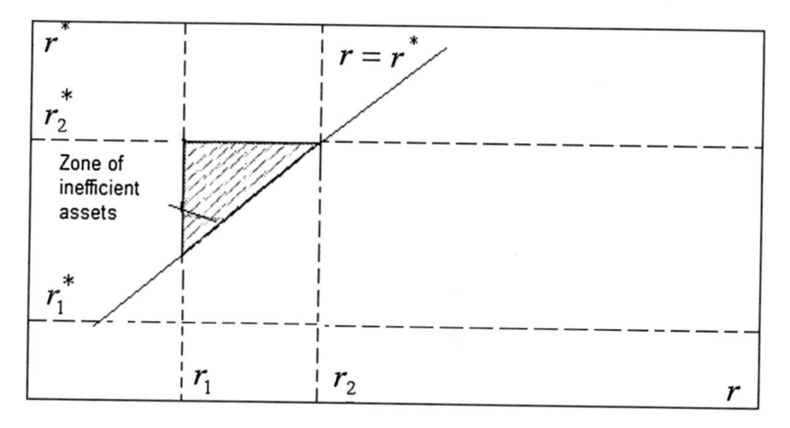

Fig. 9.2 Phase space (r, r^*)

$$S_\alpha = \begin{cases} 0, \; if \; r_1 \geq r_2^* \\ \frac{(r_2^* - r_1)^2}{2}, \; if \; r_2^* > r_1 \geq r_1^*; \; r_2 \geq r_2^* \\ \frac{(r_1^* - r_1) + (r_2^* - r_1)}{2} \cdot (r_2^* - r_1^*), \; \text{if} \; r_1 < r_1^*, \; r_2 > r_2^* \\ (r_2^* - r_1^*)(r_2 - r_1) - \frac{(r_2 - r_1^*)^2}{2}, \; if \; r_1 < r_1^* \leq r_2; \; r_2 < r_2^* \\ (r_2^* - r_1^*)(r_2 - r_1), \; if \; r_2 \geq r_1^* \end{cases} . \tag{9.4}$$

Since all realizations (r, r^*) at set membership level $\varphi(\alpha)$ are equally possible, so the degree of inefficiencies risk $\varphi(\alpha)$ is geometrical probability of event to drop into any point (r, r^*) in the zone of inefficient distribution of the capital [4, 5]:

$$\varphi(\alpha) = \frac{S_\alpha}{(r_2^* - r_1^*)(r_2 - r_1)}$$

Then total value of risk level of portfolio inefficiency equal to:

$$\beta = \int_0^{\alpha_1} \varphi(a)\partial\alpha \tag{9.5}$$

When the criterion of investor is defined as crisp level r^* limiting transition at $r_2^* \to r_1^* \to r^*$ gives:

$$\varphi(\alpha) = \begin{cases} 0, \ \mathit{if} \ r^* < r_1 \\ \frac{(r^* - r_1)}{(r_2 - r_1)}, \ \mathit{if} \ r_1 \le r^* \le r_2; \ \alpha \in [0; 1]. \\ 1, \ \mathit{if} \ r^* > r_2 \end{cases} \tag{9.6}$$

Then expected value risk degree of a portfolio is defined so [4]:

$$\beta = \begin{cases} 0, \ \mathit{if} \ r^* < r_{\min} \\ R\left(1 + \frac{1-\alpha_1}{\alpha_1}\ln(1-\alpha_1)\right), \ \mathit{if} \ r_{\min} \le r^* \le \tilde{r} \\ 1 - (1-R)\left(1 + \frac{1-\alpha_1}{\alpha_1}\ln(1-\alpha_1)\right), \ \mathit{if} \ \tilde{r} \le r^* \le r_{\max} \\ 1, \ \mathit{if} \ r^* \ge r_{\max} \end{cases}, \tag{9.7}$$

where

$$R = \begin{cases} \frac{r^* - r_{\min}}{r_{\max} - r_{\min}}, \ \mathit{if} \ r^* < r_{\max} \\ 1, \qquad \mathit{if} \ r^* \ge r_{\max} \end{cases}. \tag{9.8}$$

$$\alpha_1 = \begin{cases} 0, & \mathit{if} \ r^* < r_{\min} \\ \frac{r^* - r_{\min}}{\tilde{r} - r_{\min}}, & \mathit{if} \ r_{\min} \le r^* < \tilde{r} \\ 1, & \mathit{if} \ r^* = \tilde{r} \\ \frac{r_{\max} - r^*}{r_{\max} - \tilde{r}}, & \mathit{if} \ \tilde{r} < r^* < r_{\max} \\ 0, & \mathit{if} \ r^* \ge r_{\max} \end{cases}$$

Taking into account also that profitableness of a portfolio is equal to:

$$r = \left(r_{\min} = \sum_{i=1}^{N} x_i r_{1i}; \bar{r} = \sum_{i=1}^{N} x_i \bar{r}_i; r_{\max} = \sum_{i=1}^{N} x_i r_{2i}\right)$$

where $(r_{1i}, \bar{r}_i, r_{2i})$ is the profitableness of ith security, we obtain the following direct optimization problem (9.9)—(9.11) [4, 7]:

$$\bar{r} = \sum_{i=1}^{N} x_i \bar{r}_i \to \max \tag{9.9}$$

$$\beta = const \tag{9.10}$$

$$\sum_{i=1}^{N} x_i = 1,\ x_i \geq 0,\ i = \overline{1,N} \tag{9.11}$$

At a risk level variation β 3 cases are possible. Consider in detail each of them.

1. $\beta = 0$

   From (9.7) it is evident, that this case is possible when $r^* < \sum_{i=1}^{N} x_i r_{1i}$.

   The following problem of linear programming is obtained:

$$\bar{r} = \sum_{i=1}^{N} x_i \bar{r}_i \to \max \tag{9.12}$$

$$\sum_{i=1}^{N} x_i r_{1i} > r^*, \tag{9.13}$$

$$\sum_{i=1}^{N} x_i = 1,\ x_i \geq 0,\ i = \overline{1,N} \tag{9.14}$$

   The solution of the problem (9.12)–(9.14)—vector $x = \{x_i\}\ i = \overline{1,N}$ is a required structure of the optimum portfolio for the given risk level.

2. $\beta = 1$. This case practically is not accepted and is considered only for completeness of analysis. From (9.7) it follows, that this case is possible when $r^* \geq \sum_{i=1}^{N} x_i r_{i2}$.

   Then we get the following problem

$$\bar{r} = \sum_{i=1}^{N} x_i \bar{r}_i \to \max$$

$$\sum_{i=1}^{N} x_i r_{2i} \leq r^*$$

$$\sum_{i=1}^{N} x_i = 1,\ x_i \geq 0,\ i = \overline{1,N}$$

Found result of the problem (9.9)–(9.11) is a solution vector $x = \{x_i\}\ i = \overline{1,N}$ which determines a required structure of an optimum portfolio for the given risk level.

3. $0 < \beta < 1$

From (9.7) it is evident, that this case is possible when $\sum_{i=1}^{N} x_i r_{1i} \leq r^* \leq \sum_{i=1}^{N} x_i \bar{r}_i$ or when $\sum_{i=1}^{N} x_i \bar{r}_i \leq r^* \leq \sum_{i=1}^{N} x_i r_{i2}$.

- Assume $\sum_{i=1}^{N} x_i r_{i1} \leq r^* \leq \sum_{i=1}^{N} x_i \bar{r}_i$. Then using (9.7)–(9.8) the problem (9.9)–(9.11) is reduced to the following nonlinear programming problem:

$$\bar{r} = \sum_{i=1}^{N} x_i \bar{r}_i \rightarrow \max \tag{9.15}$$

$$\frac{1}{\sum_{i=1}^{N} x_i r_{i2} - \sum_{i=1}^{N} x_i r_{i1}} \left(\left(r^* - \sum_{i=1}^{N} x_i r_{i1} \right) + \left(\sum_{i=1}^{N} x_i \bar{r}_i - r^* \right) * \right.$$
$$\left. * \ln \left(\frac{\sum_{i=1}^{N} x_i \bar{r}_i - r^*}{\sum_{i=1}^{N} x_i \bar{r}_i - \sum_{i=1}^{N} x_i r_{i1}} \right) \right) = \beta \tag{9.16}$$

$$\sum_{i=1}^{N} x_i r_{i1} \leq r^*, \tag{9.17}$$

$$\sum_{i=1}^{N} x_i \bar{r}_i > r^* \tag{9.18}$$

$$\sum_{i=1}^{N} x_i = 1,\ x_i \geq 0,\ i = \overline{1,N} \tag{9.19}$$

- Let $\sum_{i=1}^{N} x_i \bar{r}_i \leq r^* \leq \sum_{i=1}^{N} x_i r_{i2}$. Then the problem (9.9)–(9.11) is reduced to the following nonlinear programming problem:

$$\bar{r} = \sum_{i=1}^{N} x_i \bar{r}_i \to \max \tag{9.20}$$

$$\left(r^* - \sum_{i=1}^{N} x_i r_{i1}\right) - \left(r^* - \sum_{i=1}^{N} x_i \bar{r}_i\right) * \ln\left(\frac{r^* - \sum_{i=1}^{N} x_i \bar{r}_i}{\sum_{i=1}^{N} x_i \bar{r}_{i2} - \sum_{i=1}^{N} x_i r_{i1}}\right) *, \\ * \frac{1}{\sum_{i=1}^{N} x_i r_{i2} - \sum_{i=1}^{N} x_i r_{i1}} = \beta \tag{9.21}$$

$$\sum_{i=1}^{N} x_i r_{i2} > r^*, \tag{9.22}$$

$$\sum_{i=1}^{N} x_i \bar{r}_i \leq r^*, \tag{9.23}$$

$$\sum_{i=1}^{N} x_i = 1, \ x_i \geq 0, \ i = \overline{1,N}. \tag{9.24}$$

The R-algorithm of minimization of not differentiated functions is suggested to find the solution of problems (9.15)–(9.19) and (9.20)–(9.24). Let both problems: (9.15) –(9.19) and (9.20)–(9.24) be solvable. Then to the structure of a required optimum portfolio will correspond such vector—$x = \{x_i\}$ $i = \overline{1,N}$ the solution that of problems (9.15)–(9.19), (9.20)–(9.24) the criterion function value of which will be greater.

9.4 Fuzzy-Sets Approach with Bell-Shaped Membership Functions

In case of using bell-shaped membership functions (MF) we should solve the problem of portfolio optimization where parameter $r_i = (r_{1i}, \bar{r}_i, r_{2i})$, the profitability of the ith asset, is a fuzzy number with bell-shaped form $\mu(x) = \frac{1}{1+\left(\frac{x-a}{c}\right)^2}$, as shown in Fig. 9.3.

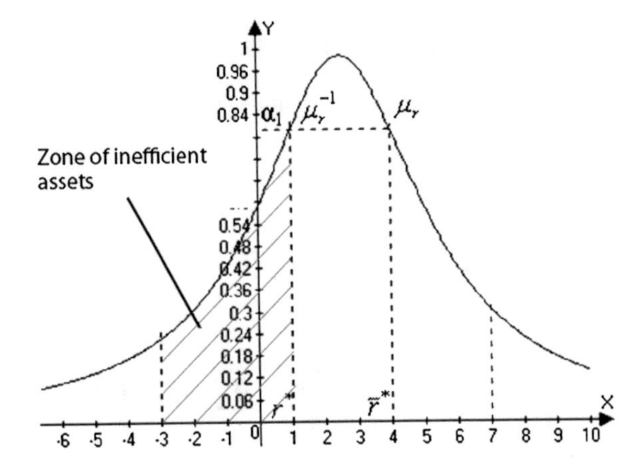

Fig. 9.3 Efficiency criterion for the bell-shaped MF

In this case:

$$\alpha_1 = \begin{cases} \dfrac{1}{1+\left(\dfrac{r^* - \bar{r}}{r_{\max} - r_{\min}}\right)^2} & where\; r_{\min} < r^* < r_{\max} \\ 0, & if\; r_{\min} > r^*\; and\; r^* > r_{\max} \\ 1, & if\; r^* = \bar{r} \end{cases} \tag{9.25}$$

$$\beta = \begin{cases} 0,\; if\; r^* < r_{\min} \\ \frac{1}{2}\alpha_1 + \dfrac{1}{2}\dfrac{r^* - \bar{r}}{r_{\max} - r_{\min}} * L,\; if\; r_{\min} \le r^* \le \bar{r} \\ 1 - \left(\frac{1}{2}\alpha_1 - \dfrac{1}{2}\dfrac{r^* - \bar{r}}{r_{\max} - r_{\min}} * L)\right),\; if\; \bar{r} \le r^* < r_{\max} \\ 1,\; if\; r^* > r_{\max} \end{cases} \tag{9.26}$$

where we can find α_1 from (9.25), $L = \arcsin\left(\sqrt{\alpha_1}\right) - \left(\sqrt{\alpha_1 - \alpha_1^2}\right)$

In order to determine the structure of the portfolio, which will provide maximum return with the given level of risk, we need to solve the problem (9.3) where α_1 and β are determined from formulas (9.27), (9.28) [8].

$$\beta = \begin{cases} 0, & if\; r^* \le r_{\min} \\ \frac{1}{2}(\alpha_1 - \alpha_0) + \dfrac{1}{2}\dfrac{(r^* - \bar{r})}{(r_{\max} - r_{\min})} * L + \dfrac{(r^* - r_{\min})}{(r_{\max} - r_{\min})} * \alpha_0, & if\; r_{\min} \le r^* \le \bar{r} \\ 1 - \left(\frac{1}{2}(\alpha_1 - \alpha_0) - \dfrac{1}{2}\dfrac{(r^* - \bar{r})}{(r_{\max} - r_{\min})} * L + \dfrac{(r_{\max} - r^*)}{(r_{\max} - r_{\min})} * \alpha_0\right), & if\; \bar{r} < r^* \le r_{\max} \\ 1, & if\; r^* > r_{\max} \end{cases} \tag{9.27}$$

where

$$\alpha_1 = \begin{cases} \dfrac{1}{1+\left(\dfrac{\bar{r}-r_{\min}}{r_{\max}-r_{\min}}\right)^2} & where\, r_1 \le r^* \le \bar{r} \\ \dfrac{1}{1+\left(\dfrac{r_{\max}-\bar{r}}{r_{\max}-r_{\min}}\right)^2} & where\, \bar{r} < r^* \le r_2 \end{cases} \tag{9.28}$$

$$L = \arcsin(\sqrt{\alpha_1}) - \arcsin(\sqrt{\alpha_0}) - \left(\left(\sqrt{\alpha_1 - \alpha_1^2}\right) - \left(\sqrt{\alpha_0 - \alpha_0^2}\right)\right) \tag{9.29}$$

Optimization problem consists of tasks (9.3), (9.25)–(9.29).

9.5 Fuzzy-Sets Approach with Gaussian Membership Functions

In case of using Gaussian MF we should solve the portfolio optimization problem where parameter $r_i = (r_{1i}, \bar{r}_i, r_{2i})$, the profitability of the ith stock is a fuzzy number with Gaussian form $\mu(x) = e^{-\frac{1}{2}\left(\frac{x-a}{c}\right)^2}$, as shown in Fig. 9.4.

In this case [8]:

$$\alpha_1 = \begin{cases} \exp\left\{-\dfrac{1}{2}\left(\dfrac{r^*-\bar{r}}{r_{\max}-r_{\min}}\right)^2\right\} \\ 0, \;\; if\;\; r^* < r_{\min}\;\; and\, r^* > r_{\max} \\ 1, \;\; if\;\; r^* = \bar{r} \end{cases} \tag{9.30}$$

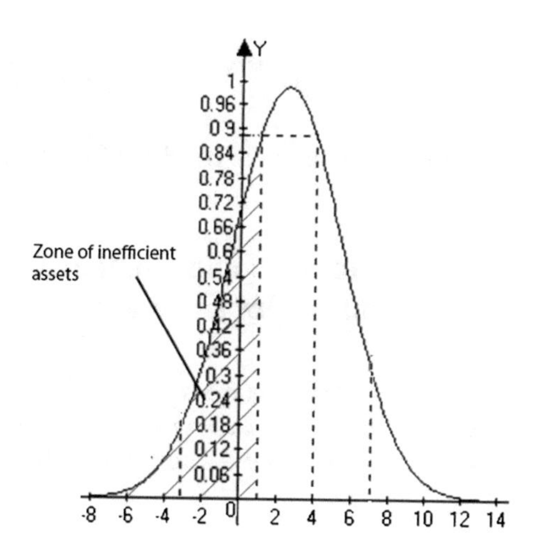

Fig. 9.4 Efficiency criterion for the Gaussian MF

$$\beta = \begin{cases} 0, & if\ r^* \le r_{\min} \\ \frac{1}{2}(\alpha_1 - \alpha_0) + \frac{\sqrt{\pi}}{2\sqrt{2}} R\left(\Phi(\sqrt{\frac{1}{\alpha_0}}) - \Phi\left(\sqrt{\ln \frac{1}{\alpha_1}}\right)\right), & if\ r_{\min} \le r^* \le \bar{r} \\ 1 - \left(\frac{1}{2}(\alpha_1 - \alpha_0) - \frac{\sqrt{\pi}}{2\sqrt{2}} R\left(\Phi(\sqrt{\frac{1}{\alpha_0}}) - \Phi\left(\sqrt{\ln \frac{1}{\alpha_1}}\right)\right)\right), & if\ \bar{r} < r^* \le r_{\max} \\ 1, & if\ r^* > r_{\max} \end{cases} \tag{9.31}$$

where α_0 is very small, α_1 we can find from (9.30),

$$\text{R is} \qquad R = \frac{r^* - \bar{r}}{r_{\max} - r_{\min}} \tag{9.32}$$

$\Phi(x)$ is the Laplace function: $\Phi(x) = \frac{2}{\sqrt{\pi}} \int\limits_0^x e^{-t^2} dt$

In order to determine the structure of the portfolio, which will provide maximum return with the given level of risk, we need to solve the problem (9.3) where β and α_1 are determined from formulas (9.30), (9.31).

$$\beta = \begin{cases} 0, & if\ r^* \le r_{\min} \\ \frac{1}{2}(\alpha_1 - \alpha_0) + \frac{\sqrt{\pi}}{2\sqrt{2}} R\left(\Phi(\sqrt{\frac{1}{\alpha_0}}) - \Phi\left(\sqrt{\ln \frac{1}{\alpha_1}}\right)\right) + \frac{r^* - r_{\min}}{r_{\max} - r_{\min}} * \alpha_0, & if\ r_{\min} \le r^* \le \bar{r} \\ 1 - \left(\frac{1}{2}(\alpha_1 - \alpha_0) - \frac{\sqrt{\pi}}{2\sqrt{2}} R\left(\Phi(\sqrt{\frac{1}{\alpha_0}}) - \Phi\left(\sqrt{\ln \frac{1}{\alpha_1}}\right)\right) + \frac{r_{\max} - r^*}{r_{\max} - r_{\min}} * \alpha_0\right), & if\ \bar{r} < r^* \le r_{\max} \\ 1, & if\ r^* > r_{\max} \end{cases} \tag{9.33}$$

where

$$\alpha_0 = \begin{cases} \exp\left\{-\frac{1}{2}\left(\frac{\bar{r} - r_{\min}}{r_{\max} - r_{\min}}\right)^2\right\}, & where\ r_{\min} \le r^* \le \bar{r} \\ \exp\left\{-\frac{1}{2}\left(\frac{r_{\max} - \bar{r}}{r_{\max} - r_{\min}}\right)^2\right\}, & where\ \bar{r} < r^* \le r_{\max} \end{cases} \tag{9.34}$$

Optimization problem consists of tasks (9.3), (9.30), (9.31)–(9.33).

9.6 The Analysis and Comparison of the Results Received by Markovitz and Fuzzy-Sets Model

For the comparative analysis of investigated methods of a share portfolio optimisation real data on share prices of the companies RAO » EES (EERS2) and Gazprom (GASP), were taken from February, 2000 till May, 2006 [4, 7].

In Markovitz model expected profitableness of a share is calculated as a mean $m = M\{r\}$ and risk of an asset is considered as a dispersion of the profitableness value $\sigma^2 = M\left[(m-r)^2\right]$, i.e. level of variability of expected incomes.

In the fuzzy-sets model obtained from a situation at the share market we conclude:

- shares profitableness of EERS2 lies in a settlement corridor [−1.0: 3.9], the most expected value of profitableness is 2.1 %
- shares profitableness of GASP lies in a settlement corridor [−4.1: 5.7], the most expected value of profitableness is 4.8 %

Let critical profitableness of a portfolio be 3.5 % i.e. portfolio investments which bring the income below 3.5 %, are considered as the inefficient.

Expected profitableness of the optimum portfolios received by Markovitz model, is higher, than profitableness of optimum portfolios, received by means of fuzzy-set model because in Markovitz model the calculation of expected share profitableness is based on indicators for the preceding periods and the situation in the share market at the moment of decision-making by the investor is not accounted. As profitableness of shares EERS2 and GASP till July 2006 was much more higher than at the present moment, Markovitz model gives unfairly high estimate.

In the fuzzy-set model the profitableness of each asset is a fuzzy number. Its expected value is calculated not from statistical data, but from analysis of the market at the moment of decision-making by the investor. Thus, in the considered case, the expected profitableness of a portfolio is not too high.

The structures of an optimum portfolio which are obtained by both methods for the same risk levels are quite different too. To understand the reason of this consider following dependences obtained for both models (Fig. 9.5) [3, 4].

Dependence of expected profitableness on risk degree of the portfolio is presented on Fig. 9.5.

Dependences of expected profitableness on degree of risk of the portfolio, received by the above specified methods, are practically opposite. The reason of such result is the various understanding of a portfolio risk.

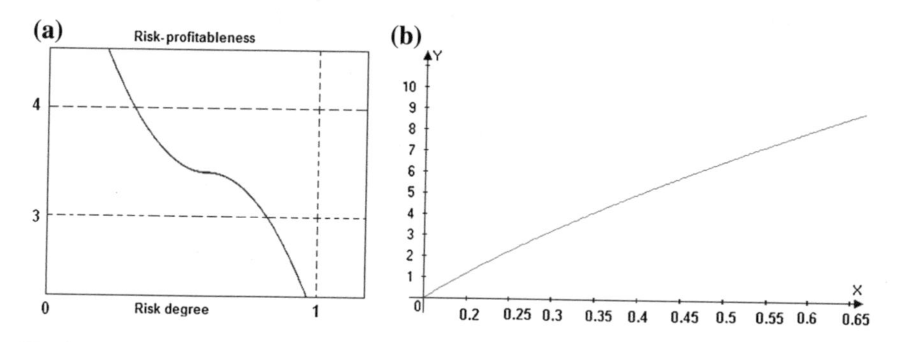

Fig. 9.5 Dependence of the expected profitableness on risk degree for considered models. **a** By fuzzy-set method **b** by Markovitz model

In the fuzzy-set method the risk is understood as a situation when expected profitableness of a portfolio falls below the set critical level, so with decrease of expected profitableness risk of portfolio investments to be less than the critical value, increases.

In Markovitz model the risk is considered as the degree of expected income variability of a portfolio, in both cases of smaller and greater income that contradicts common sense. The various understanding of portfolio risk level is also the reason of difference of a portfolio structure, received by different methods.

For share EERS2, with the growth of portion of low profitable securities in a portfolio, even in spite of the fact that the settlement corridor for EERS2 is narrower, rather than a settlement corridor for GASP, expected profitableness of a portfolio in general falls and the risk of an inefficiency portfolio grows.

Level of volatility of expected incomes for shares EERS2 obtained from data 2000–2006 is much more lower, than for shares GASP. Therefore, in Markovitz model which consider it as risk of portfolio investments, with the increase of portion of share EERS2 the risk of a portfolio decreases.

From the point of view of the fuzzy-set approach, the greater is the portion of GASP shares in a portfolio, the less is the risk of that efficiency of share investments will appear below the critical level which is in our case 3.5 %.

From the point of view of Markovitz model, average mean deviation from average value for GASP shares is great enough, therefore with growth of this share the risk of a portfolio increases. It leads to that portion of highly profitable assets in the share portfolio received by Markovitz model, is unfairly small.

According to Markovitz model, thanks to correlation between assets it is possible to receive a portfolio with a risk level less than volatility of the least risk security.

In this task after investing 96 % of the capital in EERS2 shares and 4 % in GASP shares, the investor received portfolio with expected profitableness of 2.4 % and degree of risk 0.19. However investments with expected profitableness of 2.4 % in the fuzzy-set model are considered as the inefficient. If to set critical value of expected portfolio profitableness equal to 2.4 % the risk of inefficient investments will decrease, too.

9.7 The Dual Portfolio Optimization Problem

9.7.1 Problem Statement

The initial portfolio optimization problem which is naturally to be called as direct has the following form [8]: optimize the expected portfolio profitableness

$$\bar{r} = \sum_{i=1}^{N} \bar{r}_l x_i \rightarrow \max \tag{9.35}$$

under constraints on risk

$$0 \leq \beta \leq 1 \tag{9.36}$$

$$\sum_{i=1}^{N} x_i = 1, \quad x_i \geq 0 \tag{9.37}$$

Let's consider the case when the criterion value r^* satisfies the conditions

$$\sum_{i=1}^{N} x_i r_{i1} \leq r^* \leq \sum_{i=1}^{N} \bar{r}_l x_i = \bar{r} \tag{9.38}$$

Then the risk value is equal

$$\beta(x) = \frac{1}{\sum_{i=1}^{N} x_i r_{i2} - \sum_{i=1}^{N} x_i r_{i1}} [\left(r^* - \sum_{i=1}^{N} x_i r_{i1}\right) + \left(\sum_{i=1}^{N} x_i \bar{r}_l - r^*\right) * \\ * \ln\left(\frac{\sum_{i=1}^{N} x_i \bar{r}_l - r^*}{\sum_{i=1}^{N} x_i \bar{r}_l - \sum_{i=1}^{N} x_i r_{i1}}\right)] \tag{9.39}$$

Now consider the portfolio optimization problem dual to the problem (9.35)–(9.37).

$$\text{To minimize} \qquad \beta(x) \tag{9.40}$$

under conditions

$$\bar{r} = \sum_{i=1}^{N} x_i \bar{r}_l \geq r_{given} = r^* \tag{9.41}$$

$$\sum_{i=1}^{N} x_i = 1, \quad x_i \geq 0 \tag{9.42}$$

9.7.2 *Optimality Conditions for Dual Fuzzy Portfolio Problem*

As it was proved in [9] function $\beta(x)$ is convex thefore the dual portfolio problem (9.40)–(9.42) is a convex programming problem. Taking into account that constraints (9.38) are linear compose Lagrangian function:

$$L(x, \lambda, \mu) = \beta(x) + \lambda(r^* - \sum_{i=1}^{N} x_i \bar{r}_l) + \mu\left(\sum_{i=1}^{N} x_i - 1\right) \quad (9.43)$$

The optimality conditions by Kuhn-Tucker are such [3, 8]

$$\frac{\partial L}{\partial x_i} = \frac{\partial \beta(x)}{\partial x_i} - \lambda \bar{r}_l + \mu \geq 0, \quad 1 \leq i \leq N, \quad (9.44)$$

$$\frac{\partial L}{\partial \lambda} = -\sum_{i=1}^{N} x_i \bar{r}_l + r^* \leq 0, \quad \frac{\partial L}{\partial \mu} = \sum_{i=1}^{N} x_i - 1 = 0 \quad (9.45)$$

and conditions of complementary slackness

$$\frac{\partial L}{\partial x_i} x_i = 0, \ 1 \leq i \leq N,$$

$$\frac{\partial L}{\partial \lambda} \lambda = \lambda\left(-\sum_{i=1}^{N} x_i \bar{r}_l + r^*\right) = 0, \ x_i \geq 0, \ \lambda \geq 0 \quad (9.46)$$

where $\lambda \geq 0$ and μ are indefinite Lagrange multpliers.

This problem may be solved by standard methods of convex programming, for example method of feasible directions or method of penalty functions.

9.8 The Application of FGMDH for Stock Prices Forecasting in the Fuzzy Portfolio Optimization Problem

The profitableness of leading companies at NYSE in the period from 03.09.2013 to 17.01.2014 were used as the input data in experimental investigations. The companies included: Canon Inc. (CAJ), McDonald's Corporation (MCD), PepsiCo, Inc (PEP), The Procter & Gamble Company (PG), SAP AG (SAP). The corresponding data is presented in the Table 9.1 [8]:

For forecasting we have used the Fuzzy GMDH method with triangular membership functions, linear partial descriptions, training sample of 70 %, forecasting for 1 step.

The fuzzy GMDH allows to construct forecasting model using experimental data automatically without participation of an expert. Besides, it may work under uncertainty conditions with fuzzy input data or data given as intervals.

This method is considered and explored in the Chap. 6. The next profitableness values on date 17.01.2014 were obtained (Table 9.2):

Table 9.1 The profitableness of shares, %

Companies / Dates	CAJ	MCD	PEP	PG	SAP
06.09.13	0.510	−1.841	1.172	0.772	2.692
13.09.13	−0.880	−0.933	−1.184	−1.139	−0.945
20.09.13	0.990	0.829	−0.889	0.961	−1.509
27.09.13	1.110	0.164	1.012	2.611	−0.582
04.10.13	0.060	1.569	−0.151	−0.569	0.649
11.10.13	−0.410	−0.403	−2.239	−3.741	−3.190
18.10.13	−1.230	−0.507	−2.368	−0.851	−0.477
25.10.13	2.300	−0.201	−1.190	−1.304	−3.979
01.11.13	2.620	−1.961	0.059	0.185	2.494
08.11.13	0.190	0.308	−1.754	−1.451	−0.879
15.11.13	−3.600	0.175	−0.679	−3.136	−1.675
22.11.13	−0.180	−0.635	0.140	−0.449	−0.329
29.11.13	−1.830	1.567	1.066	1.393	−0.510
06.12.13	1.950	−0.300	0.657	−1.416	−0.636
13.12.13	1.650	1.337	2.128	2.843	1.566
20.12.13	−0.060	−1.111	−1.000	−0.184	−2.247
27.12.13	1.560	−0.633	−1.038	−0.861	−1.132
03.01.14	0.880	0.484	0.808	1.890	2.655
10.01.14	1.770	0.052	−1.483	0.422	1.042
17.01.14	−1.270	−0.105	0.206	0.162	0.843

Table 9.2 The profitableness of shares on date 17.01.2014, %

Companies	Profitableness				MAPE test sample	MSE test sample
	Real value	Low bound	Forecasted value	Upper bound		
CAJ	−1.270	−1.484	−1.246	−1.008	2.2068	0.0295
MCD	−0.105	−0.347	−0.118	0.111	2.5943	0.0091
PEP	0.206	0.001	0.242	0.483	3.0179	0.0177
PG	0.162	0.041	0.170	0.299	1.6251	0.0197
SAP	0.843	0.675	0.867	1.059	2.3065	0.0164

Thus, as the result of application of FGMDH the shares profitableness values were forecasted to the end of 20th week (17.01.2014) as follows [8]:

- Profitableness of CAJ shares lies in the calculated corridor [−1.484; −1.008], the expected value is—1.246 %;
- Profitableness of MCD shares lies in the calculated corridor [−0.347; 0.111], the expected value is—0.118 %;

- Profitableness of PEP shares lies in the calculated corridor [0.001; 0.483], the expected value is 0.242 %;
- Profitableness of PG shares lies in the calculated corridor [0.041; 0.299], the

Table 9.3 Distribution of components of the optimal portfolio for triangular MF with critical level $r^* = 0.7\,\%$

CAJ	MCD	PEP	PG	SAP
0.05482	0.00196	0.0027	0.00234	0.93818
0.06145	0.00113	0.00606	0.0039	0.92746
0.0698	0.00577	0.00235	0.00219	0.91989
0.06871	0.00228	0.0057	0.00244	0.92087
0.07567	0.00569	0.00106	0.00094	0.91664
0.07553	0.00002	0.0029	0.00208	0.91947
0.06774	0.00121	0.006	0.00234	0.92271
0.0764	0.001	0.00612	0.00464	0.91184
0.09072	0.00849	0.00655	0.0039	0.89034

Table 9.4 Parameters of the optimal portfolio for triangular MF with critical level $r^* = 0.7\,\%$

Low bound	Expected profitableness	Upper bound	Risk level
0.55133	0.74591	0.94049	0.2
0.53462	0.72954	0.92446	0.25
0.51544	0.71084	0.90624	0.3
0.51894	0.71431	0.90968	0.35
0.5045	0.70018	0.89587	0.4
0.50877	0.70425	0.89973	0.45
0.522	0.71731	0.91262	0.5
0.50197	0.69752	0.89308	0.55
0.46358	0.66014	0.8567	0.6

expected value is 0.17 %;
- Profitableness of SAP shares lies in the calculated corridor [0.675; 1.059], the expected value is 0.867 %.

In this way the portfolio optimization system stops to be dependent on factor of expert subjectivity. Besides, we can get data for this method automatically, without expert's estimates.

Let the critical profitableness level be 0.7 %. Varying the risk level we obtain the following results at the end of 2nd week (17.01.2014) for triangular MF. The results are presented in the Tables 9.3, 9.4 and the Fig. 9.6.

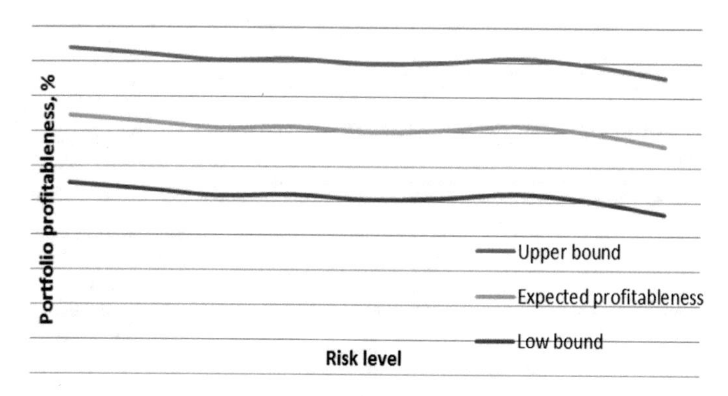

Fig. 9.6 Dependence of the expected portfolio profitableness versus risk level for triangular MF

As we can see in Fig. 9.6, the dependence profitableness—risk has descending type, the greater is the risk the lesser is profitableness which is opposite to classical probabilistic method. It may be explained so that at fuzzy approach by risk is meant

Table 9.5 Distribution of components of the optimal portfolio for bell-shaped MF with critical level $r^* = 0.7\%$

CAJ	MCD	PEP	PG	SAP
0.00129	0.0026	0.00279	0.00316	0.99016
0.00264	0.00249	0.0028	0.0037	0.98837
0.00249	0.00268	0.00246	0.00238	0.98999
0.00209	0.00148	0.00039	0.00024	0.9958
0.00102	0.00105	0.0021	0.00222	0.99361
0.00425	0.00325	0.00343	0.0032	0.98587
0.00128	0.00125	0.00207	0.002	0.9934
0.00084	0.0022	0.0018	0.00163	0.99353
0.00165	0.00105	0.00198	0.00137	0.99395

Table 9.6 Parameters of the optimal portfolio for bell-shaped MF with critical level $r^* = 0,7\%$

Low bound	Expected profitableness	Upper bound	Risk level
0.66568	0.85777	1.04986	0.2
0.66252	0.85464	1.04676	0.25
0.66372	0.8559	1.04808	0.3
0.6612	0.85335	1.04551	0.35
0.6594	0.85145	1.0435	0.4
0.65816	0.85044	1.04272	0.45
0.63574	0.82782	1.0199	0.5
0.62921	0.82131	1.01341	0.55
0.59079	0.78291	0.97504	0.6

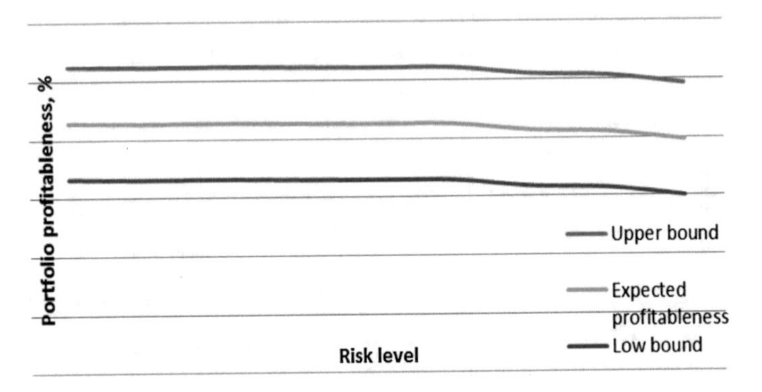

Fig. 9.7 Dependence of expected portfolio profitableness versus risk for bell-shaped MF

the situation when the expected profitableness happens to be less than the given criteria level. When the expected profitableness decreases, the risk grows.

The profitableness of the real portfolio is 0.7056 %. This value falls in calculated corridor of profitableness for optimal portfolio [0.5346, 0.7295, 0.9245] built with application of forecasting, indicating the high accuracy of the forecast.

Now consider the same portfolio using bell-shaped MF (Tables 9.5 and 9.6 and Fig. 9.7).

The profitableness of the real portfolio is 0.8339 %. This value also falls in the calculated corridor of profitableness for optimal portfolio [0.6657; 0.8578; 1.0499] [8].

Now consider the same portfolio using Gaussian MF (results are presented in Tables 9.7 and 9.8 and Fig. 9.8).

The profitableness of the real portfolio is 0.8316 %. This value also falls in calculated corridor of profitableness for optimal portfolio [0.6833; 0.8756; 1.0677].

In the above presented results the optimal portfolio corresponds to the first row of tables. As it can be seen from these data, the profitableness obtained using

Table 9.7 Distribution of components of the optimal portfolio for Gaussian MF with critical level $r^* = 0.7\%$

CAJ	MCD	PEP	PG	SAP
0.0028	0.00277	0.00221	0.0021	0.99012
0.0009	0.00126	0.00153	0.00162	0.99469
0.00028	0.00189	0.00232	0.00213	0.99338
0.00193	0.00243	0.00284	0.00278	0.99002
0.00144	0.00096	0.00088	0.00138	0.99534
0.00083	0.001	0.00225	0.00144	0.99448
0.00223	0.0024	0.003	0.00209	0.99028
0.0013	0.00124	0.00129	0.0019	0.99427
0.00261	0.00191	0.00204	0.00239	0.99105

Table 9.8 Parameters of the optimal portfolio for Gaussian MF with critical level $r^* = 0.7\,\%$

Low bound	Expected profitableness	Upper bound	Risk level
0.6833	0.87551	1.06772	0.2
0.66972	0.86178	1.05384	0.25
0.66955	0.86161	1.05368	0.3
0.66468	0.85682	1.04896	0.35
0.64944	0.8415	1.03356	0.4
0.65975	0.85185	1.04394	0.45
0.63439	0.8266	1.0188	0.5
0.63184	0.82389	1.01594	0.55
0.62452	0.81666	1.0088	0.6

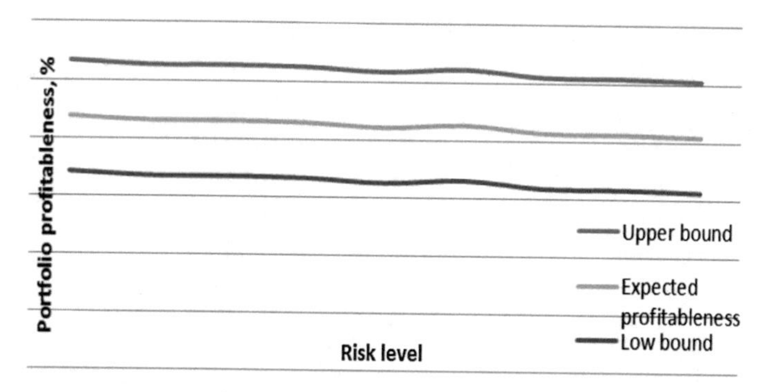

Fig. 9.8 Dependence of expected portfolio profitableness versus risk for Gaussian MF

Gaussian and bell-shaped MF is higher than the profitableness obtained using triangular MF. The reason is the form of used MF curves. The bell-shaped and Gaussian function are more convex, so an area of inefficient assets is bigger, and the risk of getting into this area is higher.

The optimal portfolios obtained with different MF actually have the same structure, the main portion belongs to the company SAP shares, due to high rates of return as compared with other companies.

Let's consider the results obtained by solving the dual problem using triangular MF. In this case, the investor sets the rate of return and the problem is to minimize the risk.

The optimal portfolio is presented in Tables 9.9 and 9.10, and Fig. 9.9:

From these results one may see that the curve "dependence risk—given critical level of profitability" has a ascending character, because with the growth of the critical value of profitability increases the probability that the expected return would be lower than a given critical value [8].

Table 9.9 Distribution of components of the optimal portfolio (dual task)

CAJ	MCD	PEP	PG	SAP
0.01627	0.02083	0.02226	0.02231	0.91833
0.01112	0.02085	0.02391	0.02383	0.92029
0.00333	0.01992	0.02517	0.02476	0.92682
0.0021	0.01579	0.02457	0.02344	0.9341
0.00004	0.00921	0.02423	0.02135	0.94517
0.00224	0.00144	0.01825	0.01095	0.96712
0.00044	0.00682	0.02508	0.02058	0.94708
0.0011	0.00917	0.02448	0.02039	0.94486
0.00294	0.01206	0.02533	0.02154	0.93813

Table 9.10 Parameters of the optimal portfolio (dual task)

Low bound	Expected profitableness	Upper bound	Risk level	Critical rate of return
0.58944	0.78264	0.97584	0.00025	0.6
0.59846	0.79141	0.98437	0.01468	0.65
0.61478	0.80735	0.99991	0.04973	0.7
0.6229	0.81531	1.00772	0.13347	0.75
0.63606	0.82822	1.02037	0.26399	0.8
0.64945	0.84181	1.03417	0.49937	0.85
0.63712	0.82933	1.02153	0.72631	0.86
0.63382	0.82612	1.01843	0.8333	0.87
0.62559	0.81805	1.01052	0.91214	0.88

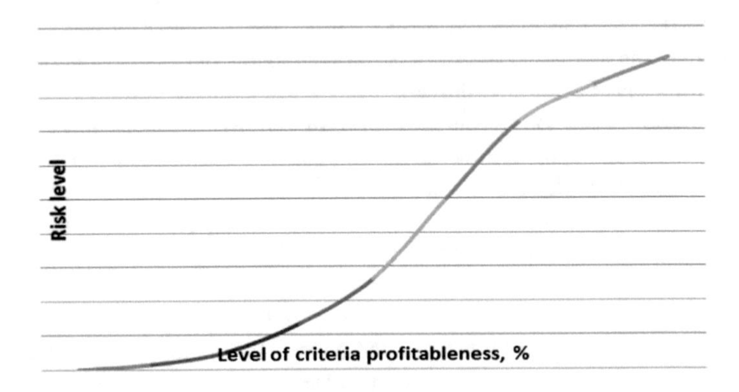

Fig. 9.9 Dependence of the risk level on a given critical return

9.9 Conclusion

The problem of optimization of the investment portfolio under uncertainty is considered in this chapter. The fuzzy-set approach for solving the direct and dual portfolio optimization problems was suggested and explored. In the direct problem triangular, bell-shaped and Gaussian membership functions were used. The results of solving the problems were presented. The optimal portfolios for the five assets at NYSE stock market were constructed and analyzed.

The problem of stock prices forecasting for portfolio optimization was also investigated. The fuzzy GMDH was proposed for its solution and its experimental investigations are presented. The fuzzy GMDH was applied for stocks profitableness forecasting at NYSE stock market in the problem of fuzzy portfolio optimization. The application of fuzzy GMDH enabled to decrease risk of the wrong decisions concerning portfolio content.

After analysis of experiments it was detected that the dependence "profitableness-risk" has descending type, the greater risk the lesser is profitableness that is opposite to classical probabilistic methods.

The dependence "risk versus given critical level of profitability" has ascending type, because with the growth of the critical level of profitability increases the probability that the expected return appears to be lower than a given critical value.

As the main result of this research the fundamentals of theory of fuzzy portfolio optimization under uncertainty have been developed.

References

1. Zadeh, L.A.: Fuzzy sets. Inf. Control **8**(3), 338–353 (1965)
2. Zadeh, L.A.: Fuzzy sets as a basis for a theory of possibility. Fuzzy Sets Syst. **N1**, 3–28 (1978)
3. Zaychenko, Y.P.: Fuzzy models and methods in intellectual systems, 354 p. Publ. House, Slovo, Kiev (2008). Zaychenko, Y.: Fuzzy group method of data handling under fuzzy input data. In: System research and information technologies, №. 3, pp. 100–112 (2007) (in Russ.)
4. Zaychenko, Y.P., Maliheh Esfandiyarifard. Analysis and comparison results of invest portfolio optimization using Markovitz model and fuzzy sets method. In: XIII-th International Conference KDS-2007, vol. 1, pp. 278–286, Sofia (2007). (in Russ.)
5. Nedosekin, A.O.: System of portfolio optimization of Siemens services. Banking Technol. № 5 (2003). (in Russ.) http://www.finansy.ru/publ/fin/004.htm
6. Nedosekin, A.O.: Corporation business portfolio optimization (in Russ.). Аудит сайте: http://sedok.narod.ru/s_files/2003/Art_070303.doc
7. Zaychenko, Y.P.: Maliheh Esfandiyarifard. Analysis of invest portfolio for different membership functions. Syst. Res. Inf. Technol. 1(2) 59–76 (2008). (in Russ.)
8. Zaychenko, Y.: Inna Sydoruk. Direct and dual problem of investment portfolio optimization under uncertainty. Int. J. Inf. Technol. Knowl. **8**(3), 225–242 (2014)
9. Zgurovsky, M.Z., Zaychenko, Y.P.: Models and methods of decision-making under uncertainty, 275 c. Publ. House, Naukova Dumnka, Kiev (2011). (in Russ.)

Index

M.Z. Zgurovsky and Y.P. Zaychenko, *The Fundamentals of Computational Intelligence: System Approach*, Studies in Computational Intelligence 652,
DOI 10.1007/978-3-319-35162-9

Printed in the United States
By Bookmasters